Applications of Covalent Organic Frameworks in Environmental Remediation

This book provides a comprehensive overview of covalent organic frameworks (COFs) and their growing significance in environmental applications. With their highly ordered, porous structures and tunable properties, COFs offer innovative solutions for challenges such as water purification, air filtration, and pollutant degradation. This book introduces the fundamentals of COF synthesis, structure, and properties while highlighting recent research advances and practical applications. Emphasizing both theoretical insights and real-world relevance, it explores how molecular design and structural engineering can enhance COF performance. This resource is ideal for researchers, students, and professionals focused on sustainable materials and environmental remediation technologies.

Features:

- Highlights the use of COFs in water purification, air filtration, and pollutant degradation.
- Explores both fundamental mechanisms and practical challenges in environmental applications of COFs.
- Combines recent research, synthesis methods, and real-world case studies in an accessible format.
- Serves as a valuable resource for researchers, students, and professionals in environmental and materials science.

Applications of Covalent Organic Frameworks in Environmental Remediation

Edited by

Pallavi Jain and Sapna Raghav

CRC Press
Taylor & Francis Group
Boca Raton London New York

CRC Press is an imprint of the
Taylor & Francis Group, an **informa** business

Designed cover image: Pallavi Jain and Sapna Raghav

First edition published 2026
by CRC Press
2385 NW Executive Center Drive, Suite 320, Boca Raton FL 33431

and by CRC Press
4 Park Square, Milton Park, Abingdon, Oxon, OX14 4RN

CRC Press is an imprint of Taylor & Francis Group, LLC

Library of Congress Cataloging-in-Publication Data
Names: Jain, Pallavi (Professor of chemistry) editor | Raghav, Sapna editor
Title: Applications of covalent organic frameworks in environmental remediation / edited by Pallavi Jain and Sapna Raghav
Description: First edition. | Boca Raton : CRC Press, 2026. | Includes bibliographical references and index. |
Identifiers: LCCN 2025024622 (print) | LCCN 2025024623 (ebook) | ISBN 9781032996677 hbk | ISBN 9781032996684 pbk | ISBN 9781003605386 ebk
Subjects: LCSH: Pollution prevention—Materials | Organic compounds—Environmental aspects | Mesoporous materials—Environmental aspects | Chemical bonds
Classification: LCC TD192.2. A67 2026 (print) | LCC TD192.2 (ebook)
LC record available at https://lccn.loc.gov/2025024622
LC ebook record available at https://lccn.loc.gov/2025024623

ISBN: 978-1-032-99667-7 (hbk)
ISBN: 978-1-032-99668-4 (pbk)
ISBN: 978-1-003-60538-6 (ebk)

DOI: 10.1201/9781003605386

Typeset in Times
by codeMantra

Contents

Preface

Environmental sustainability is one of the most pressing challenges of our time, necessitating innovative materials and technologies to combat pollution, resource scarcity, and climate change. Among the advancements in materials science, covalent organic frameworks (COFs) have emerged as a particularly promising class of materials. COFs are distinguished by their crystalline structures and strong covalent bonds, which form highly ordered, porous networks. Their unique properties make them ideal for addressing a range of environmental issues. This book, *Applications of Covalent Organic Frameworks in Environmental Remediation*, seeks to explore the multifaceted potential of COFs, focusing specifically on their roles in environmental remediation and sustainability.

These frameworks have a high structural diversity and versatility. Formed from organic building blocks linked by covalent bonds, these frameworks exhibit structures with large surface areas, tunable porosity, and well-defined, stable structures. This book introduces the basic principles of COFs, including their synthesis, properties, and the tremendous progress of the last decade. By considering molecular design and the characterization of COFs, this book offers a deep understanding of how these materials are engineered for specific environmental applications.

COF environmental applications is a prime focus of this book. These chapters deal with the exploitation of COFs in domains such as water purification, air filtration, and the degradation of pollutants. In this regard, this book examines the role that COFs could play in these critical areas, bringing out their capacity to provide effective solutions to real-world environmental challenges. The mechanisms of COF materials in terms of applications toward environmental health and sustainability are expounded upon through detailed discussions supported by recent research findings and practical insights. This book was motivated by increasing awareness that environmental challenges require advanced materials. Although much effort has been made to investigate COFs, to date there remains a lack of comprehensive resources dedicated to their environmental applications. This book aims to address that gap by offering a single, up-to-date reference that systematically covers recent developments, synthesis methods, and practical applications of COFs. Thus, this book is intended to be an invaluable reference for researchers, professionals, and students involved in materials science, chemistry, and environmental engineering by offering an inclusive and accessible account of the subjects.

Besides practical applications, this book also covers the level of mechanisms governing COF efficiency in environment-related applications. It is intended to provide readers with an overview of design strategies that enhance COF performance, such as how structural and compositional engineering can modulate efficiency and selectivity. Challenges in scaling up COF technology for industrial applications will also be discussed, offering a balanced views on both theoretical and practical aspects of COF research.

Applications of Covalent Organic Frameworks in Environmental Remediation seeks to be a spur for new developments in environmental science by placing COFs at the forefront of this research. It is hoped that state-of-the-art developments and insights into future trends will inspire further research and collaboration. Through focused, in-depth analyses, it contributes to scientific progress while offering practical solutions for building a more sustainable and resilient future.

About the Editors

Pallavi Jain has a distinguished academic and professional career. She currently heads the Student Welfare Department at the SRM Institute of Science and Technology, Delhi-NCR Campus, Modinagar, Ghaziabad, India. She received her Ph.D. in Chemistry from Banasthali Vidyapith, Rajasthan, India, in 2017. She has approximately 19 years of extensive teaching experience. Her research interest focuses on the development of transition metal complexes incorporating Schiff base ligands for their biological importance in drug delivery, as well as the synthesis of deep eutectic solvents. She has authored and co-authored over 80 publications in reputed international journals and contributed more than 65 book chapters. Her work has been cited more than 1,250 times. She has also edited four books.

Sapna Raghav, born in Haryana, India, is an accomplished academic and researcher in the field of Chemistry. She earned her Ph.D. in 2019 from the Department of Chemistry at Banasthali University, Rajasthan. Her primary research interests lie in Environmental and Materials Chemistry, with a focus on water decontamination, sensing technologies, and material science. She has contributed over 25 research articles in reputed international journals. In addition, she has authored more than 50 book chapters with renowned publishers. She has actively participated in numerous national and international conferences and has served as a Women Scientist in Intellectual Property Rights under the DST-TIFAC scheme in New Delhi. She currently holds the position of State Vice President of the Haryana Geographical Indication (GI) Tagging Council under WICCI. With four years of teaching experience, she is presently working as Head of the Department of Chemistry at Shri Jagdishprasad Jhabarmal Tibrewala University, Jhunjhunu, Rajasthan.

Contributors

Amjad Ali
Ram Sewak Singh Mahila College
Sitamarhi, Bihar, India
and
A Constituent Unit of B.R.A. Bihar University, Muzaffarpur
Muzaffarpur, Bihar, India

Satyesh Raj Anand
Carbon Nanomaterials Research Laboratory, Faculty of Engineering & Technology
Jagannath University
Jaipur, Rajasthan, India

Rahul Badru
Department of Chemistry
Sri Guru Granth Sahib World University
Fatehgarh Sahib, Punjab, India

Deepti Chaudhary
Department of Chemistry
Shri Ram Murti Samarak College of Engineering and Technology
Bareilly, Uttar Pradesh, India

Vishwajit Chavda
Chemical Engineering Department
Gujarat Technological University
Valsad, Gujarat, India

Neha Choudhary
Department of Chemistry
School of Sciences, IFTM University
Moradabad, Uttar Pradesh, India

Meenu Devi
Department of Chemistry
Banasthali Vidyapith
Jaipur, Rajasthan, India

Serajul Haque Faizi
L.S. College
A Constituent Unit of B.R.A. Bihar University, Muzaffarpur
Muzaffarpur, Bihar, India

Sushmita Gajurel
Department of Chemistry
University Centre for Research and Development, Chandigarh University
Mohali, Punjab, India

Mansi Gandhi
Department of Chemistry
Gujarat Technological University
Chandkheda, Ahmedabad, Gujarat, India

Piyush Gupta
Department of Chemistry, Faculty of Engineering and Technology
SRM Institute of Science and Technology, Delhi NCR Campus
Modinagar, Ghaziabad, Uttar Pradesh, India

Mohamed M. Hammam
Faculty of Science
Ain Shams University
Cairo, Egypt

Pallavi Jain
Department of Chemistry, Faculty of Engineering and Technology
SRM Institute of Science and Technology, Delhi NCR Campus
Modinagar, Ghaziabad, Uttar Pradesh, India

Bhawana Jangir
Department of Chemistry
JECRC University
Jaipur, Rajasthan, India

Jasvinder Kaur
Department of Chemistry
School of Sciences, IFTM University
Moradabad, Uttar Pradesh, India

Tarnveer Kaur
Department of Basic & Applied Sciences
Sri Guru Granth Sahib World University
Fatehgarh Sahib, Punjab, India

Vishnu Kumar Khandelwal
Department of Chemistry
JECRC University
Jaipur, Rajasthan, India

Rajesh Kumar
R.D.S. College
A Constituent Unit of B.R.A. Bihar University, Muzaffarpur
Muzaffarpur, Bihar, India

Rajesh Kumar Meena
Kalindi College
University of Delhi
Delhi, India

Amrutha V. Nambiar
Department of Chemistry
Banasthali Vidyapith
Jaipur, India
and
Amity University Dubai
Dubai, UAE

Amani F.H. Noureldeen
Faculty of Science
Ain Shams University
Cairo, Egypt

Ramadan M. Ramadan
Faculty of Science
Ain Shams University
Cairo, Egypt

Shashi Ranga
Chemical Engineering Department
Government Engineering College-Valsad
Valsad, Gujarat, India

Manisha Rawat
Department of Chemistry
Banasthali Vidyapith
Jaipur, Rajasthan, India

Shobhit Saini
Department of Chemistry
SRM Institute of Science and Technology, Delhi NCR Campus
Modinagar, Ghaziabad, Uttar Pradesh, India

Aruna Sharma
Department of Chemistry
JECRC University
Jaipur, Rajasthan, India

Rekha Sharma
Department of Chemistry
Banasthali Vidyapith
Jaipur, Rajasthan, India

Vaibhav Sharma
Department of Chemistry
SRM Institute of Science and Technology, Delhi NCR Campus
Modinagar, Ghaziabad, Uttar Pradesh, India

Priyanshi Shukla
Department of Chemistry
Banasthali Vidyapith
Jaipur, Rajasthan, India

Avtar Singh
Research and Development
Molekule Inc.
Tampa, Florida
and
Department of Chemistry
Sri Guru Tegh Bahadur Khalsa College
Anandpur Sahib, Punjab, India

Shubhangi Singhal
Department of Chemistry
Banasthali Vidyapith
Jaipur, Rajasthan, India

Shivanshi Tandon
Department of Chemistry
School of Sciences, IFTM University
Moradabad, Uttar Pradesh, India

Khushboo Verma
Department of Chemistry
Sri Guru Granth Sahib World University
Fatehgarh Sahib, Punjab, India

Anjali Yadav
Department of Chemistry
JECRC University
Jaipur, Rajasthan, India

1 Introduction of Covalent Organic Frameworks

Mansi Gandhi, Vishwajit Chavda, and Shashi Ranga

1.1 INTRODUCTION

Materials science research focuses on tailoring porous frameworks from basic molecular building components. Scientifically developed, permeable, and chemically strong nanostructured components contain enormous potential for enhancing the effectiveness of electrical equipment (M. Zeng et al. 2018), chemical sensors (Varghese et al. 2015), catalysts (R. Dong et al. 2015), energy storage gadgets (Sheberla et al. 2017), and many more (Chavda, Hirpara, and Kumar 2022; Chavda et al. 2023; Chavda, Hirpara, et al. 2024; Chavda, Patel, et al. 2024; Shah et al. 2024). Reticular chemistry is a novel branch that arises from the controlled joining of molecular building fragments by strong bonds, producing prolonged crystalline structures. Metal-organic frameworks (MOFs) and covalent organic frameworks (COFs) are synthesized using this method (Mishra et al. 2024, 2025). The exact control at the molecular level allows for the construction of open frameworks in COFs and MOFs.

A type of porous crystalline material termed COFs is formed whenever organic components are attached via covalent bonds to each other in a specific pattern, forming unique two-dimensional (2D) and three-dimensional (3D) polymeric frameworks. Strong covalent bonds connect light atoms like C, B, N, and O to form the backbone structure of COFs (Raghav et al. 2025). In 2005, Yaghi et al. postulated the reticular production of COFs (Côté et al. 2005). Their modular design makes them extremely adaptable materials for various uses by providing exact control over pore size, topology, and performance.

There are many remarkable points about COFs (Guan et al. 2020), including their large surface area, numerous pores, low display density due to their light components, and high durability. COFs are better than MOFs in several ways, such as having a lower density, a unique mix of chemical modularity, structural diversity, stability, and thermal stability (Ma and Scott 2018). Large pore sizes up to 4.7 nm, 4,210 $m^2 g^{-1}$ surface area, 0.17 g cm^{-3} ultralow density, and 8.1 cm^2 Vs high charge carrier mobility have all been demonstrated by the frameworks created by rationalizing the design of the COFs using the principles of reticular chemistry and improving control over the modifications of their pore measurements (Diercks and Yaghi 2017). These attributes render suitable material significant potential for applications such as the delivery of medicines (Mitra et al. 2017), gas storage (Zhongping Li et al. 2014), gas separation (Kang et al. 2016), catalytic processes (Zhi et al. 2017), and lithium-ion batteries (Du et al. 2016).

DOI: 10.1201/9781003605386-1

The synthesis of COFs concentrates on postsurface functionalization, structural regularity, and permeability to modify unique qualities for particular applications (Evans et al. 2018; Wang et al. 2018; Liu et al. 2019). The most significant difficulty with building an extremely crystalline, consistent network of porous polymeric material is maintaining atomic-level regulation of its structure and composition via appropriate synthetic pathways. The regulation of pore dimensions is another constraint in the synthesis of COFs. The two categories of the COFs are 2D and 3D. Which fundamental building pieces are used affects whether the COF's structure is 3D or 2D. For instance, the tetrahedral building blocks may make the 3D COFs, whereas the linear building blocks can produce the 2D COFs (Uribe-Romo et al. 2009; Lin et al. 2018). The exfoliated COF structures are like graphene, with flexible atomic frameworks and periodic channels (Wang et al. 2017). The potential for charge-carrier transferability, organized porous structure, and enormous surface area has made 2D COFs new candidate materials for various applications (Zhan, Chen, and Zhang 2017). Bulk COFs can be exfoliated to make COF nanosheets (CONs) (Y. Peng et al. 2017).

In the energy industry, 2D COF materials exhibit potential for novel nanomaterials. For instance, the shuttle effect of ions in common electrode materials results in low cycle life and battery capacity loss. A conceivable way to get around the ion shuttle effect is to develop ionic sieves using 2D COFs with uniform and controlled nanopores (Zhan, Chen, and Zhang 2017). 2D materials can alleviate the trade-off between permeability and selectivity in separation applications. There is a chance to create a flawless membrane with improved flux and separation efficiency because of the COFs' great compatibility with the polymers. Unlike graphene, the inherent nanopores in 2D COFs function as molecular transport channels (Peng et al. 2017). Organic layers are stacked to form 2D COFs with desired characteristics. These materials can store gases like NH_3 and H_2 because of their large surface areas and diverse pore diameters (Huang et al. 2021). The material's conductive or semiconductive behavior is caused by the electronic interactions between the various sheets developed by the 2D COFs because of their small interlayer distance and π-conjugated (Furukawa and Yaghi 2009; J. Li et al. 2019).

CONs are a novel type of thin, porous 2D nanostructures that are readily functionalized and designed. Layer restacking, problematic post-synthetic alterations, and poor dispersion are the main obstacles to the scale production of CON thin films. Moreover, the exfoliated CONs have low dispersion stability in various solvents, which makes it more difficult to fabricate thin films using drop casting, dip coating, and spin coating (Wan et al. 2011). The high specific area of 3D COFs draws attention to them and 2D COFs. However, the collapsing 3D porous network makes it difficult to synthesize 3D COFs, especially new ones (Wan et al. 2008; Khayum et al. 2016; Li et al. 2017). Porphyrin linkages (Lin et al. 2017; Yahiaoui et al. 2018), silicate (Fang et al. 2015), boroxine (Messner et al. 2013; Baldwin et al. 2016), and imides (C. Wu et al. 2018) are employed for the development of claimed 3D COFs (Figure 1.1).

However, compared to 2D-structured COFs, which can be created by self-assembly and straightforward chemical-based processes, the development of 3D-structured COFs remains limited since it primarily takes place through elevated temperature solvothermal techniques (X. Guan et al. 2018). Since any reaction may be readily fixed, synthesizing COFs via dynamic covalent chemistry (DCC) leads to the development

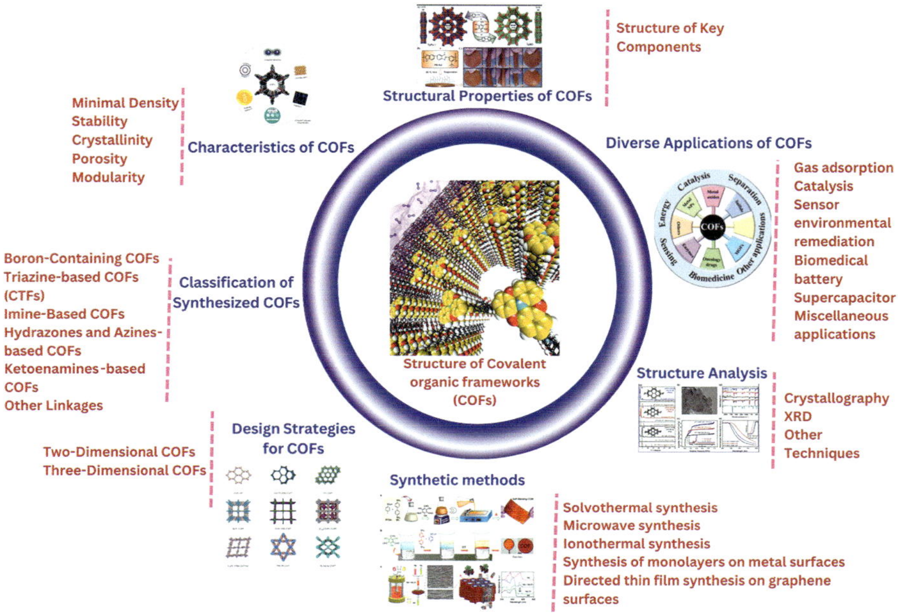

FIGURE 1.1 Overview of COFs—structure, properties, synthesis, and applications.

of arranged and crystalline structures. COFs may also be readily functionalized and combined with various materials (Feng et al. 2021). The COFs are classified into three types based on the linkages: (a) COFs connected to boronate ester, (b) COFs linked to imines and hydrazones, and (c) frameworks based on covalent triazine that includes cyano (Qian et al. 2018).

According to building units and structural characteristics, COFs may be classified as imine, triazine, or boron-based (Ding and Wang 2013; Qian et al. 2018). The C–N connections are one of the main groups of COFs. Developing COF materials using imine bonding has become a flexible approach. Aldehyde-amine co-condensation and hydrazone-linked aldehyde-hydrazide formation can create imine-based COFs (Yue et al. 2015). By condensing organic building blocks of hydrazine and aldehyde, Uribe and the team created two novel COFs, COF-43 and COF-42, which result in hydrazone-linked structures. The hydrazone connections' pH-dependent reversibility regulated these COFs' crystallinity. These COFs have great chemical stability, persistent porosity, and a wide surface area (Uribe-Romo et al. 2011).

Nitrogen-containing heterocycles consist of COFs based on triazine, which are also COFs containing nitrogen. The triazine functionalized, hexagonally organized covalent organic polymer was created by Gomes and team using the Schiff base condensation process between 1,3,5-tris-(4-aminophenyl)triazine and terephthaldehyde. The surface area of the produced triazine-based COF is $716\,m^2g^{-1}$ (Gomes, Bhanja, and Bhaumik 2015). The B-O links are identified as belonging to a different class of COFs.

A COF was developed by Yaghi and team via the condensation methods of phenyl diboronic acid $C_6H_4[B(OH)_2]_2$ and hexahydroxytriphenylene $[C_{18}H_6(OH)_6]$

(Côté et al. 2005). B-O connections are highly common and widespread in generating COFs, and various building blocks are accessible to vary their structure (M.-X. Wu and Yang 2017). The COF may be constructed using a wide variety of links in addition to the ones listed above, including the triazine lineage (Guo et al. 2013), olefin lineage (Zhuang et al. 2016), polyimide (Zhongping Li et al. 2014), benzobisoxazole family (Pyles et al. 2016), and squaraine type (Nagai et al. 2013). Although many other links may generate COFs, azine, imine, phenazine, squaraine, and hydrazone-based connections are the most frequently employed (Huang, Wang, and Jiang 2016a).

However, stability will be minimally influenced by reversibility if the reaction scenarios are low. Because of their chemical and thermal breakdown during the process, stable COFs will only have limited functional groups bound to them if they are synthesized at a higher temperature. To get around this conundrum and create a stable, crystalline COF with the requisite functions, some design approach is thus needed. The stability and crystallinity of COFs have been enhanced by several novel synthetic techniques that allow for functionalities incompatible with reversible processes. Reversible reactions, single reaction routes, and pre-orientation of building blocks are the three methods that may be used to create crystalline COFs. A crystalline COF can be obtained by conjugation using techniques (Haase and Lotsch 2020).

This chapter delves into the structural designs, functional modifications, and advancements of COFs, highlighting their key properties and applications in gas separation and storage, catalysis, optoelectronics, energy storage and conversion, drug-based delivery, sensing, and water purification.

1.2 STRUCTURAL PROPERTIES OF COFs

In past centuries, polymer-based materials have been developed based on kinetically regulated procedures, resulting in irreversible covalent connections. The crystallization of connected organic polymers is challenging when using irreversible methods. Conversely, reversible covalent bonds that may be produced, broken, and regenerated are advanced by DCC (Rowan et al. 2002). Unlike the typical formation of covalent bonds, DCC is thermodynamically controlled, resulting in a reversible process with "proofreading" and "error checking" characteristics that create thermodynamically stable structures. COF synthesis uses the DCC principle concurrently with the crystallization process to create a polymer skeleton. By reducing structural anomalies, self-healing signals enable the creation of an ordered structure (Feng, Ding, and Jiang 2012a). The product has a well-defined crystalline structure and is very thermodynamically stable. Two major factors need to be considered when developing and producing COFs: the structure of the essential components and the synthetic process to preserve thermodynamic control in reversible processes involving reaction conditions and reaction medium (Rowan et al. 2002).

1.2.1 Structure of Key Components

The structure of the essential elements must meet two fundamental requirements to have a crystalline and controlled COF (Feng, Ding, and Jiang 2012a):

1. There must be no irreversible reaction in the production of COF.
2. In COF, the form of important components must be carefully preserved.

Key components must have groups that are reactive enough to initiate the formation of dynamic covalent bonds; in other words, there must be no irreversible side reactions, and, in the presence of thermodynamic circumstances, the reaction system should only include exchangeable monomers, polymers, and oligomers. Building blocks must be rigid and conformational, and bond formation should follow a certain orientation. Several reversible procedures have been studied to satisfy the first requirement for the effective synthesis of COF. Most of the known structures have a boron-based chemistry. Dialcohols and boric acids self-condense to generate boronate ester linkages with five and six members, respectively (Côté et al. 2005). A reaction between boronic acids and acetonide-protected catechols produced COFs in the presence of the Lewis acid catalyst BF_3OEt_2 (Spitler and Dichtel 2010). Avoiding oxidation and solubility problems with the building elements—especially large aromatic dialcohols—is made easier with this in situ deprotection method. Like boroxine or boronate esters, COFs have been made by forming borosilicate linkages by tetraboronic acid condensation with tert-butylsilanetriol (Hunt et al. 2008). Boron-based COFs are susceptible to impact and can react with airborne water vapors, even though they can withstand heat (Lanni et al. 2011).

New COF connection chemistries were created by Yaghi's group (Uribe-Romo et al. 2011). The cyclical trimerization of cyano groups produced covalent triazine-based frameworks (CTFs) under ionothermal conditions (Kuhn, Antonietti, and Thomas 2008). Because of the restricted reversibility of the trimerization reaction, CTFs often have reduced crystallinity even when they have a high degree of conjugation and outstanding thermal, mechanical, and chemical stabilities (Bojdys et al. 2010).

1.3 CHARACTERISTICS OF COFs

The building blocks' size, shape, and effectiveness may be controlled to produce the conceptual model and the atomic layer structure. In topology diagrams, the characteristics of COFs might be resolved by using appropriate knot and binder structures (Figure 1.2). COFs produced possess unique characteristics distinctively absent from all other materials (Song et al. 2019).

1.3.1 Minimal Density

In addition to having light components, COFs store energy and guest molecules with exceptional gravimetric performance. COF-108 has a density of 0.17 g cm^3, lower than any other crystalline solid (El-Kaderi et al. 2007).

1.3.2 Stability

Recent developments (low amine bond polarity, hydrogen bond contacts, keto-enol tautomerization, Michael addition–elimination/benzoxazole mechanism) have

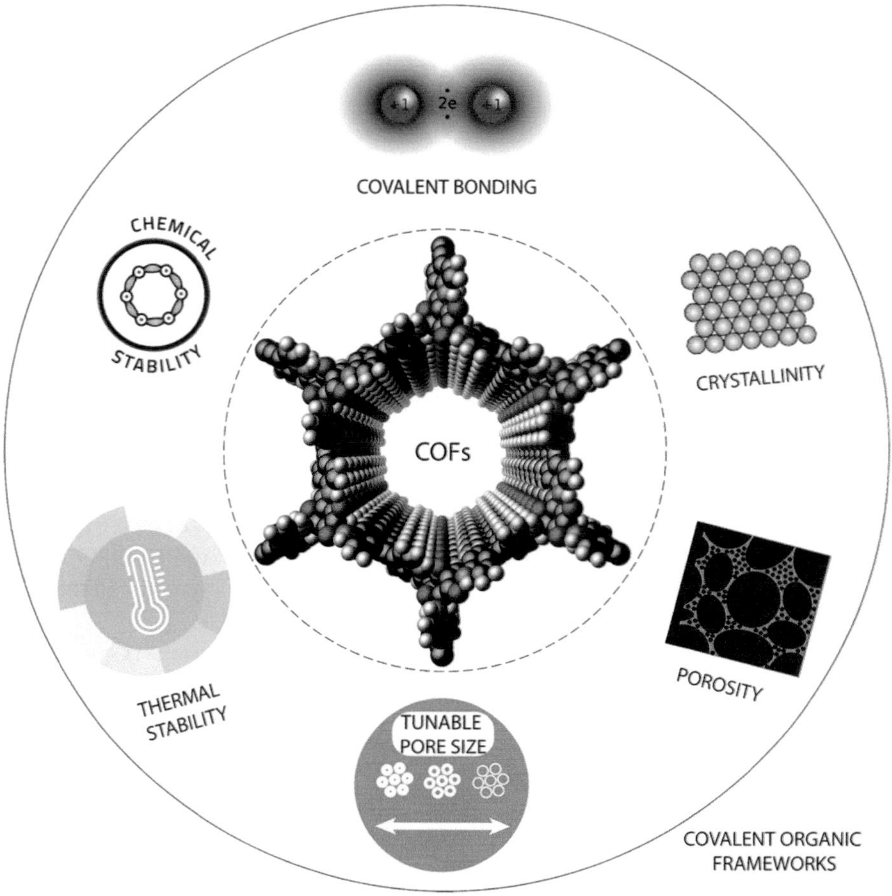

FIGURE 1.2 The most important properties of COFs are decisive for the potential of these materials in versatile applications. (Reproduced with permission from Zhao, Pachfule, and Thomas (2021). Copyright 2024. Royal Society of Chemistry.)

reinforced COF structures (Xinle Li et al. 2018), making them more stable than most MOFs due to their strong covalent connections. These methods of stabilizing COFs result in hydrophobic materials that can withstand a wide range of pH and are suitable for use in reductive and oxidative conditions—all of which have never been shown in a MOF before.

1.3.3 Crystallinity

Since they are crystalline materials, COFs provide highly controlled and consistent functional group organization. This enables the use of diffraction methods for characterization and defines the structural connectivity. Catalysis and optoelectronic devices may potentially benefit from this structural consistency (Song et al. 2019).

1.3.4 Porosity

The regular and systematic permeability of COFs allows for greater gas separation and catalysis performance, where a thorough approach to pores is usually required. Surface areas over 3,000 and 5,000 m^2g^{-1} were measured for 2D and 3D COFs, respectively (Song et al. 2019).

1.3.5 Modularity

With careful building unit selection, the flexible properties of COFs might be changed before synthesis. This allowed scientists to control the composition and structure of a porous and crystalline material while retaining control over the active site's temporal organization, chemical functionality, and mass/volume. New strategies, including metal coordination, integrating monomers with decreased symmetry, and various bond formation processes, have made it possible to manage COF structures more intricately (Zhang et al. 2018). COFs may be built at three structural levels: skeletal model, pore design, and complementary skeleton and pore model, depending on the necessary features (Mellah et al. 2018). For directed molecule manufacturing, which has still not been fully examined, COFs offer appealing principles.

1.4 CLASSIFICATION OF SYNTHESIZED COFs

COF-based materials have been developed utilizing various building blocks and ideal reaction conditions. According to their structural features, the reported COFs are separated into three categories (Table 1.1) (Ding and Wang 2013).

TABLE 1.1
Overview of Various Types of COFs with Their Building Blocks, Synthesis Methods, and Applications

Type of COFs	Building Blocks	Synthesis Methods	Applications	Ref.
Two-dimensional COFs	Planar aromatic linkers	Solvothermal, mechanochemical	Sensing, gas storage, catalysis	Kang et al. (2016)
Three-dimensional COFs	Rigid 3D organic linkers	Ionothermal, solvothermal	Drug delivery, energy storage	Kuhn, Antonietti, and Thomas (2008)
π-Conjugated COFs	π-extended linkers (e.g., porphyrins)	Polymerization under solvothermal conditions	Optoelectronics, photocatalysis, conductive materials	G. Lin et al. (2017)
Hydrazone-linked COFs	Hydrazone (-C=N-N=C-) bonds	Schiff base condensation	Chemical sensing, water purification	Stegbauer, Schwinghammer, and Lotsch (2014)
Imine-linked COFs	Imine (-C=N-) bonds	Dynamic covalent chemistry, solvothermal	Gas separation, heterogeneous catalysis	Pandey et al. (2010)

(Continued)

TABLE 1.1 (*Continued*)
Overview of Various Types of COFs with Their Building Blocks, Synthesis Methods, and Applications

Type of COFs	Building Blocks	Synthesis Methods	Applications	Ref.
β-Ketoenamine COFs	β-Ketoenamine (-C=C-C=O) linkages	Acid-catalyzed synthesis	Energy storage, supercapacitors	DeBlase et al. (2013)
Boroxine/Boronate Ester COFs	Boroxine/ Boronate ester (-B-O-B-) linkages	Thermally assisted reversible reactions	CO_2 capture, gas adsorption	Hunt et al. (2008)
Thioether-linked COFs	Sulfur-containing linkages (-S-)	Thiol-ene click chemistry	Heavy metal adsorption, ion exchange	S.-Y. Ding et al. (2016)
Triazine-based COFs	Triazine (-C_3N_3) core	Ionothermal, microwave-assisted synthesis	CO_2 reduction, electrocatalysis	Gomes, Bhanja, and Bhaumik (2015)
Azine-linked COFs	Azine (-C=N-N=C-) bonds	Schiff base reaction under solvothermal conditions	Molecular separation, adsorption	Dalapati et al. (2013)
Porphyrin-linked COFs	Porphyrin-based macrocycles	Condensation reactions under solvothermal conditions	Photocatalysis, artificial photosynthesis	Liao et al. (2016); Cui et al. (2019)
Aminophenol-linked COFs	Aminophenol (-NH_2 and -OH) functional groups	Oxidative polymerization	CO_2 adsorption, gas separation	F.-Z. Cui et al. (2019)
Amide-linked COFs	Amide (-CONH-) bonds	Thermally assisted polymerization	Biomedical applications, gas separation	Zhan et al. (2017)

1.4.1 Boron-Containing COFs

Following the discovery of COF-5 and COF-1 by Yaghi and colleagues, boron-containing COFs, which are produced via boronate anhydride or boronate ester synthesis, have garnered much attention from researchers (Côté et al. 2005). The synthesis of these COFs places them into two groups. The 1,4-benzene boronic acid reaction, which yielded 2D COF-1, is an example of a class resulting from mono-building block self-condensation (Côté et al. 2005). The co-condensation of boronic acid and diol produces the second class, a five-membered C_2O_2B ring formed by dehydration (Wan et al. 2009). Layered eclipsed structures frequently occur in 2D COFs, but tetrahedral building blocks can create complex 3D COFs. By altering the combinations of boronic acid and diol, this co-condensation technique's adaptability allows the production of COFs with various features.

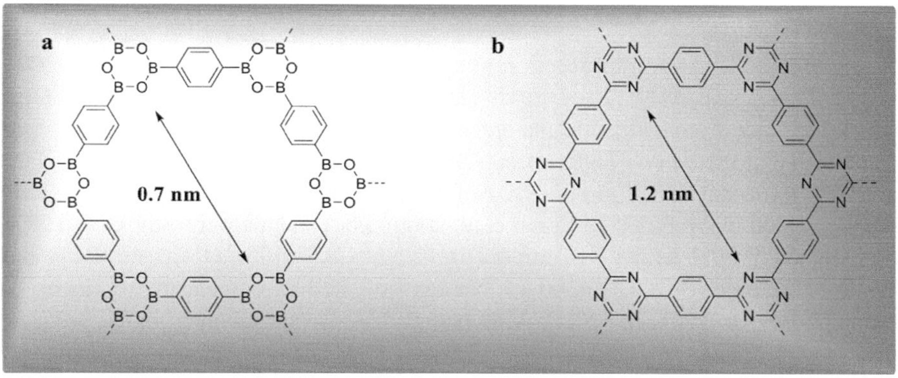

FIGURE 1.3 (a) Boron-containing COF. (b) Triazine-based COF (CTF). (Reproduced with permission from S.-Y. Ding and Wang (2013). Copyright 2024. Royal Society of Chemistry.)

For example, monomer co-condensation generated the 2D HHTP-DPB COF, which has a 4.7 nm pore size (Spitler et al. 2011). COF-202, which was produced by co-condensing tetra(4-dihydroxyborylphenyl)methane and tert-butyl silane triol [t-BuSi$(OH)_3$] to produce borosilicate-linked structures, also demonstrates that functional tuning is feasible (Lanni et al. 2011). In borosilicate-linked COFs, every –$B(OH)_2$unit interacts with distinct triols, unlike boronic acid dehydration, which forms C_2O_2B five-membered rings (Bunck and Dichtel 2012). It is also possible to use ^1H NMR spectroscopy to study the integration of monomers in COF materials (Figure 1.3).

1.4.2 Triazine-Based COFs (CTFs)

CTFs are a class of COFs identified by Thomas and associates (Bunck and Dichtel 2012). At 400°C, nitrile building blocks undergo cyclotrimerization with $ZnCl_2$ to develop CTFs. For example, the CTF-1 material has a 1.2 nm pore size and a 791 $m^2 g^{-1}$ Brunauer–Emmett–Teller (BET) surface area. It is created by cyclotrimerizing 1,4-dicyanobenzene (DCB). Like COF-1, the shape of CTF-1 is a 2D hexagonal structure, according to the powder X-ray diffraction (PXRD) design. An amorphous polymer with a larger surface area (1,123 $m^2 g^{-1}$) is synthesized by a higher $ZnCl_2$ (10:1) ratio. 2,6-dicyanonaphthaline was also used to effectively manufacture the supercrystalline CTF-2 material, which had a surface area of 90 $m^2 g$ BET (Bojdys et al. 2010). The material layers' irregular arrangement might explain the lower surface area of CTF-2. CTF-2's surface area increased when the $ZnCl_2$ ratios increased (Palkovits et al. 2009). Despite having superior chemical and thermal stability, CTFs are less crystalline than COFs containing boron. There is potential for using CTFs as catalyst supports because of their high nitrogen concentration (Chan-Thaw et al. 2011).

1.4.3 Imine-Based COFs

Yaghi and colleagues introduced imine-based COFs, a new category of COFs (Uribe-Romo et al. 2009). They are divided into two types: hydrazone-linked COFs, which are made from hydrazide and aldehyde, and Schiff base COFs, which originate from amine-aldehyde co-condensation. With a 3D diamond-like structure, a surface area of 1,360 m^2g^{-1} BET, and a pore size of 7.8 Å, COF-300, the first imine-based COF, was described in 2009 (Uribe-Romo et al. 2009). For catalytic purposes, COF-LZU1 was created in 2011 with 1,4-diaminobenzene and 1,3,5-triformylbenzene (Ding and Wang 2013). Since then, COF-366 has been investigated for optoelectronics, incorporating porphyrin (Wan et al. 2011). After being created using solvothermal techniques, hydrazone-linked COFs, namely COF-42 and COF-43, showed 2D-eclipsed structures with BET surface areas of 710 m^2g^{-1} and 620 m^2g^{-1}, respectively. A large amount of H_2 bonding contributes to their stability (Schwab et al. 2009; Uribe-Romo et al. 2011).

Crystallinity is equivalent to COFs containing boron, and imine-based COFs perform better in terms of functional consistency than CTFs. Their future uses will be expanded by the metal coordination made possible by their nitrogen framework (Pandey et al. 2010).

Amorphous porous polymers with Schiff base chemistry have surfaced despite the scarcity of imine-based COFs. To demonstrate the significance of reaction optimization for synthesizing new COFs, Nguyen and colleagues synthesized such polymers under changed circumstances utilizing a 1,3,5-triformylbenzene co-condensation technique (Schwab et al. 2009; Ding and Wang 2013).

1.4.4 Hydrazones and Azine-Based COFs

Yaghi's team used the reversible condensation of hydrazides with aldehydes to make hydrazones for the first time in 2011 when synthesizing COF. Using 2,5-die thoxyterephthalohydrazide and TFB, they created COF-42 and then used the same hydrazide with an extended aldehyde linker, TFPB, to create COF-43 (Uribe-Romo et al. 2011). In sensing (Das et al. 2015; Ding et al. 2016) and catalysis (Stegbauer, Schwinghammer, and Lotsch 2014; Liu et al. 2017), hydrazone-linked COFs have been extensively studied despite being a small family. Dalapati et al. initially showed that hydrazine may create azine linkages when it combines with aldehydes in the Py-Azine COF synthesis from hydrazine and 1,3,5,8-tetrakis(4-formylphenyl)pyrene. This COF's azine units and rhombic holes (~2 nm) act as docking sites for guest molecules, improving its sensing capabilities (Liu et al. 2017). Different geometries are made possible by azine-linked COFs, in contrast to conventional hexagonal pores.

For example, researchers produced triangular holes as tiny as 1 nm using hydrazine and hexaphenylbenzene aldehyde (Alahakoon et al. 2016). Tritopic linkers with selective elongation at one binding site were incorporated into HP-COF-1 and HP-COF-2 to create a dual-pore structure (Youlong Zhu et al. 2015). Using TPG as an aldehyde and 4-aminobenzohydrazide and 4-amino-2-hydroxybenzohydrazide as amine/hydrazine-functionalized building blocks, Mitra et al. created COFs that integrated ketoenamine and hydrazone linkages (Mitra et al. 2017). Furthermore, in situ deprotection of aldehyde dimethyl acetals was used to create hydrazine and azine-linked COFs (Seoa Kim et al. 2025). Since TPG is the aldehyde linker in imine-based networks, enol-keto tautomerization has also been used for hydrazone and azine COFs (Dalapati et al. 2013; Plas et al. 2016).

1.4.5 Ketoenamine-Based COFs

In COF applications, stability improvement is a critical problem. TPG was used as an aldehyde linker in a 2012 technique by Banerjee et al. to create COFs resistant to strong bases and acids. This building block permits the production of imine-linked COFs, which subsequently go through irreversible enol-keto tautomerization to create ketoenamine connections since it includes hydroxyl groups next to formyl groups. These β-ketoenamine COFs are very stable chemically, even against strong bases. However, the irreversible procedure that prevents lattice error correction frequently results in their crystallinity being lower than that of imine COFs (Biswal et al. 2013).

In addition, Banerjee's group created a structural series that included various amines in β-ketoenamine-based COFs. These could be mechanochemically delaminated simply by a grinding mortar, resulting in nanosheets that were just a few layers thick (Chandra et al. 2013). Low crystallinity ketoenamine COFs were also produced via mechanochemical synthesis using linkers and TPG grinding (Kandambeth et al. 2012). Their surface areas were quite small (~70 m^2g^{-1}), but solvothermal synthesis yielded much larger surface areas (>300 m^2g^{-1}) (Figure 1.4).

Numerous uses have been made possible by the exceptional stability of ketoenamine COFs, such as proton-conducting electrolytes (Chandra et al. 2014; Chandra et al. 2016; Ma et al. 2016; Shinde et al. 2016) and reversible redox reactions in acidic electrolytes for energy storage (DeBlase et al. 2013, 2015). We have recently shown that β-ketoenamine COFs may be used in lactic acid adsorption by undergoing nitro-to-amino group reduction in an acidic environment (Lohse et al. 2016). A 3D β-ketoenamine COF was created by Yan et al. using 1,3,5,7-tetraaminoadamantane (Fang et al. 2014). To produce intramolecular hydrogen-bond-stabilized β-ketoenamines, Rao et al. presented an alternate method that uses Michael addition–elimination to condense amines with β-ketoenols. The ketoenamine linkage motif may be synthesized directly in a single step using this technique (Rao et al. 2017).

1.4.6 Other Linkages

Additional links have been created beyond the COF coupling chemistries previously addressed. A 3D COF was created by Yaghi's group in 2008 using borosilicate bonding between tert-butylsilane triol and tetrahedral boronic acid (Kahveci et al. 2013). Later, Kuhn et al. developed ionothermal nitrile trimerization, which turns dicyanobenzenes into crystalline triazine frameworks (Spitler and Dichtel 2010; Ding et al. 2012; Colson et al. 2015). Most CTFs are amorphous, although they are highly stable and have uses in gas separation (Puthiaraj et al. 2016; Ying et al. 2016), Li–S battery polysulfide sorption (Ghazi et al. 2016), and catalysis.

Borazine-linked COFs (thermal breakdown of aminophenylbenzene-borane) (Jackson, Reich, and El-Kaderi 2012), amine-linked TriCF (melamine + cyanuric

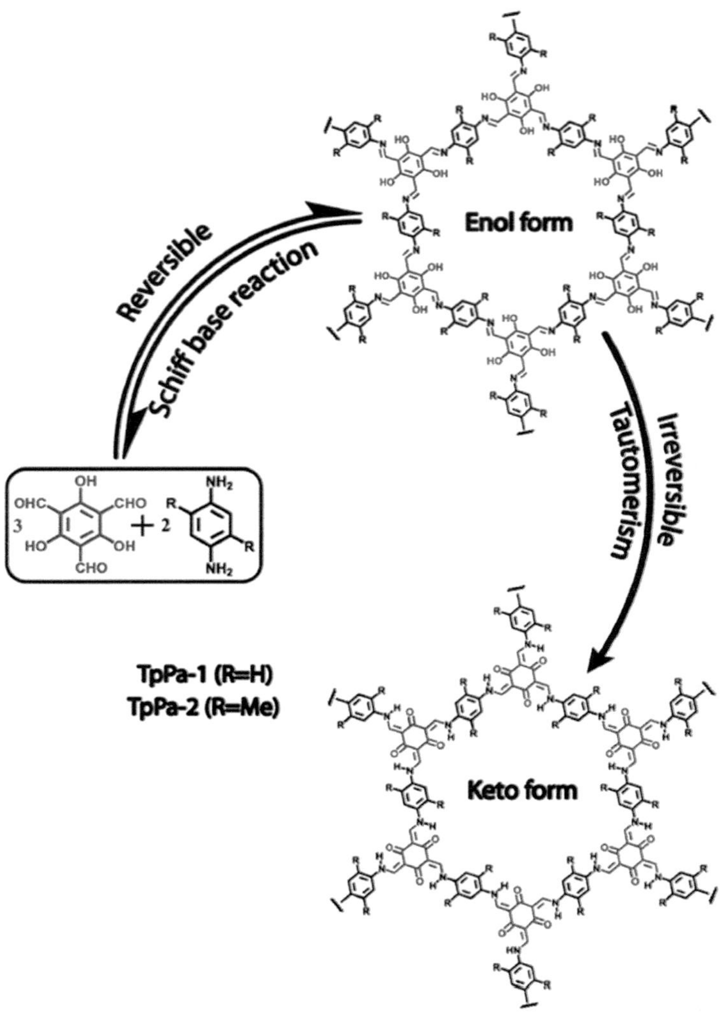

FIGURE 1.4 Schematic illustration of β-ketoenamine COF synthesis through enol-keto tautomerization. (Reproduced with authorization Kandambeth et al. (2012). Copyright 2012, American Chemical Society.)

chloride) (Zhang et al. 2023), and squarine-linked COFs (squaric acid + aromatic amines) (Nagai et al. 2013; Yuanzhi Zhu et al. 2017) are further links. Benzimidazole-linked COFs were created from diaminobenzidine and TPG (Nandi et al. 2016), whereas benzobisoxazole-linked COFs were formed via trigonal aldehyde condensation (Nath et al. 2016).

Azole-linked COFs (self-polymerized tetrakis(arylnitroso)porphyrin) (Nath et al. 2016), spiroborate-linked COFs (diol-functionalized macrocycles + trimethylborate) with high gas storage and Li-ion transport properties, and phenazine-linked CS-COFs (triphenylene hexamine + tert-butylpyrene tetraone) (Guo et al. 2013) are some other noteworthy developments. Moreover, the olefin-linked COF 2DPPV (phenylene diacetonitrile + TFPB) was generated via Knoevenagel condensation (Zhuang et al. 2016).

These growing connection patterns serve catalysis, stability, and electrical characteristics. Ketenamines improve durability despite decreased crystallinity, while imines, azines, and hydrazones provide strong stability. Despite being understudied, phenols, benzimidazoles, benzobisoxazoles, and olefins provide conjugation. Linkages that coordinate metals make catalysis easier, whereas nitrogen-rich backbones enhance CO_2 sorption. The structural configuration and linker choice determine the COF characteristics.

1.5 DESIGN STRATEGIES FOR COFs

By directing the expansion of the polymer backbone, the topology design diagram acts as a guide to produce covalent bonds. Rigid backbones with reactive sites organized in certain geometries are necessary for monomers to retain a specified covalent bond orientation. This idea is demonstrated in the block model, where the shape of each monomer dictates the location of the reactive site. The covalent bonding pattern determines the alignment of monomers, which guarantees that chain development adheres to the specified topology. In the x-y plane, planar monomers produce intrinsic nanopores and regularly arranged open polygonal areas by restricting the synthesis of polymer backbones to a 2D plane. While π-systems allow for extra non-covalent interactions that form distinct high-order structures, the strong skeleton of 2D covalent polymers guarantees structural stability.

In 2D covalent polymers, π-units encourage layered stacking because of their rigid structure, where precise interlayer interactions control alignment. This stacking creates completely ordered π-arrays and one-dimensional (1D) open channels with careful monomer engineering and selection, allowing for structural customization of 2D COFs (Figure 1.5).

1.5.1 Two-Dimensional COFs

Polygon skeletons for irregular (anisotropic) and regular (isotropic) topologies have been designed for COF design, allowing 2D COFs to be built with various topologies. The structure of the lattice is highly ordered, and the pore is distinct in every case. The structural diversity of 2D COFs is based on the topological diversity. Massive COF members with various linkers and construction components support each topology.

- **Symmetrical Topologies:** Diverse monomer combinations can create COFs with various skeletons and pore architectures thanks to the topology map, which results in unique framework orderings. For example, $[C_3+C_2]$ produces hexagonal 2D COFs, but $[C_2+C_2+C_2]$ and $[C_3+C_3]$ produce hexagonal COFs with different π-orderings and pore diameters. $[C_4+C_2]$ and $[C_4+C_4]$ combinations give rise to tetragonal COFs (Feng, Ding, and Jiang 2012a; Dalapati et al. 2015).

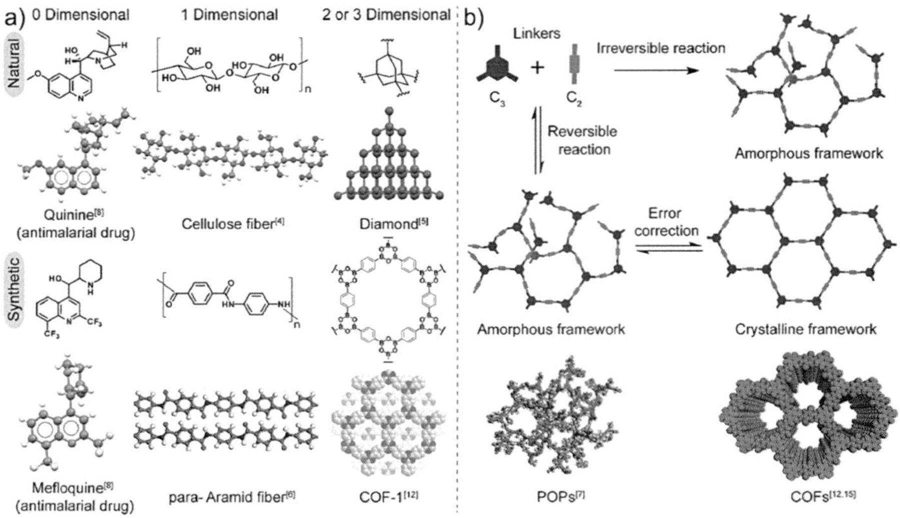

FIGURE 1.5 (a) Examples of natural and synthetic covalent structures in different dimensions. (b) Schematic representation of crystalline and amorphous COF formation by reversible and irreversible reactions. (Reproduced with permission from Côté et al. (2005). Copyright 2024.)

The topology determines the pore size:

1. Microporous COFs are created via the combinations $[C_2+C_2+C_2]$ and $[C_4+C_4]$.
2. There are three types of mesoporous COFs: $[C_3+C_2]$, $[C_3+C_3]$, and $[C_4+C_2]$.

π-density is also affected by topology. Creating microporous COFs with small pore diameters and high π-density is feasible using the $[C_6+C_2]$ structure. Triangular pores are more appropriate for semiconducting and molecular separation applications than tetragonal and hexagonal COFs because of their significantly reduced π-unit density of $2/\sqrt{3}$ L-2 and pore size of $L/\sqrt{3}$ (Feng, Dong, and Jiang 2013; Huang, Wang, and Jiang 2016a; Xinyi Chen et al. 2020). Moreover, dual-pore triangular lattices, such as kagome COFs with one dodecagonal and six triangular pores, may be formed thanks to C_3-symmetric vertices. Another important topology produces two structures, $[C_2+C_2]$: Uniform pore size in rhombic skeletons (Ascherl et al. 2016). Kagome frameworks with six trigonal micropores on the periphery and one core hexagonal mesopore. Topology is determined by the knots and linkers used. Different C_2-symmetric knots generate kagome topologies, but bulky knots with strong π–π interactions prefer rhombic forms. The docking effect of twisted linkers compensates for weak interlayer π–π interactions in kagome COFs, guaranteeing stability (Xiong Chen et al. 2014; Zhou et al. 2014).

- **Asymmetric Topologies:** A knot and linker alternate to generate regular polygons and isotropic tiling patterns in [1 + 1] two-component systems. Nevertheless, this limits COF variation [1 + 1] combination; for example, only 10 COFs may be built with one knot and ten linkers (W. Gao et al.

2015; N. Huang et al. 2016). This restriction can be addressed by employing a multiple-component (MC) approach, which uses two or three distinct linkers in the COF design (Huang, Wang, and Jiang 2016a). A C_3-symmetric knot with 10 C_2-symmetric linkers may make 210 hexagonal COFs, while a [1 + 2] approach produces two separate hexagonal COFs depending on the molar ratio of linkers (1/2 or 2/1).

Increasing structural complexity, the [1 + 3] method produces hexagonal COFs with three edge pairings. The anisotropic tiling and unique hexagonal pores of MC-based COFs contrast with the isotropic tiling of [1 + 1] COFs. This strategy's capacity to combine multiple donor–acceptor units allows for remarkable conductivity beyond a simple sum of [1 + 1] COFs and strong intercolumnar interactions (Pang et al. 2016). Asymmetrically tiled pores are introduced using [1 + 2] systems in tetragonal COFs, where the MC method is extended. A total of 45 tetragonal COFs with unique edges have been created using phthalocyanine knots with 10 linkers. Furthermore, asymmetric hexagonal COF design is possible in a [2 + 1] system with two C_3-symmetric knots of varying sizes and a C_2-symmetric linker (Huang, Wang, and Jiang 2016b; Xinyi Chen et al. 2020).

The creation of HP-COF-1 and HP-COF-2—COFs with two hexagonal holes of varied sizes—by desymmetrized knots, like 5-(4-formylphenyl) isophthalaldehyde, represents a milestone in multi-pore COF building. In TPE-based COFs, the MC approach has also been used to create dual-pore structures in kagome topologies (Spitler and Dichtel 2010; Pachfule et al. 2014; Fang et al. 2015; Li et al. 2019). By condensing TPE knots (4,4',4'',4'''-(Ethene-1,1,2,2-tetrayl)tetrabenzaldehyde (ETTA)) with two dialdehydes in a 1/1 ratio, triple-pore COFs such as SIOC-COF-1 and SIOC-COF-2 are established (Zhu et al. 2015; Crowe, Baldwin, and McGrier 2016).

1.5.2 Three-Dimensional COFs

To direct polymer backbone extension into a covalently connected 3D network, 3D COFs require at least one Td or orthogonal unit, in contrast to 2D COFs (Kang et al. 2016), which require planar construction units. C_1, C_2, C_3, C_4, and Td mix with different Td or orthogonal knots to create interpenetrated polymer chains with unique skeletons and pores. Based on topology, 3D COFs are classified into networks like ctn, bor, dia, srs, rra, and pts (Spitler and Dichtel 2010). Large surface areas and noninterpenetrated skeletons result from the [Td + C_3] diagram, giving birth to the ctn or bor network. Moreover, [Td + Td] provides access to the CTN network. With multifold-interpenetrated 1D microporous channels (0.7–1.5 nm), the dia network, the biggest 3D COF family, is based on the [Td + C_2] diagram (Feng, Ding, and Jiang 2012a; Matsumoto et al. 2018).

SiCOF-5 is the only two-fold interpenetrated example of the srs net from [Td + C_3]. Four reactive sites are needed for two-fold interpenetration in the pts network, which is the outcome of [Td + C_2] or [Td + C_4] (Zhan, Chen, and Zhang 2017; Lan et al. 2018). The interwoven 3D helical COFb-505 is formed by an orthogonal Cu(II)–phenanthroline complex with a C_2-symmetric linker. We have also investigated an uncommon [C_2+C_3] diagram for 3D COFs containing genetic structural units (Lan et al. 2018). Because the chemical factors governing folding are yet unknown, interpenetration makes it impossible for the topology diagram to predict folding independently. 3D COFs are more challenging to predesign or artificially regulate than 2D COFs (Royuela et al. 2019; Li et al. 2022). Despite being primarily microporous, they frequently show lower-than-expected porosity because of multifold interpenetration, underscoring the necessity of improving folding and interpenetration control (Trewin and Cooper 2009; Lan et al. 2018).

1.6 SYNTHESIS METHODS

Forming highly structured covalent networks requires controlling the thermodynamic equilibrium during covalent bond creation. The reaction media and circumstances (temperature, pressure, and template presence or absence) should be considered explicitly to develop thermodynamically stable polymeric crystalline structures. Reaction media, molten metal salts, and mixed solvent systems have been created to offer solvothermal and ionothermal conditions for synthesizing COFs, respectively. COFs have been synthesized rapidly using microwave processes in solvothermal conditions. Unlike these heavy techniques, COF films or monolayers have been investigated by reactions on substrates like graphene sheets and metal surfaces (Figure 1.6).

1.6.1 Solvothermal Synthesis

A common solvothermal technique for the preparation of COF entails putting monomers and a mixture of solvents in a Pyrex tube, degassing it using cycles of freeze-pump-thaw, closing the tube, and heating it to a certain temperature for a predetermined amount of time. The precipitate is collected, cleaned, and vacuum-dried to produce COF powder. The crystal nucleation, reaction rate, solubility, self-healing, and growth characteristics are significant considerations (Figure 1.7) (Côté et al. 2005).

- **Solvent Mixtures and Ratios:** They impact crystallization and framework development. While dioxane–toluene is utilized for borosilicate COFs, THF–methanol, DMAc-o-dichlorobenzene, and dioxane–mesitylene are common solvent combinations for boroxine and boronate ester COFs. Mesitylene–dioxane–aqueous acetic acid is advantageous for hydrazone-based COFs, whereas dioxane–aqueous acetic acid is necessary for imine-linked COFs (Ding et al. 2011; Uribe-Romo et al. 2011). Thermodynamic control relies on maintaining the proper concentration of building blocks. The ability of fully soluble or entirely insoluble precursors to generate COFs effectively requires more research.
- **Temperature Control:** Reaction reversibility, essential for COF crystallization, is guaranteed by an appropriate temperature (Spitler et al. 2011). Reaction temperatures typically vary between 85°C and 120°C, depending

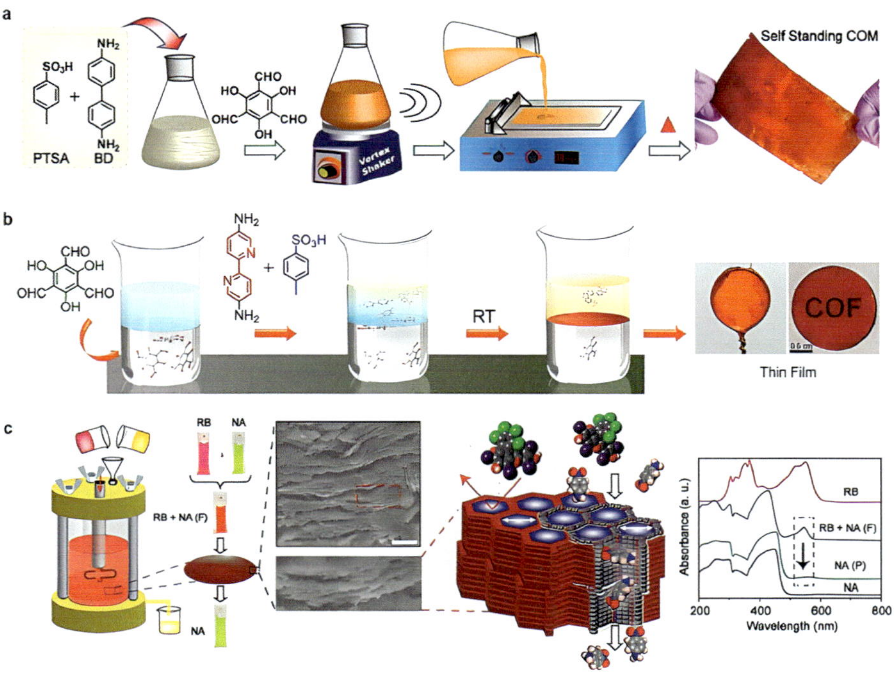

FIGURE 1.6 (a) Diagram depicting the formation of a COF membrane. (b) Schematic representation of interfacial crystallization in COF thin films. (c) Process of selective molecular separation using COF membranes and thin films. A graph illustrating the selective separation of Rose Bengal (RB) from a mixture of RB and nitroaniline (NA). (Adapted with permission from Kandambeth et al. (2017). Copyright 2017 American Chemical Society.)

on the reactivity of the building blocks. Water molecules can affect the reverse reaction; hence, a closed reaction system is required to keep them in place (Tilford et al. 2006).

1.6.2 Microwave Synthesis

Cooper and his team developed a rapid, high-throughput method for dynamic covalent reactions using a microwave reactor to synthesize boronate ester-linked COFs efficiently. Compared to solvothermal techniques, microwave synthesis offers several benefits.

- COFs are quickly produced using microwave synthesis. Large-scale synthesis is, therefore, feasible.
- The synthesis using microwaves does not require a sealed vessel.
- The microwave solvent extraction method improves porosity by eliminating residues and contaminants trapped in the frameworks more effectively. The microwave procedure may offer a potent substitute for the solvothermal approach (Campbell et al. 2009; Dogru et al. 2011).

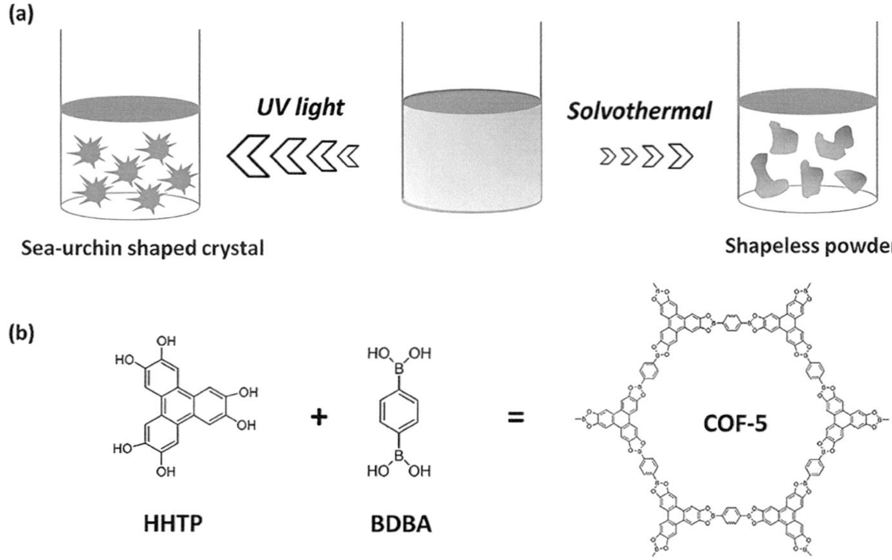

FIGURE 1.7 (a) Schematic of two alternative COF-5 synthesis techniques; UV-COF-5 was synthesized with a sea-urchin-like morphology in a few hours. Conventional th-COF-5 was produced in 72 hours without homogenous morphology. (b) The molecular structure of COF-5, which was produced by an HHTP-BDBA condensation process. (Reprinted with permission from Soyoung Kim et al. (2017). Copyright, 2017, John Wiley.)

1.6.3 Ionothermal Synthesis

To create CTFs, Thomas and colleagues investigated an ionothermal synthesis technique. Crystalline-conjugated CTFs with strong thermal and chemical stabilities are produced by cyclotrimerizing aromatic nitrile building blocks (such as DCB) in molten $ZnCl_2$ at 400°C. For the partly reversible trimerization process, $ZnCl_2$ is the solvent and the catalyst. Building block supply is, however, constrained by such severe reaction conditions. Amorphous materials without long-range molecular order constitute most of the made CTFs (Kuhn, Antonietti, and Thomas 2008; Bojdys et al. 2010).

1.6.4 Synthesis of Monolayers on Metal Surfaces

Building blocks can condense onto a metal surface to create monolayers of COF-1 and COF-5, instead of the bulky production of COFs. SCOF-2 and SCOF-1 nanoarchitectures are covalently bonded when precursors are sublimated onto a pristine Ag surface under extremely high vacuum. Along with little irregular five-, seven-, and eight-membered pores, scanning tunneling microscopy (STM) shows hexagonal pores. Selecting the right single-crystal metal substrate for improved alignment, enhancing precursor purity, and adjusting reaction conditions may all be necessary to produce defect-free monolayers (Feng, Dong, and Jiang 2013; Xiong Chen et al. 2014; Ascherl et al. 2016).

Monolayer COF synthesis may also be conducted on a highly ordered pyrolytic graphite (HOPG) surface. THF solutions were used to deposit 9,9-dihexylfluorene-2,7-diboronic acid, 1,4-benzene diboronic acid (BDBA), and biphenyldiboronic acid onto HOPG. The reactions were then heated to 150°C in a sealed autoclave for one hour. As a water reservoir, $CuSO_4 \cdot 5H_2O$ dramatically increased monolayer coverage, rising from around 7% to 98%. Released water molecules control equilibrium during heating to promote defect repair and create an ordered COF network. $CuSO_4$ reabsorbs water during cooling, preventing SCOF breakdown and guaranteeing structural integrity (Zwaneveld et al. 2008).

1.6.5 Directed Thin-Film Synthesis on Graphene Surfaces

These methods produce COFs that are either insoluble monolayers or powders that are unreliable for incorporation into actual devices or interface to electrodes. Because of this, the process of creating thin films of COF on a substrate is highly technologically relevant and of general scientific interest. On single-layer graphene (SLG on SiO_2) surfaces, Dichtel and colleagues synthesized and characterized layered-orientated 2D COF films (Spitler et al. 2011). However, solvothermal reactions oriented thin films of COF developed on the SLG when SLG/SiO_2 was added. Many COF films have been effectively produced on graphene, including COF-5, TP-COF, NiPc-COF, HHTP-DPB-COF, and ZnPc-PPE-COF. The thickness of these films may be adjusted by varying the reaction time. Grazing incidence XRD studies have shown the thin films' strong crystallinity and vertically aligned layers (Colson et al. 2011).

1.7 STRUCTURE ANALYSIS

1.7.1 Crystallography

The structures of 3D imine-based COFs, including COF-300, its hydrated form, COF-303, LZU-111, and LZU-79, have been determined using single-crystal X-ray diffraction (SXRD) at 0.85 Å resolution. Atomic locations, geometric characteristics (bond angles and lengths), and unit-cell parameters of the space group are among the exact crystal properties carefully determined by SXRD studies. This presents a pathway to a thorough mechanism analysis of COFs by providing a vast opportunity to investigate the structure-related characteristics and the absorption of the guest molecules (Uribe-Romo et al. 2011; Ma et al. 2018).

1.7.2 Diffraction of Powder X-Rays and Theoretical Simulation

Strong signals in 2D or 3D COFs imply significant structural periodicity, which is the primary method used to evaluate the crystallinity of COFs using PXRD (Jin et al. 2013). COF structures may be determined using structural simulations and PXRD pattern matching. By comparing sorption behavior with actual data, grand canonical Monte Carlo simulations confirm porous structures, whereas computational techniques such as density functional tight binding (DFTB) and density functional theory forecast structures. Other methods that track polymerization and crystallization in real time include small-angle and wide-angle X-ray scattering. According to in situ XRD research, COFs break down in hot conditions but hold up well under pressures up to 100 bar (Sun et al. 2020).

DFTB calculations evaluate stacking patterns in 2D COFs, where hexagonal honeycomb structures exhibit AA-eclipsed or AB-staggered stacking. Intermediate slipped AA-stacking modes are often more stable. Stacking energy varies with slip distance, and self-consistent charge-density-functional-based tight binding (SCC-DFTB) calculations help identify the most stable structures (Jin et al. 2013; Xiong Chen et al. 2015). Zigzag and armchair shifts (1.4 Å) optimize Coulomb interactions, while PES calculations suggest a 1.7 Å offset for certain COFs. DFTB with Lennard-Jones dispersion predicts 0.8–1.0 Å offsets. AB-staggered structures exhibit lower stabilization energy due to weaker π–π interactions, sometimes leading to mismatches in simulated and experimental XRD patterns. Comparing PXRD profiles with Pawley refinement ensures accurate space group assignments (Tylianakis, Klontzas, and Froudakis 2011; Koo, Dichtel, and Clancy 2012).

1.7.3 Other Techniques for Structural Assessment of COFs

Along with PXRD, techniques like 3D electron diffraction tomography and 3D rotation electron diffraction using Transmission Electron Microscopy (TEM) or High-Resolution Transmission Electron Microscopy (HR-TEM) enable single-crystal structural analysis of sub-micrometer- or micrometer-sized COF crystals (Peng et al. 2017; Li et al. 2018). Unlike traditional HR-TEM, which can potentially cause structural damage, low-dose TEM with direct-detection electron-counting cameras exposes honeycomb structures by recording structural characteristics at 4 Å resolution. Further supporting the lattice identification transform of TEM images is the Fast Fourier Transform (Figure 1.8) (Colson et al. 2011).

Grazing incidence X-ray diffraction provides information on the preferred orientation and crystallinity of COF films, while STM characterizes monolayer COFs on various substrates. Gas sorption isotherm measurements, fourier-transform infrared spectroscopy (FT-IR), nuclear magnetic resonance (NMR), elemental analysis, X-ray photoelectron spectroscopy (XPS), field emission scanning electron microscopy (FE-SEM), atomic force microscopy (AFM), thermogravimetric analysis (TGA), thermal analysis (TA) spectroscopy, BET surface area analysis, and elemental analysis are other characterization methods. COF particle size in polymerization systems is determined by dynamic techniques such as dynamic light scattering, and the turbidity of the reaction system aids in evaluating the rates of COF generation and the sensitivity of polymerization (S.-L. Cai et al. 2014; Smith and Dichtel 2014; DeBlase et al. 2015).

1.8 APPLICATIONS OF COFs

High crystallinity, a large BET surface area, adjustable pore diameters, and remarkable durability in harsh environments and solvents are just a few of the distinctive physical and chemical characteristics of COFs that set them apart from other porous organic structures. They have great promise for various scientific, technical, and

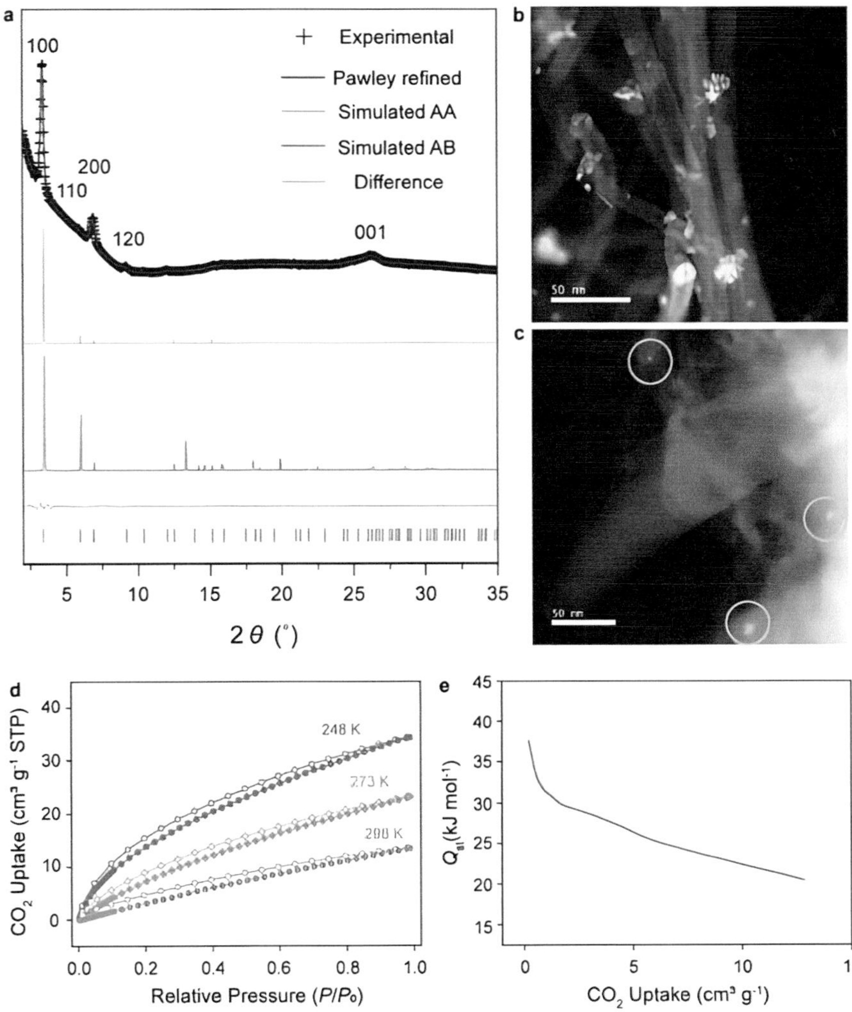

FIGURE 1.8 Characterization of the Tf-DHzOPr COF and COF-Ru@CNT cathodes. (a) Tf-DHzOPr COF indexed PXRD pattern, simulated AA and AB stacking. (b) A dark-field STEM picture of Ru@CNT. (c) A dark-field STEM image of COF-Ru@CNT. (Ru nanoparticles are shown with yellow circles). (c) CO_2 sorption isotherms of COF-Ru@CNT at 248, 273, and 298 K. (e) Adsorption curve for COF-Ru@CNT with isosteric heat. (Reprinted with permission from Xing Li et al. (2019). Copyright, 2024, John Wiley.)

industrial applications due to their adaptability. Many COF structures have been created or altered in less than ten years for sensing, drug release, gas storage, separation, catalysis, selective extraction, ion exchange, photoconductivity, energy storage, electron transport, proton conductivity, and pollutant adsorption. These applications focus on the links between structure and activity.

1.8.1 Adsorption of Gas

Due to their high porosity, low density, and light-element composition, which allowed for good adsorption, COFs were first investigated for gas storage. They have been studied for storing H_2, CO_2, CH_4, and NH_3 throughout the last ten years. COFs provide interesting options for hydrogen storage, which is still a problem. Due to their high surface area and free volume, 3D nature COFs outperform 2D ones. COF-102, for example, rivaled the best MOFs with a 77 mg g^{-1} H_2 uptake at 35 bar and 77 K. Adsorption is further improved by metal doping, as seen in Li-COF-105 (68.4 mg g^{-1}) and Sc/Ti-doped COFs (70.1 mg g^{-1}). Compared to their nonfunctionalized counterparts, functionalized COFs (like COF-320) increased absorption by 35% (Song et al. 2019; Yaghi, Kalmutzki, and Diercks 2019).

Another important use is CO_2 collection, which has been driven by environmental concerns. Surface area and uptake are correlated, with COF-102 reaching 1,180 mg g^{-1} at 55 bars. Stable structures have been created, such as COFs based on imines, azine, and triazine, with ILCOF-1 exhibiting absorption of 1,289 mg g^{-1}. COFs containing nitrogen, such as Cz-COF and Tz-COF, improve CO_2 adsorption, reaching 110 mg g^{-1} and 154 mg g^{-1}, respectively. COFs have also been investigated for collecting SO_2, CH_4, and NH_3. HOOC-COF (containing Sr^{+2}, Ca^{+2}, and Mn^{+2}) showed significant adsorption, whereas COF-10 showed NH_3 absorption of 255 mg g^{-1}. Meso-COF-3's remarkable iodine adsorption (4,000 mg g^{-1} at 75°C) highlighted the adaptability of COFs in gas capture (Xiong Chen et al. 2014; Zhou et al. 2014; Huang, Wang, and Jiang 2016b; Chen et al. 2020).

1.8.2 COFs for Water and Soil Remediation

Soil and water quality are threatened by pollution from industry, agriculture, and human activities. There are serious concerns about inorganic contaminants (heavy metals including chromium and mercury) and organic pollutants (VOCs, PFASs, insecticides, and medications). Despite several treatments, COFs are notable for their stability, porosity, and adaptability. COFs display outstanding adsorption capabilities. The significance of BET surface area (857 m^2g^{-1}) was highlighted by PAF-111, which absorbed 1,666 mg g^{-1} of Rhodamine B, whereas CS-COF-1 removed 1,691 mg g^{-1} of methylene blue, exceeding known porous materials. COFs have outstanding recyclability and effectively remove colors like methyl orange and Congo red (Shengjun Liu et al. 2025).

Using sulfur-rich groups, TPB-DMTP-COF-SH was able to remove 4,350 mg g^{-1} of Hg(II) from hazardous metals. Under five minutes, Ethylenediaminetetraacetic acid (EDTA)-functionalized COFs eliminated Fe^{+3}, Ni^{+2}, Cu^{+2}, and Pd^{+2} with 85% effectiveness. Organic contaminants are another target of COFs. $TpBD(CF_3)_2$ COF adsorbed 119 mg g^{-1} ibuprofen at neutral pH, but an amine-functionalized COF eliminated 90% of GenX. Sulfonate-functionalized adsorbents outperformed commercial COFs, which removed 770 mg g^{-1} diclofenac. Sulfamerazine, furfural, and biotoxins are among the contaminants that COFs efficiently collect (Feng, Ding, and Jiang 2012b; Ascherl et al. 2016; N. Huang et al. 2016).

Microporosity and structural robustness are used in composite COF membranes to improve adsorption and filtration. Polyacrylonitrile with COFs incorporated into it promotes better pollution removal. In addition to expanding wastewater treatment applications, COF composites, including MOFs, carbon-based materials, and electrospun nanofibers, make COFs attractive options for environmental remediation.

1.8.3 COF-Based Heterogeneous Catalysis

Solid catalysts provide "active sites" for reactions at the interface of several phases in heterogeneous catalysis (Cui et al. 2019). COFs have drawn interest in this sector due to their high porosity, stability, and tunability (Cui et al. 2019; Jhulki et al. 2020). Pd (II)-containing imine-type COFs were among the first to be explored; they had a high degree of recyclability and could convert up to 98% in Suzuki–Miyaura reactions (Ding et al. 2011). Additional C-C coupling processes like Sonogashira and Heck have also been investigated (Pachfule et al. 2014). COF-NHC was created by Yang et al., allowing for 99% conversion at room temperature and palladium coordination (Xie et al. 2020).

In addition to C-C coupling, COFs catalyze epoxidation; sales-based, sulfonic acid-modified, and molybdophosphoric acid-based COFs have shown excellent efficacy. Moreover, they promote cyanation, oxidation, reduction, and CO_2 cycloaddition through hydroxyl, hydroxylamine, and ionic liquid-modified COFs. Chiral COFs facilitate the selective synthesis of enantiomers in asymmetric catalysis, as demonstrated by [(S)-Py]x-TPB-DMTP-COF, which produced a significant enantiomeric excess in Michael additions. Other uses include Diels-Alder reactions, aldol condensation, and Steglich rearrangement (Li et al. 2019; Shengjun Liu et al. 2025).

COFs are also exceptionally good at photocatalysis and electrocatalysis. They make visible-light oxidative hydroxylation of arylboronic acids (99% conversion) possible and aid in oxygen evolution and reduction. COFs are also utilized in pollutant degradation, heavy metal removal, and CO_2 reduction (Pachfule et al. 2014; Gonçalves et al. 2016). COFs containing azine, ketoenamine, and triazine connections have been investigated for sustainable hydrogen generation. Biswal et al. produced TpDTz COF, outperforming earlier systems with a high turnover number and steady hydrogen evolution (941 $\mu mol\ h^{-1} g^{-1}$) (Wei et al. 2018).

1.8.4 COF-Based Chemical Sensing

Like catalysis, COFs are being investigated extensively for chemical sensing because of their inherent photoluminescence from stiff aromatic structures, π-electron delocalization, and conjugated double bonds, which allow for selective photodetection (Vishnoi, Kaleeswaran, and Murugavel 2018). By complexing with guest species, COFs can activate new channels or quench fluorescence.

When IMDEA-COF-1, an imine-linked COF produced via TFB and PyDA condensation, interacted with dyestuffs such as MB, it quenched its fluorescence, which allowed it to detect contaminants in aqueous solutions with selectivity (Albacete et al. 2019). In addition, COFs may identify elements such as metals (Chen et al. 2019; Cui et al. 2020), organic compounds (Gomes and Bhaumik 2016; Q. Gao et al. 2018; X. Wu et al. 2019; Liang et al. 2020; Q. Wang et al. 2020; H. Yuan et al. 2020; C. Zhang et al. 2020; S. Zhang et al. 2020; X. Lu et al. 2021; Shengjun Liu et al. 2025), biomolecules (P. Wang et al. 2014; Ni et al. 2018; Z. Cao et al. 2019; J. Dong et al. 2019;

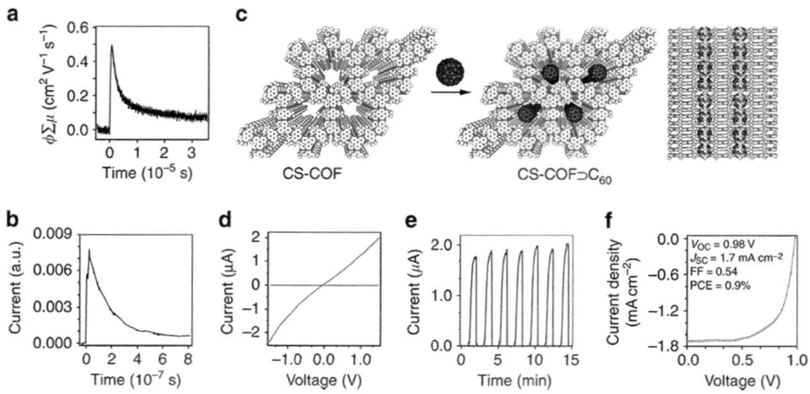

FIGURE 1.9 Carrier mobility and photoelectric properties of CS-COF: (a) Conductivity transients recorded using flash photolysis time-resolved microwave conductivity techniques after excitation with a 355 nm laser pulse. (b) Time-of-flight transient current integration for CS-COF under a 355 nm laser pulse at 34 mJ cm^{-2} per pulse. (c) Schematic representation of CS-COF-C60 synthesis via sublimed fullerene crystallization in open 1D channels (white: carbon, red: nitrogen, purple: fullerene), along with a side-view image of CS-COF-C60. (d) I–V characteristics of a 50 nm thick CS-COF-C60/PMMA film placed between Al and Au electrodes, measured at bias voltages from –1.5 to 1.5 V in air at 25°C. (e) Photocurrent switching behavior at a 1.5 V bias voltage under repeated light on–off cycles for the 50 nm thick CS-COF-C60/PMMA film in air at 25°C. (f) J–V curve of a photovoltaic cell under air mass 1.5 irradiation. (Recreated with permission from Guo et al. 2013. Copyright 2013, Nature Publishing Group.)

Yan et al. 2019; J. Yang et al. 2021), gases (F.-Z. Cui et al. 2019; Jhulki et al. 2020; W. Ma et al. 2020), water/humidity (Yuwei Zhang et al. 2016; Xie et al. 2020; Zhentao Li et al. 2022), and pH levels (Yuwei Zhang et al. 2016; Xie et al. 2020) (Figure 1.9).

TFPPy and a non-planar aldehyde were used to create Py-TPE-COF, which demonstrated high porosity (1.1 nm width, 0.77 $cm^3 g^{-1}$ volume, and 987 $m^2 g^{-1}$ BET surface area) and a 21% absolute photoluminescence quantum yield for explosive detection. In the presence of TNP, it quenched selectively (X. Lu et al. 2021). By using oxygen–nitrogen–oxygen chelation, hydrazone-linked COFs have been designed to detect Fe^{+3} (W.-R. Cui et al. 2020), while Bpy-COF nanosheets, which are made from TFP and [2,2′-bipyridine]-5,5′-diamine, showed increased fluorescence when exposed to Al^{+3} (Yan et al. 2019; W.-R. Cui et al. 2020). Peng and team created a 2D imine-linked COF for biosensing (BET 1,136.5 $m^2 g^{-1}$, pore volume 0.89 $cm^3 g^{-1}$, pore size 1.5 nm). Fluorescently tagged DNA probes stuck to the pore walls of exfoliated nanosheets, which quenched the fluorescence and served as DNA biosensors. For quantitative detection, fluorescence was restored upon target DNA binding (J. Yang et al. 2021).

1.8.5 Electronic Applications

Solid-state ionic batteries and fuel cells may use COFs because of their promising ionic conductivity and electrochemical energy storage (Xing Li and Loh 2019).

Numerous aromatic COFs have been created and saturated with acidic substances like sulfonic and phosphoric acids to improve proton conductivity (Shinde et al. 2016; Sasmal et al. 2018; Z. Meng, Aykanat, and Mirica 2019; Ranjeesh et al. 2019). Using a delayed crystallization technique for improved porosity and stability, Banerjee and the team created a self-standing COF-based membrane with high proton conductivity (7.8×10^{-2} S cm^{-1} at 80°C, 95% RH) (Sasmal et al. 2018). At 298 K and 98% RH, NUS-10, another COF created via mechanochemical grinding, had a high proton conductivity of 3.96×10^{-2} S cm^{-1} (Y. Peng et al. 2016). In water at 353 K, a mixed-matrix COF-PVDF membrane obtained 1.58×10^{-2} S cm^{-1}. In addition, COFs have been investigated for lithium (Q. Xu et al. 2018) and lithium-sulfur batteries (Y. Meng et al. 2018; Royuela et al. 2019; W. Feng et al. 2020), as well as more environmentally friendly sodium (Patra et al. 2018) and potassium-based substitutes (S.-Y. Li et al. 2019).

For organic photovoltaic applications, their stability and semiconducting qualities make them appealing (Dogru and Bein 2014). According to Wu et al., 3D imine-linked COFs were created to improve perovskite solar cells; SP-3D-COF-2 increased power conversion efficiency by 18% (C. Wu et al. 2018). As evidence of its enormous potential for future technologies, COFs are extensively researched for electrical and energy storage applications (S. K. Das et al. 2018; Halder et al. 2018; Khayum M et al. 2018; Haldar et al. 2019; Shuai Liu et al. 2019; M. Wang et al. 2019; J. Xu et al. 2020) and photovoltaics (Dogru et al. 2013; Yao et al. 2020).

1.8.6 Biomedical Applications

Current studies demonstrate the promise of COFs for biological uses, such as cancer treatment and medication delivery (Du et al. 2016; Bhunia, Deo, and Gaharwar 2020; Scicluna and Vella-Zarb 2020). Drug loading and controlled release are made possible by their high porosity, surface area, and customizable pore walls. Covalent reversibility of COFs also permits targeted biodegradation once they have served their purpose in the body.

For regulated camptothecin release, Oliveira et al. created an amino-functionalized COF-5 that increases cytotoxic action at lower dosages and improves drug stability in water (Scicluna and Vella-Zarb 2020). Combining a photosensitizer (BODIPY-2I) with a CD44-targeting chemical (GAG), Guan et al. created a COF nanocomposite ($CaCO_3$@COF-BODIPY-2I@GAG) for synergistic tumor treatment that uses Ca^{2+} overload and singlet oxygen stress to kill tumor cells (Bai et al. 2016). These results and various documentation [39, 50, 139, 197–206] validate COFs as viable options for targeted cancer and medication delivery (Figure 1.10).

1.8.7 Optoelectronic Applications of COFs

Excellent light-harvesting capabilities are demonstrated by COFs with π-conjugated systems, which makes them interesting for use in optoelectronic devices (Sick et al. 2018). Low charge carrier mobility restricts their application in high-performance photodetectors, even though their structure enables fine adjustment of photoelectric characteristics. Integrating COFs with graphene, which improves charge mobility, speed,

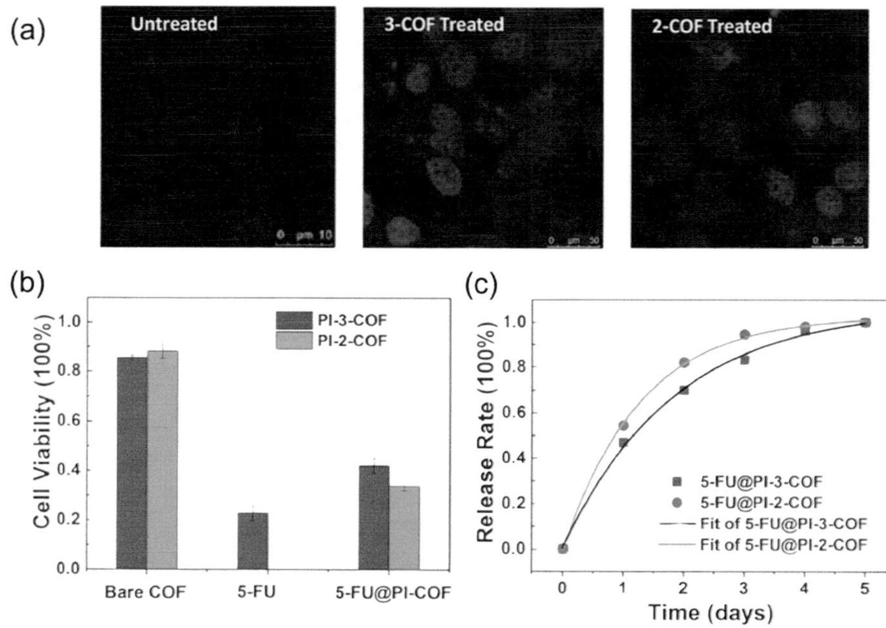

FIGURE 1.10 (a) Confocal images of MCF-7 cells before and after treatment with drug-loaded COFs. The blue tint represents the release of fluorescent 5-FU in the cells. (b) Quantitative MTT measurement of cell viability after incubation with naked COFs, 5-FU, and 5-FU-loaded COFs for 24 hours. The concentrations are 100, 16, and 100 μg mL^{-1}. (b) Release profiles and fitting curves for two COFs loaded with 5-FU at 100 μg mL^{-1}. (Reproduced with permission from Bai et al. (2016). Copyright 2016, Royal Society of Chemistry.)

and responsivity, can be avoided (L. Cao et al. 2022). Moreover, graphene's hydrophobic properties encourage COF crystallinity (Colson et al. 2011). Graphene-COF hybrids enhance photodetection performance by enabling effective charge separation (L. Cao et al. 2022). It was shown by Cao et al. that these hybrids' π-π interactions improve charge transfer (L. Xu et al. 2013). Tetraphenylethylene-based COFs also accomplished photoresponsivity (~3.2×10^7 A W^{-1} at 473 nm) and a quick reaction time (1.14 ms) by forming highly organized donor–acceptor complexes (Xiong et al. 2020). COF integration with 2D materials has potential applications in next-generation optoelectronic devices, such as solar cells and photodetectors (Figure 1.11).

1.8.8 CO_2 Capture Applications of COFs

With rising CO_2 emissions, COFs offer a highly porous, tunable solution for carbon capture (Y. Zeng, Zou, and Zhao 2016). Guan et al. (2019) synthesized a 3D azine-linked COF (3D-HNU5), showing high crystallinity, stability, and selective CO_2 adsorption (123.1 mg g^{-1}) due to nitrogen-enhanced Lewis's acid–base interactions (P. Guan et al. 2019). It retained adsorption capacity over five cycles and remained stable up to 450°C, though alkaline conditions weakened its C=N linkages (Figure 1.12).

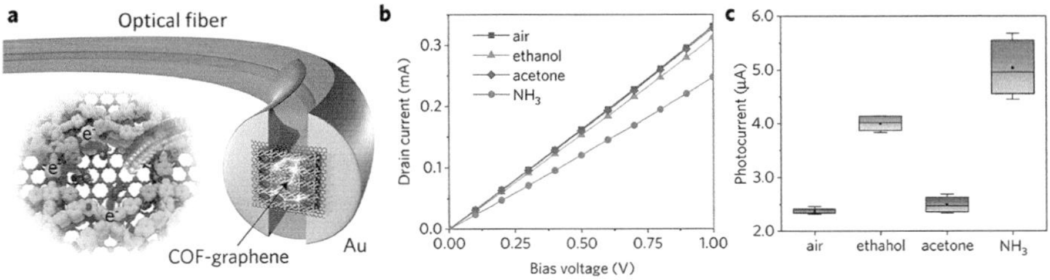

FIGURE 1.11 Target molecules control optoelectronic characteristics. (a) A schematic representation of gas molecule absorption and charge transfer at the surface of the COFETBC-TAPT-graphene film. b) IDS-VDS characteristic curves in the dark for several gas molecules (air, 1% ethanol vapor in air, 1% acetone vapor in air, and 1% NH_3). (b) Photocurrent production in various gas atmospheres. (Reprinted with permission from Xiong et al. (2020). Copyright, 2020, John Wiley.)

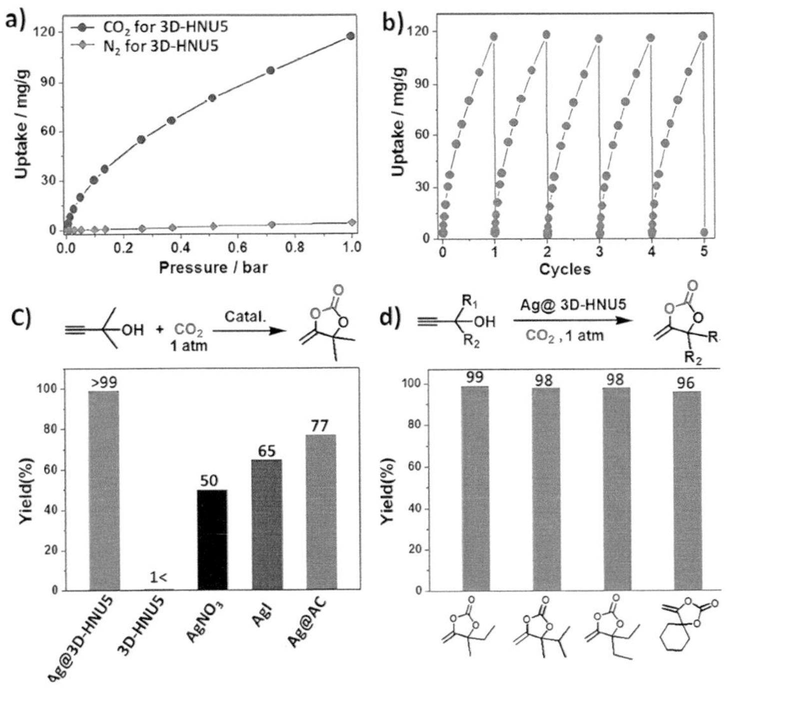

FIGURE 1.12 (a) 3D-HNU5's CO_2 and N_2 isotherms at 273 K. (b) 3D-HNU5 adsorption of recycled CO_2 at 273 K. (c) A number of catalysts' catalytic activity in the synthesis of cyclic carbonates. (d) Using Ag@3D-HNU5 as a catalyst, related propargyl alcohols react with CO_2 to produce various cyclic carbonates. Propargylic alcohols (1.0 mmol), catalyst (0.1 mol%, depending on Ag loading), DBU (1.0 mmol), CH_3CN (3 ml), CO_2 (99.999%, balloon), RT, and 12 hours were the reaction conditions. Utilizing ^1H NMR spectroscopy, the yields were determined. (Reproduced with permission P. Guan et al. (2019). Copyright, 2024, Royal Society of Chemistry.)

Additionally, COFs can support nano-Ag catalysts, enabling CO_2 conversion into carbonates via cycloaddition reactions. Nitrogen-rich COFs further enhance CO_2 capture through strong molecular interactions (Y. Zeng, Zou, and Zhao 2016). While promising, further research is needed to optimize efficiency, stability, and scalability for industrial use.

1.8.9 COF-Based Miscellaneous Applications

1.8.9.1 Membranes and Chromatographic Separations

With their adjustable pore size, low density, and large surface area, crystalline COFs are excellent for membrane applications in wastewater treatment, gas separation, and desalination (S. Yuan et al. 2019). COF membranes are an effective and energy-efficient way to extract CO_2 (Haszeldine 2009; L. Ding et al. 2018), and their potential for water purification is also encouraging (Werber, Osuji, and Elimelech 2016; S. Zeng et al. 2025). MOFs (Rodenas et al. 2015; X. Cui et al. 2016), ZIFs (Baig 2020), graphene (Koenig et al. 2012), and zeolites (J.-T. Chen et al. 2014) have all been investigated for membranes; ultra-microporous MOFs are particularly good at selective gas separation (Kang et al. 2014; Yuanyuan Zhang et al. 2015). COF thin-film membranes allow for effective nanofiltration. In their development of COFs, Kandambeth et al. created M-TpTD and M-TpBD, the latter of which has a 2.5× greater acetonitrile permeability than polyamide membranes and the former of which selectively rejects dyes (Kandambeth et al. 2017). Additionally, functionalized 2D COFs exhibit desalination potential (Kang Zhang et al. 2017). COF-poly(ether sulfone) (COF-PES) membranes were presented by Matsumoto et al. and demonstrated better solute rejection than PES separately (Valentino et al. 2017; Matsumoto et al. 2018), highlighting the potential of COFs in membrane technology.

1.8.9.2 Chromatographic Separations

Chromatographic methods such as capillary electrochromatography (Bao et al. 2016; Niu et al. 2016; Kong, Bao, and Chen 2017), solid-phase microextraction (Y. Chen and Chen 2017), and industrial chromatography (C.-X. Yang et al. 2015; Qian, Yang, and Yan 2016) are being advanced by COFs. A high-performance stationary phase for HPLC, BtaMth COF, was synthesized by Huang et al. and effectively separated positional and cis-trans isomers (Kai Zhang et al. 2017). When attached to silica capillaries, BtaMth COF efficiently separated alcohols, aromatic isomers, and alkanes while continuing to function for months. Nitrotoluene stability tests validated long-term effectiveness (2–8 months). Using chromatography and a column efficiency of 4,058 plates m^{-1}, TpBD COF, which was made from benzidine (BD) and 1,3,5-triformylphloroglucinol (Tp), was able to separate octane with great resolution (C.-X. Yang et al. 2015). Thermodynamic tests highlighted the potential of COFs in chromatography, which verified that all separations were spontaneous. These results indicate the adaptability of COFs in chromatographic and membrane applications, improving separation effectiveness (Figure 1.13).

1.8.9.3 Supercapacitors

The high-power density, extended cycle stability, and superior rate capability of micro-supercapacitors (MSCs) make them attractive for portable electronics (P. Zhang et al. 2018). Many active materials have been investigated, such as metal

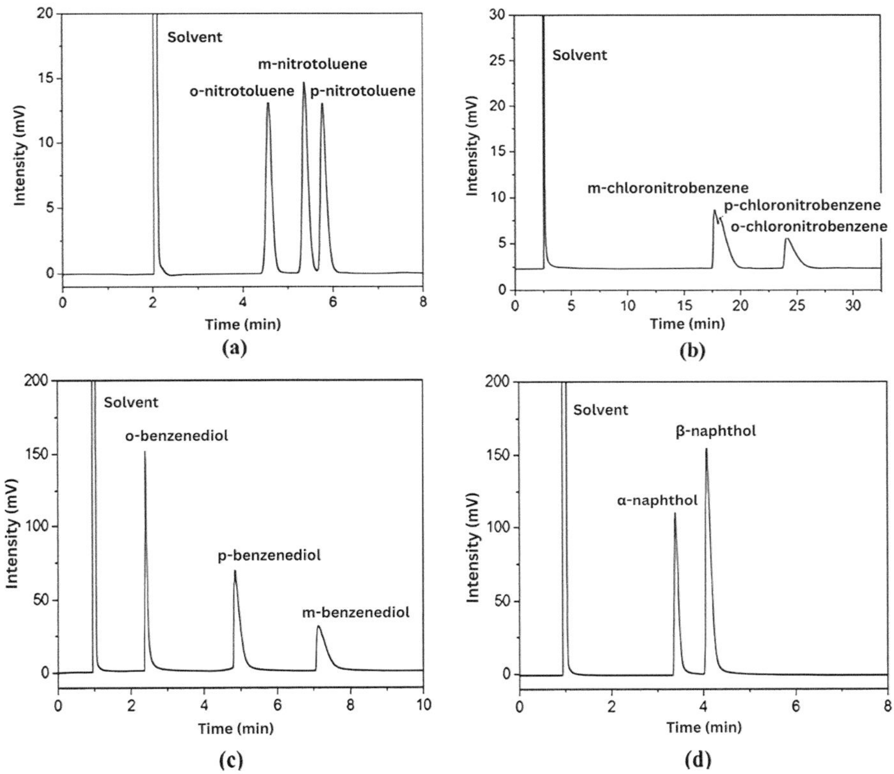

FIGURE 1.13 Using FID for detection, aromatic positional isomers were separated using a BtaMth-bound capillary column (30 m×0.32 mm). The gas chromatograms revealed that the N_2 flow rates of the following isomers were (a) 1.2 mL min^{-1} for nitrotoluene isomers, (b) 1.0 mL min^{-1} for chloronitrobenzene isomers, (c) 3.5 mL min^{-1} for benzenediol isomers, and (d) 3.5 mL min^{-1} for naphthol isomers under 230°C. (Reprinted with permission granted from Kai Zhang et al. (2017). Copyright, 2017, Elsevier.)

compounds (Jun Feng et al. 2011), carbon-based materials (Y.-Y. Peng et al. 2016), and olefin-linked COFs (J. Xu et al. 2019). Porous architectures, high packing density, and stability in acidic/basic solutions are characteristics of 2D MOFs produced from 1,4,5,8,9,12-Hexaazatriphenylene-2,3,6,7,10,11-hexacarbonitrile (HAB) (D. Feng et al. 2018). For MSCs, COFs are being made despite their inflexible shape. By using a conjugated triazine core for π-electron delocalization, Xu and the team developed g-$C_{34}N_6$-COF via Knoevenagel condensation. Up to 500°C, this COF demonstrated thermal stability and a large surface area of 1,003 m^{2}g^{-1} (J. Xu et al. 2020). The COF-based MSC showed an outstanding cycling stability (93.1% capacitance retention after 5,000 cycles), in addition to areal capacitances of 15.2 mF cm^{-2} and energy densities of up to 7.3 mWh cm^{-3} when coupled with carbon nanotubes onto a thin film. The remarkable degree of flexibility of these supercapacitors allows them to maintain electrochemical stability even when deformed (Figure 1.14).

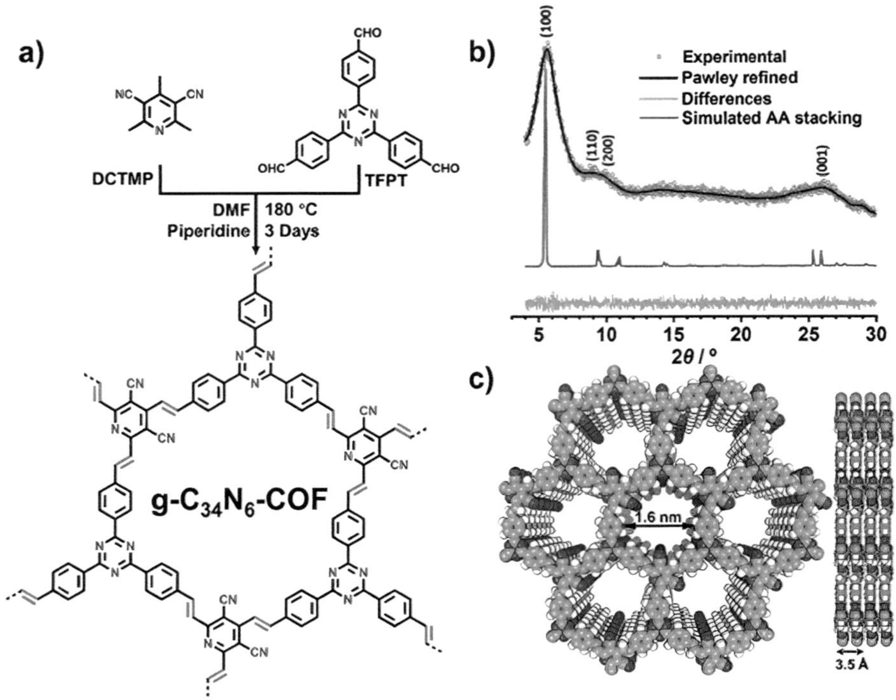

FIGURE 1.14 (a) The synthesis and structure of g-$C_{34}N_6$-COF. (b) PXRD patterns of g-$C_{34}N_6$-COF, showing a comparison between the experimental pattern (red circles), Pawley refined profile (black), the simulated pattern for the eclipsed AA-stacking mode (blue), and refinement differences (orange). (c) The eclipsed AA-stacking model of g-$C_{34}N_6$-COF, presented in both the top view (left) and side view (right). (Reprinted with permission from J. Xu et al. (2020). Copyright, 2024, John Wiley.)

1.8.9.4 Battery Applications

For rechargeable batteries, such as Zn–air (C.-X. Yang et al. 2015), lithium-ion (Luo et al. 2018), Li–CO_2 (Xing Li et al. 2019), lithium-sulfur (Liao et al. 2016), and potassium-ion batteries (F. Cai, Hu, and Chou 2018), COFs are being studied. Energy storage and CO_2 sequestration are two advantages of Li–CO_2 batteries (F. Cai, Hu, and Chou 2018). COFs improve Li_2CO_3 breakdown and CO_2 collection, which increases energy capacity by facilitating gas flow in CO_2-cathode materials (F. Cai, Hu, and Chou 2018; Xiudong Chen et al. 2019). Li et al. created hydrazine/hydrazone-linked COFs that were hybridized with Ru@CNT, resulting in a strong catalytic activity and short ion diffusion routes. The resultant Li–CO_2 battery demonstrated stability over 200 cycles at 1 A g^{-1}, a low cutoff potential (1.24 V), and an ultrahigh capacity of 27,348 mAh g^{-1} (M. Lu et al. 2022). COFs are interesting options for upcoming rechargeable battery applications due to their open channels and functionalization capability (Sun et al. 2020) (Figure 1.15).

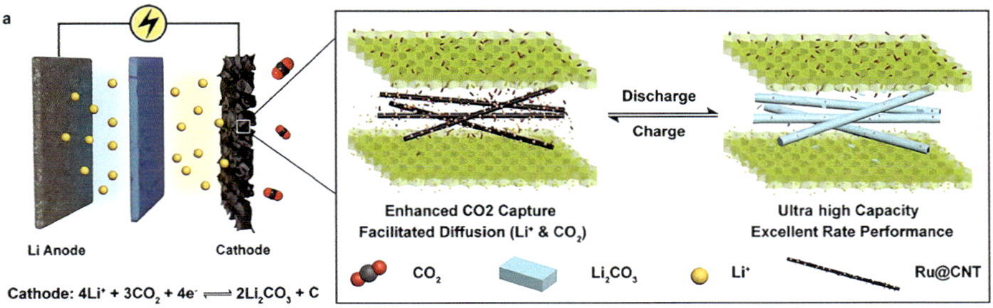

FIGURE 1.15 A schematic representation of COF acting as a CO_2 collector and gas-ion diffusion channels for Li–CO_2 battery cathode. (Reprinted with permission from Xing Li et al. (2019). Copyright, 2024, John Wiley.)

1.9 CHALLENGES AND FUTURE PERSPECTIVES

COFs are a promising class of porous, crystalline materials with high surface areas, tunable structures, and exceptional chemical stability, making them ideal for various applications. In the future, COFs are expected to play a transformative role in energy storage and conversion, including advanced batteries, supercapacitors, and hydrogen storage, as well as in photocatalytic water splitting for clean energy production. Their ability to selectively capture and separate gases, such as CO_2 and methane, positions them as key materials for addressing environmental challenges like greenhouse gas emissions and energy storage. Additionally, COFs show immense potential in catalysis, where their well-defined structures can host active sites for efficient chemical reactions, and in sensing, where they can detect specific molecules for environmental monitoring, medical diagnostics, and security applications. COFs are being explored for targeted drug delivery and biomedical imaging in the biomedical field due to their tunable properties and biocompatibility. They also promise environmental remediation by removing pollutants and heavy metals, such as water purification and air filtration. Despite their potential, several challenges must be overcome to realize the complete application of COFs. One major issue is the complexity of their synthesis, which often requires precise control over reaction conditions, making large-scale production difficult. Stability under harsh conditions, such as elevated temperatures or acidic environments, is another concern, as it can limit long-term usability. Functionalizing COFs for specific applications without compromising their structural integrity remains a significant hurdle. Additionally, the high cost of raw materials and energy-intensive synthesis processes pose economic challenges. Addressing these issues through innovative synthesis techniques, improved material stability, and cost-effective production methods will be critical to unlocking the full potential of COFs. Overcoming these challenges will enable COFs to significantly impact industries, from energy and environmental sustainability to healthcare and advanced materials.

1.10 CONCLUSION

COFs represent a groundbreaking class of porous, crystalline materials constructed from organic building blocks via strong covalent bonds. Their unique structural precision, high surface area, and tunable porosity have positioned them as versatile materials with applications spanning gas storage, catalysis, sensing, energy storage, and drug delivery. The ability to design COFs with specific functionalities by selecting appropriate molecular precursors and linkage types underscores their potential for tailored applications. Recent advancements in synthetic strategies have improved their crystallinity, stability, and scalability, addressing earlier challenges related to their practical implementation. Furthermore, integrating COFs with other functional materials has expanded their utility in hybrid systems, enhancing their performance in complex environments. Despite these successes, challenges remain, such as achieving large-scale production, improving mechanical robustness, and understanding long-term stability under operational conditions. Future research should focus on developing greener synthesis methods, exploring novel topologies, and expanding their applications in emerging fields like photocatalysis and environmental remediation. As the field progresses, COFs are poised to be pivotal in addressing global energy, sustainability, and healthcare challenges, making them a cornerstone of modern materials science. Their continued development promises to unlock new possibilities for innovation and technological advancement.

REFERENCES

Alahakoon, Sampath B., Christina M. Thompson, Amy X. Nguyen, Gino Occhialini, Gregory T. McCandless, and Ronald A. Smaldone. 2016. "An Azine-Linked Hexaphenylbenzene Based Covalent Organic Framework." *Chemical Communications* 52 (13): 2843–45. doi:10.1039/C5CC10408D.

Albacete, Pablo, Alejandro López-Moreno, Sofía Mena-Hernando, Ana E. Platero-Prats, Emilio M. Pérez, and Félix Zamora. 2019. "Chemical Sensing of Water Contaminants by a Colloid of a Fluorescent Imine-Linked Covalent Organic Framework." *Chemical Communications* 55 (10): 1382–85. doi:10.1039/C8CC08307J.

Ascherl, Laura, Torben Sick, Johannes T. Margraf, Saul H. Lapidus, Mona Calik, Christina Hettstedt, Konstantin Karaghiosoff, et al. 2016. "Molecular Docking Sites Designed for the Generation of Highly Crystalline Covalent Organic Frameworks." *Nature Chemistry* 8 (4): 310–16. doi:10.1038/nchem.2444.

Bai, Linyi, Soo Zeng Fiona Phua, Wei Qi Lim, Avijit Jana, Zhong Luo, Huijun Phoebe Tham, Lingzhi Zhao, Qiang Gao, and Yanli Zhao. 2016. "Nanoscale Covalent Organic Frameworks as Smart Carriers for Drug Delivery." *Chemical Communications* 52 (22): 4128–31. doi:10.1039/C6CC00853D.

Baig, Nadeem. 2020. "Contaminants in Our Water: Identification and Remediation Methods." In *Recent Progress on the Development of Superhydrophobic and Superoleophilic Meshes for Oil and Water Separation: A Review*, pp. 175–96. ACS Publication. doi:10.1021/bk-2020-1352.ch010.

Baldwin, Luke A., Jonathan W. Crowe, David A. Pyles, and Psaras L. McGrier. 2016. "Metalation of a Mesoporous Three-Dimensional Covalent Organic Framework." *Journal of the American Chemical Society* 138 (46): 15134–37. doi:10.1021/jacs.6b10316.

Bao, Tao, Pingxiu Tang, Deying Kong, Zhenkun Mao, and Zilin Chen. 2016. "Polydopamine-Supported Immobilization of Covalent-Organic Framework-5 in Capillary as Stationary Phase for Electrochromatographic Separation." *Journal of Chromatography A* 1445 (May): 140–48. doi:10.1016/j.chroma.2016.03.085.

Bhunia, Sukanya, Kaivalya A. Deo, and Akhilesh K. Gaharwar. 2020. "2D Covalent Organic Frameworks for Biomedical Applications." *Advanced Functional Materials* 30 (27): 2002046. doi:10.1002/adfm.202002046.

Biswal, Bishnu P., Suman Chandra, Sharath Kandambeth, Binit Lukose, Thomas Heine, and Rahul Banerjee. 2013. "Mechanochemical Synthesis of Chemically Stable Isoreticular Covalent Organic Frameworks." *Journal of the American Chemical Society* 135 (14): 5328–31. doi:10.1021/ja4017842.

Bojdys, Michael J., Jekaterina Jeromenok, Arne Thomas, and Markus Antonietti. 2010. "Rational Extension of the Family of Layered, Covalent, Triazine-Based Frameworks with Regular Porosity." *Advanced Materials* 22 (19): 2202–5. doi:10.1002/adma.200903436.

Bunck, David N., and William R. Dichtel. 2012. "Internal Functionalization of Three-Dimensional Covalent Organic Frameworks." *Angewandte Chemie International Edition* 51 (8): 1885–89. doi:10.1002/anie.201108462.

Cai, Fengshi, Zhe Hu, and Shu-Lei Chou. 2018. "Progress and Future Perspectives on Li(Na)–CO_2 Batteries." *Advanced Sustainable Systems* 2 (8–9). doi:10.1002/adsu.201800060.

Cai, Song-Liang, Yue-Biao Zhang, Andrew B. Pun, Bo He, Jinhui Yang, Francesca M. Toma, Ian D. Sharp, et al. 2014. "Tunable Electrical Conductivity in Oriented Thin Films of Tetrathiafulvalene-Based Covalent Organic Framework." *Chemical Science* 5 (12): 4693–4700. doi:10.1039/C4SC02593H.

Campbell, Neil L., Rob Clowes, Lyndsey K. Ritchie, and Andrew I. Cooper. 2009. "Rapid Microwave Synthesis and Purification of Porous Covalent Organic Frameworks." *Chemistry of Materials* 21 (2): 204–6. doi:10.1021/cm802981m.

Cao, Li, Xiaowei Liu, Digambar B. Shinde, Cailing Chen, I-Chun Chen, Zhen Li, Zongyao Zhou, Zhongyu Yang, Yu Han, and Zhiping Lai. 2022. "Oriented Two-Dimensional Covalent Organic Framework Membranes with High Ion Flux and Smart Gating Nanofluidic Transport." *Angewandte Chemie International Edition* 61 (6). doi:10.1002/anie.202113141.

Cao, Zhengming, Fenghe Duan, Xiaoyu Huang, Yang Liu, Nan Zhou, Lei Xia, Zhihong Zhang, and Miao Du. 2019. "A Multiple Aptasensor for Ultrasensitive Detection of MiRNAs by Using Covalent-Organic Framework Nanowire as Platform and Shell-Encoded Gold Nanoparticles as Signal Labels." *Analytica Chimica Acta* 1082 (November): 176–85. doi:10.1016/j.aca.2019.07.062.

Chandra, Suman, Sharath Kandambeth, Bishnu P. Biswal, Binit Lukose, Shrikant M. Kunjir, Minakshi Chaudhary, Ravichandar Babarao, Thomas Heine, and Rahul Banerjee. 2013. "Chemically Stable Multilayered Covalent Organic Nanosheets from Covalent Organic Frameworks via Mechanical Delamination." *Journal of the American Chemical Society* 135 (47): 17853–61. doi:10.1021/ja408121p.

Chandra, Suman, Tanay Kundu, Kaushik Dey, Matthew Addicoat, Thomas Heine, and Rahul Banerjee. 2016. "Interplaying Intrinsic and Extrinsic Proton Conductivities in Covalent Organic Frameworks." *Chemistry of Materials* 28 (5): 1489–94. doi:10.1021/acs.chemmater.5b04947.

Chandra, Suman, Tanay Kundu, Sharath Kandambeth, Ravichandar BabaRao, Yogesh Marathe, Shrikant M. Kunjir, and Rahul Banerjee. 2014. "Phosphoric Acid Loaded Azo (−N=N−) Based Covalent Organic Framework for Proton Conduction." *Journal of the American Chemical Society* 136 (18): 6570–73. doi:10.1021/ja502212v.

Chan-Thaw, Carine E., Alberto Villa, Laura Prati, and Arne Thomas. 2011. "Triazine-Based Polymers as Nanostructured Supports for the Liquid-Phase Oxidation of Alcohols." *Chemistry – A European Journal* 17 (3): 1052–57. doi:10.1002/chem.201000675.

Chavda, Vishwajit, Brijesh Patel, Sneha Singh, Darshna Hirpara, V. Devi Rajeswari, and Sanjeev Kumar. 2023. "A Sustainable Approach for the Adsorption of Methylene Blue from an Aqueous Background: An Adsorbent Based on DES/CGS Modified GO@ZrO_2." *RSC Sustainability* 1 (8): 2038–57. doi:10.1039/D3SU00236E.

Chavda, Vishwajit, Darshna Hirpara, and Sanjeev Kumar. 2022. "GO/Ionic Surfactant Inspired Photophysical Modulation of Rhodamine B in Reline with or without Additives." *Journal of Molecular Liquids* 368 (December): 120614. doi:10.1016/j.molliq.2022.120614.

Chavda, Vishwajit, Darshna Hirpara, Vandana Rao, and Sanjeev Kumar. 2024. "Deep Eutectic Solvent/Surfactant-Engineered GO@TiO_2 Advanced Material for Sustained and Ultrafast Adsorption of Hazardous Dye from Contaminated Aqueous Environment." *ACS Applied Engineering Materials* 2 (3): 604–17. doi:10.1021/acsaenm.3c00696.

Chavda, Vishwajit, Harsh D. Patel, Avani K. Patel, Darshna Hirpara, N. K. Acharya, and Sanjeev Kumar. 2024. "Deep Eutectic Solvent Engineered GO/ZrO_2 Nanofiller for Mixed Matrix Membranes: An Efficient Hydrogen Gas Separation." *Journal of Applied Polymer Science* 141 (37). doi:10.1002/app.55938.

Chen, Gui, Hai-Hang Lan, Song-Liang Cai, Bing Sun, Xin-Le Li, Zi-Hao He, Sheng-Run Zheng, Jun Fan, Yi Liu, and Wei-Guang Zhang. 2019. "Stable Hydrazone-Linked Covalent Organic Frameworks Containing O,N,O′-Chelating Sites for Fe(III) Detection in Water." *ACS Applied Materials & Interfaces* 11 (13): 12830–37. doi:10.1021/acsami.9b02640.

Chen, Jung-Tsai, Chien-Chung Shih, Ywu-Jang Fu, Shu-Hsien Huang, Chien-Chieh Hu, Kueir-Rarn Lee, and Juin-Yih Lai. 2014. "Zeolite-Filled Porous Mixed Matrix Membranes for Air Separation." *Industrial & Engineering Chemistry Research* 53 (7): 2781–89. doi:10.1021/ie403833u.

Chen, Xinyi, Keyu Geng, Ruoyang Liu, Ke Tian Tan, Yifan Gong, Zhongping Li, Shanshan Tao, Qiuhong Jiang, and Donglin Jiang. 2020. "Covalent Organic Frameworks: Chemical Approaches to Designer Structures and Built-In Functions." *Angewandte Chemie International Edition* 59 (13): 5050–91. doi:10.1002/anie.201904291.

Chen, Xiong, Matthew Addicoat, Enquan Jin, Hong Xu, Taku Hayashi, Fei Xu, Ning Huang, Stephan Irle, and Donglin Jiang. 2015. "Designed Synthesis of Double-Stage Two-Dimensional Covalent Organic Frameworks." *Scientific Reports* 5 (1): 14650. doi:10.1038/srep14650.

Chen, Xiong, Ning Huang, Jia Gao, Hong Xu, Fei Xu, and Donglin Jiang. 2014. "Towards Covalent Organic Frameworks with Predesignable and Aligned Open Docking Sites." *Chemical Communications* 50 (46): 6161–63. doi:10.1039/C4CC01825G.

Chen, Xiudong, Hang Zhang, Chenggang Ci, Weiwei Sun, and Yong Wang. 2019. "Few-Layered Boronic Ester Based Covalent Organic Frameworks/Carbon Nanotube Composites for High-Performance K-Organic Batteries." *ACS Nano* 13 (3): 3600–3607. doi:10.1021/acsnano.9b00165.

Chen, Yuling, and Zilin Chen. 2017. "COF-1-Modified Magnetic Nanoparticles for Highly Selective and Efficient Solid-Phase Microextraction of Paclitaxel." *Talanta* 165 (April): 188–93. doi:10.1016/j.talanta.2016.12.051.

Colson, John W., Arthur R. Woll, Arnab Mukherjee, Mark P. Levendorf, Eric L. Spitler, Virgil B. Shields, Michael G. Spencer, Jiwoong Park, and William R. Dichtel. 2011. "Oriented 2D Covalent Organic Framework Thin Films on Single-Layer Graphene." *Science* 332 (6026): 228–31. doi:10.1126/science.1202747.

Colson, John W., Jason A. Mann, Catherine R. DeBlase, and William R. Dichtel. 2015. "Patterned Growth of Oriented 2 D Covalent Organic Framework Thin Films on Single-layer Graphene." *Journal of Polymer Science Part A: Polymer Chemistry* 53 (2): 378–84. doi:10.1002/pola.27399.

Côté, Adrien P., Annabelle I. Benin, Nathan W. Ockwig, Michael O'Keeffe, Adam J. Matzger, and Omar M. Yaghi. 2005. "Porous, Crystalline, Covalent Organic Frameworks." *Science* 310 (5751): 1166–70. doi:10.1126/science.1120411.

Crowe, Jonathan W., Luke A. Baldwin, and Psaras L. McGrier. 2016. "Luminescent Covalent Organic Frameworks Containing a Homogeneous and Heterogeneous Distribution of Dehydrobenzoannulene Vertex Units." *Journal of the American Chemical Society* 138 (32): 10120–23. doi:10.1021/jacs.6b06546.

Cui, Fu-Zhi, Jiao-Jiao Xie, Shu-Yan Jiang, Shi-Xian Gan, De-Li Ma, Rong-Ran Liang, Guo-Fang Jiang, and Xin Zhao. 2019. "A Gaseous Hydrogen Chloride Chemosensor Based on a 2D Covalent Organic Framework." *Chemical Communications* 55 (31): 4550–53. doi:10.1039/C9CC01548E.

Cui, Wei-Rong, Cheng-Rong Zhang, Wei Jiang, Fang-Fang Li, Ru-Ping Liang, Juewen Liu, and Jian-Ding Qiu. 2020. "Regenerable and Stable Sp2 Carbon-Conjugated Covalent Organic Frameworks for Selective Detection and Extraction of Uranium." *Nature Communications* 11 (1): 436. doi:10.1038/s41467-020-14289-x.

Cui, Xili, Kaijie Chen, Huabin Xing, Qiwei Yang, Rajamani Krishna, Zongbi Bao, Hui Wu, et al. 2016. "Pore Chemistry and Size Control in Hybrid Porous Materials for Acetylene Capture from Ethylene." *Science* 353 (6295): 141–44. doi:10.1126/science.aaf2458.

Dalapati, Sasanka, Matthew Addicoat, Shangbin Jin, Tsuneaki Sakurai, Jia Gao, Hong Xu, Stephan Irle, Shu Seki, and Donglin Jiang. 2015. "Rational Design of Crystalline Supermicroporous Covalent Organic Frameworks with Triangular Topologies." *Nature Communications* 6 (1): 7786. doi:10.1038/ncomms8786.

Dalapati, Sasanka, Shangbin Jin, Jia Gao, Yanhong Xu, Atsushi Nagai, and Donglin Jiang. 2013. "An Azine-Linked Covalent Organic Framework." *Journal of the American Chemical Society* 135 (46): 17310–13. doi:10.1021/ja4103293.

Das, Gobinda, Bishnu P. Biswal, Sharath Kandambeth, V. Venkatesh, Gagandeep Kaur, Matthew Addicoat, Thomas Heine, Sandeep Verma, and Rahul Banerjee. 2015. "Chemical Sensing in Two Dimensional Porous Covalent Organic Nanosheets." *Chemical Science* 6 (7): 3931–39. doi:10.1039/C5SC00512D.

Das, Sabuj Kanti, Kousik Bhunia, Arijit Mallick, Anirban Pradhan, Debabrata Pradhan, and Asim Bhaumik. 2018. "A New Electrochemically Responsive 2D π-Conjugated Covalent Organic Framework as a High Performance Supercapacitor." *Microporous and Mesoporous Materials* 266 (August): 109–16. doi:10.1016/j.micromeso.2018.02.026.

DeBlase, Catherine R., Katharine E. Silberstein, Thanh-Tam Truong, Héctor D. Abruña, and William R. Dichtel. 2013. "β-Ketoenamine-Linked Covalent Organic Frameworks Capable of Pseudocapacitive Energy Storage." *Journal of the American Chemical Society* 135 (45): 16821–24. doi:10.1021/ja409421d.

DeBlase, Catherine R., Kenneth Hernández-Burgos, Katharine E. Silberstein, Gabriel G. Rodríguez-Calero, Ryan P. Bisbey, Héctor D. Abruña, and William R. Dichtel. 2015. "Rapid and Efficient Redox Processes within 2D Covalent Organic Framework Thin Films." *ACS Nano* 9 (3): 3178–83. doi:10.1021/acsnano.5b00184.

Diercks, Christian S., and Omar M. Yaghi. 2017. "The Atom, the Molecule, and the Covalent Organic Framework." *Science* 355 (6328). doi:10.1126/science.aal1585.

Ding, Li, Yanying Wei, Libo Li, Tao Zhang, Haihui Wang, Jian Xue, Liang-Xin Ding, Suqing Wang, Jürgen Caro, and Yury Gogotsi. 2018. "MXene Molecular Sieving Membranes for Highly Efficient Gas Separation." *Nature Communications* 9 (1): 155. doi:10.1038/s41467-017-02529-6.

Ding, San-Yuan, and Wei Wang. 2013. "Covalent Organic Frameworks (COFs): From Design to Applications." *Chemical Society Reviews* 42 (2): 548–68. doi:10.1039/C2CS35072F.

Ding, San-Yuan, Jia Gao, Qiong Wang, Yuan Zhang, Wei-Guo Song, Cheng-Yong Su, and Wei Wang. 2011. "Construction of Covalent Organic Framework for Catalysis: Pd/COF-LZU1 in Suzuki–Miyaura Coupling Reaction." *Journal of the American Chemical Society* 133 (49): 19816–22. doi:10.1021/ja206846p.

Ding, San-Yuan, Ming Dong, Ya-Wen Wang, Yan-Tao Chen, Huai-Zhen Wang, Cheng-Yong Su, and Wei Wang. 2016. "Thioether-Based Fluorescent Covalent Organic Framework for Selective Detection and Facile Removal of Mercury(II)." *Journal of the American Chemical Society* 138 (9): 3031–37. doi:10.1021/jacs.5b10754.

Ding, Xuesong, Jia Guo, Xiao Feng, Yoshihito Honsho, Jingdong Guo, Shu Seki, Phornphimon Maitarad, Akinori Saeki, Shigeru Nagase, and Donglin Jiang. 2011. "Synthesis of Metallophthalocyanine Covalent Organic Frameworks That Exhibit High Carrier Mobility and Photoconductivity." *Angewandte Chemie International Edition* 50 (6): 1289–93. doi:10.1002/anie.201005919.

Ding, Xuesong, Xiao Feng, Akinori Saeki, Shu Seki, Atsushi Nagai, and Donglin Jiang. 2012. "Conducting Metallophthalocyanine 2D Covalent Organic Frameworks: The Role of Central Metals in Controlling π-Electronic Functions." *Chemical Communications* 48 (71): 8952. doi:10.1039/c2cc33929c.

Dogru, Mirjam, and Thomas Bein. 2014. "On the Road towards Electroactive Covalent Organic Frameworks." *Chemical Communications* 50 (42): 5531–46. doi:10.1039/C3CC46767H.

Dogru, Mirjam, Andreas Sonnauer, Andrei Gavryushin, Paul Knochel, and Thomas Bein. 2011. "A Covalent Organic Framework with 4 Nm Open Pores." *Chemical Communications* 47 (6): 1707. doi:10.1039/c0cc03792c.

Dogru, Mirjam, Matthias Handloser, Florian Auras, Thomas Kunz, Dana Medina, Achim Hartschuh, Paul Knochel, and Thomas Bein. 2013. "A Photoconductive Thienothiophene-Based Covalent Organic Framework Showing Charge Transfer Towards Included Fullerene." *Angewandte Chemie International Edition* 52 (10): 2920–24. doi:10.1002/anie.201208514.

Dong, Jinqiao, Xu Li, Shing Bo Peh, Yi Di Yuan, Yuxiang Wang, Dongxiao Ji, Shengjie Peng, et al. 2019. "Restriction of Molecular Rotors in Ultrathin Two-Dimensional Covalent Organic Framework Nanosheets for Sensing Signal Amplification." *Chemistry of Materials* 31 (1): 146–60. doi:10.1021/acs.chemmater.8b03685.

Dong, Renhao, Martin Pfeffermann, Haiwei Liang, Zhikun Zheng, Xiang Zhu, Jian Zhang, and Xinliang Feng. 2015. "Large-Area, Free-Standing, Two-Dimensional Supramolecular Polymer Single-Layer Sheets for Highly Efficient Electrocatalytic Hydrogen Evolution." *Angewandte Chemie International Edition* 54 (41): 12058–63. doi:10.1002/anie.201506048.

Du, Ya, Haishen Yang, Justin Michael Whiteley, Shun Wan, Yinghua Jin, Se-Hee Lee, and Wei Zhang. 2016. "Ionic Covalent Organic Frameworks with Spiroborate Linkage." *Angewandte Chemie International Edition* 55 (5): 1737–41. doi:10.1002/anie.201509014.

El-Kaderi, Hani M., Joseph R. Hunt, José L. Mendoza-Cortés, Adrien P. Côté, Robert E. Taylor, Michael O'Keeffe, and Omar M. Yaghi. 2007. "Designed Synthesis of 3D Covalent Organic Frameworks." *Science* 316 (5822): 268–72. doi:10.1126/science.1139915.

Evans, Austin M., Lucas R. Parent, Nathan C. Flanders, Ryan P. Bisbey, Edon Vitaku, Matthew S. Kirschner, Richard D. Schaller, Lin X. Chen, Nathan C. Gianneschi, and William R. Dichtel. 2018. "Seeded Growth of Single-Crystal Two-Dimensional Covalent Organic Frameworks." *Science* 361 (6397): 52–57. doi:10.1126/science.aar7883.

Fang, Qianrong, Junhua Wang, Shuang Gu, Robert B. Kaspar, Zhongbin Zhuang, Jie Zheng, Hongxia Guo, Shilun Qiu, and Yushan Yan. 2015. "3D Porous Crystalline Polyimide Covalent Organic Frameworks for Drug Delivery." *Journal of the American Chemical Society* 137 (26): 8352–55. doi:10.1021/jacs.5b04147.

Fang, Qianrong, Shuang Gu, Jie Zheng, Zhongbin Zhuang, Shilun Qiu, and Yushan Yan. 2014. "3D Microporous Base-Functionalized Covalent Organic Frameworks for Size-Selective Catalysis." *Angewandte Chemie International Edition* 53 (11): 2878–82. doi:10.1002/anie.201310500.

Feng, Dawei, Ting Lei, Maria R. Lukatskaya, Jihye Park, Zhehao Huang, Minah Lee, Leo Shaw, et al. 2018. "Robust and Conductive Two-Dimensional Metal−organic Frameworks with Exceptionally High Volumetric and Areal Capacitance." *Nature Energy* 3 (1): 30–36. doi:10.1038/s41560-017-0044-5.

Feng, Juanjuan, Jiaqing Feng, Xiangping Ji, Chunying Li, Sen Han, Haili Sun, and Min Sun. 2021. "Recent Advances of Covalent Organic Frameworks for Solid-Phase Microextraction." *TrAC Trends in Analytical Chemistry* 137 (April): 116208. doi:10.1016/j.trac.2021.116208.

Feng, Jun, Xu Sun, Changzheng Wu, Lele Peng, Chenwen Lin, Shuanglin Hu, Jinlong Yang, and Yi Xie. 2011. "Metallic Few-Layered VS_2 Ultrathin Nanosheets: High Two-Dimensional Conductivity for In-Plane Supercapacitors." *Journal of the American Chemical Society* 133 (44): 17832–38. doi:10.1021/ja207176c.

Feng, Wenhui, Jie Yuan, Fan Gao, Bo Weng, Wenting Hu, Yanhua Lei, Xueyan Huang, et al. 2020. "Piezopotential-Driven Simulated Electrocatalytic Nanosystem of Ultrasmall MoC Quantum Dots Encapsulated in Ultrathin N-Doped Graphene Vesicles for Superhigh H2 Production from Pure Water." *Nano Energy* 75 (September): 104990. doi:10.1016/j.nanoen.2020.104990.

Feng, Xiao, Xuesong Ding, and Donglin Jiang. 2012a. "Covalent Organic Frameworks." *Chemical Society Reviews* 41 (18): 6010. doi:10.1039/c2cs35157a.

Feng, Xiao, Xuesong Ding, and Donglin Jiang. 2012b. "Covalent Organic Frameworks." *Chemical Society Reviews* 41 (18): 6010. doi:10.1039/c2cs35157a.

Feng, Xiao, Yuping Dong, and Donglin Jiang. 2013. "Star-Shaped Two-Dimensional Covalent Organic Frameworks." *CrystEngComm* 15 (8): 1508–11. doi:10.1039/C2CE26371H.

Furukawa, Hiroyasu, and Omar M. Yaghi. 2009. "Storage of Hydrogen, Methane, and Carbon Dioxide in Highly Porous Covalent Organic Frameworks for Clean Energy Applications." *Journal of the American Chemical Society* 131 (25): 8875–83. doi:10.1021/ja9015765.

Gao, Qiang, Xing Li, Guo-Hong Ning, Kai Leng, Bingbing Tian, Cuibo Liu, Wei Tang, Hai-Sen Xu, and Kian Ping Loh. 2018. "Highly Photoluminescent Two-Dimensional Imine-Based Covalent Organic Frameworks for Chemical Sensing." *Chemical Communications* 54 (19): 2349–52. doi:10.1039/C7CC09866A.

Gao, Wenxiu, Xiuyun Sun, Huiling Niu, Xiaojing Song, Kaige Li, Hongcheng Gao, Wenxiang Zhang, Jihong Yu, and Mingjun Jia. 2015. "Phosphomolybdic Acid Functionalized Covalent Organic Frameworks: Structure Characterization and Catalytic Properties in Olefin Epoxidation." *Microporous and Mesoporous Materials* 213 (September): 59–67. doi:10.1016/j.micromeso.2015.04.009.

Ghazi, Zahid Ali, Lingyun Zhu, Han Wang, Abdul Naeem, Abdul Muqsit Khattak, Bin Liang, Niaz Ali Khan, Zhixiang Wei, Lianshan Li, and Zhiyong Tang. 2016. "Efficient Polysulfide Chemisorption in Covalent Organic Frameworks for High-Performance Lithium-Sulfur Batteries." *Advanced Energy Materials* 6 (24). doi:10.1002/aenm.201601250.

Gomes, Ruth, and Asim Bhaumik. 2016. "A New Triazine Functionalized Luminescent Covalent Organic Framework for Nitroaromatic Sensing and CO_2 Storage." *RSC Advances* 6 (33): 28047–54. doi:10.1039/C6RA01717G.

Gomes, Ruth, Piyali Bhanja, and Asim Bhaumik. 2015. "A Triazine-Based Covalent Organic Polymer for Efficient CO_2 Adsorption." *Chemical Communications* 51 (49): 10050–53. doi:10.1039/C5CC02147B.

Gonçalves, Raoni S. B., Alline B. V. de Oliveira, Haryadylla C. Sindra, Bráulio S. Archanjo, Martín E. Mendoza, Leonardo S. A. Carneiro, Camilla D. Buarque, and Pierre M. Esteves. 2016. "Heterogeneous Catalysis by Covalent Organic Frameworks (COF): $Pd(OAc)_2$ @COF-300 in Cross-Coupling Reactions." *ChemCatChem* 8 (4): 743–50. doi:10.1002/cctc.201500926.

Guan, Pengxin, Jikuan Qiu, Yuling Zhao, Huiyong Wang, Zhiyong Li, Yunlei Shi, and Jianji Wang. 2019. "A Novel Crystalline Azine-Linked Three-Dimensional Covalent Organic Framework for CO_2 Capture and Conversion." *Chemical Communications* 55 (83): 12459–62. doi:10.1039/C9CC05710B.

Guan, Xinyu, Fengqian Chen, Qianrong Fang, and Shilun Qiu. 2020. "Design and Applications of Three Dimensional Covalent Organic Frameworks." *Chemical Society Reviews* 49 (5): 1357–84. doi:10.1039/C9CS00911F.

Guan, Xinyu, Yunchao Ma, Hui Li, Yusran Yusran, Ming Xue, Qianrong Fang, Yushan Yan, Valentin Valtchev, and Shilun Qiu. 2018. "Fast, Ambient Temperature and Pressure Ionothermal Synthesis of Three-Dimensional Covalent Organic Frameworks." *Journal of the American Chemical Society* 140 (13): 4494–98. doi:10.1021/jacs.8b01320.

Guo, Jia, Yanhong Xu, Shangbin Jin, Long Chen, Toshihiko Kaji, Yoshihito Honsho, Matthew A. Addicoat, et al. 2013. "Conjugated Organic Framework with Three-Dimensionally Ordered Stable Structure and Delocalized π Clouds." *Nature Communications* 4 (1): 2736. doi:10.1038/ncomms3736.

Haase, Frederik, and Bettina V. Lotsch. 2020. "Solving the COF Trilemma: Towards Crystalline, Stable and Functional Covalent Organic Frameworks." *Chemical Society Reviews* 49 (23): 8469–8500. doi:10.1039/D0CS01027H.

Haldar, Sattwick, Rinku Kushwaha, Rahul Maity, and Ramanathan Vaidhyanathan. 2019. "Pyridine-Rich Covalent Organic Frameworks as High-Performance Solid-State Supercapacitors." *ACS Materials Letters* 1 (4): 490–97. doi:10.1021/acsmaterialslett.9b00222.

Halder, Arjun, Meena Ghosh, Abdul Khayum M, Saibal Bera, Matthew Addicoat, Himadri Sekhar Sasmal, Suvendu Karak, Sreekumar Kurungot, and Rahul Banerjee. 2018. "Interlayer Hydrogen-Bonded Covalent Organic Frameworks as High-Performance Supercapacitors." *Journal of the American Chemical Society* 140 (35): 10941–45. doi:10.1021/jacs.8b06460.

Haszeldine, R. Stuart. 2009. "Carbon Capture and Storage: How Green Can Black Be?" *Science* 325 (5948): 1647–52. doi:10.1126/science.1172246.

Huang, Ding, Cong Liang, Lining Chen, Mi Tang, Zijian Zheng, and Zhengbang Wang. 2021. "MOF Composite Fibrous Separators for High-Rate Lithium-Ion Batteries." *Journal of Materials Science* 56 (9): 5868–77. doi:10.1007/s10853-020-05559-6.

Huang, Ning, Lipeng Zhai, Damien E. Coupry, Matthew A. Addicoat, Keiko Okushita, Katsuyuki Nishimura, Thomas Heine, and Donglin Jiang. 2016. "Multiple-Component Covalent Organic Frameworks." *Nature Communications* 7 (1): 12325. doi:10.1038/ncomms12325.

Huang, Ning, Ping Wang, and Donglin Jiang. 2016a. "Covalent Organic Frameworks: A Materials Platform for Structural and Functional Designs." *Nature Reviews Materials* 1 (10): 16068. doi:10.1038/natrevmats.2016.68.

Huang, Ning, Ping Wang, and Donglin Jiang. 2016b. "Covalent Organic Frameworks: A Materials Platform for Structural and Functional Designs." *Nature Reviews Materials* 1 (10): 16068. doi:10.1038/natrevmats.2016.68.

Hunt, Joseph R., Christian J. Doonan, James D. LeVangie, Adrien P. Côté, and Omar M. Yaghi. 2008. "Reticular Synthesis of Covalent Organic Borosilicate Frameworks." *Journal of the American Chemical Society* 130 (36): 11872–73. doi:10.1021/ja805064f.

Jackson, Karl T., Thomas E. Reich, and Hani M. El-Kaderi. 2012. "Targeted Synthesis of a Porous Borazine-Linked Covalent Organic Framework." *Chemical Communications* 48 (70): 8823. doi:10.1039/c2cc33583b.

Jhulki, Samik, Austin M. Evans, Xue-Li Hao, Matthew W. Cooper, Cameron H. Feriante, Johannes Leisen, Hong Li, et al. 2020. "Humidity Sensing through Reversible Isomerization of a Covalent Organic Framework." *Journal of the American Chemical Society* 142 (2): 783–91. doi:10.1021/jacs.9b08628.

Jin, Shangbin, Ko Furukawa, Matthew Addicoat, Long Chen, Seiya Takahashi, Stephan Irle, Toshikazu Nakamura, and Donglin Jiang. 2013. "Large Pore Donor–Acceptor Covalent Organic Frameworks." *Chemical Science* 4 (12): 4505. doi:10.1039/c3sc52034j.

Kahveci, Zafer, Timur Islamoglu, Ghulam Abbas Shar, Ransheng Ding, and Hani M. El-Kaderi. 2013. "Targeted Synthesis of a Mesoporous Triptycene-Derived Covalent Organic Framework." *CrystEngComm* 15 (8): 1524–27. doi:10.1039/C2CE26487K.

Kandambeth, Sharath, Arijit Mallick, Binit Lukose, Manoj V. Mane, Thomas Heine, and Rahul Banerjee. 2012. "Construction of Crystalline 2D Covalent Organic Frameworks with Remarkable Chemical (Acid/Base) Stability via a Combined Reversible and Irreversible Route." *Journal of the American Chemical Society* 134 (48): 19524–27. doi:10.1021/ja308278w.

Kandambeth, Sharath, Bishnu P. Biswal, Harshal D. Chaudhari, Kanhu Charan Rout, Shebeeb Kunjattu H., Shouvik Mitra, Suvendu Karak, et al. 2017. "Selective Molecular Sieving in Self-Standing Porous Covalent-Organic-Framework Membranes." *Advanced Materials* 29 (2). doi:10.1002/adma.201603945.

Kang, Zixi, Ming Xue, Lili Fan, Lin Huang, Lijia Guo, Guoying Wei, Banglin Chen, and Shilun Qiu. 2014. "Highly Selective Sieving of Small Gas Molecules by Using an Ultra-Microporous Metal–Organic Framework Membrane." *Energy & Environmental Science* 7 (12): 4053–60. doi:10.1039/C4EE02275K.

Kang, Zixi, Yongwu Peng, Yuhong Qian, Daqiang Yuan, Matthew A. Addicoat, Thomas Heine, Zhigang Hu, Lincoln Tee, Zhengang Guo, and Dan Zhao. 2016. "Mixed Matrix Membranes (MMMs) Comprising Exfoliated 2D Covalent Organic Frameworks (COFs) for Efficient CO_2 Separation." *Chemistry of Materials* 28 (5): 1277–85. doi:10.1021/acs.chemmater.5b02902.

Khayum M, Abdul, Vidyanand Vijayakumar, Suvendu Karak, Sharath Kandambeth, Mohitosh Bhadra, Karthika Suresh, Nikhil Acharambath, Sreekumar Kurungot, and Rahul Banerjee. 2018. "Convergent Covalent Organic Framework Thin Sheets as Flexible Supercapacitor Electrodes." *ACS Applied Materials & Interfaces* 10 (33): 28139–46. doi:10.1021/acsami.8b10486.

Khayum, M. Abdul, Sharath Kandambeth, Shouvik Mitra, Sanoop B. Nair, Anuja Das, Samadhan S. Nagane, Rabibrata Mukherjee, and Rahul Banerjee. 2016. "Chemically Delaminated Free-Standing Ultrathin Covalent Organic Nanosheets." *Angewandte Chemie International Edition* 55 (50): 15604–8. doi:10.1002/anie.201607812.

Kim, Seoa, Won-Gwang Lim, Hyeonjung Jung, Yo Chan Jeong, Cheol-Young Park, Seung Bo Yang, Chang Hoon Lee, et al. 2025. "Protective Catalytic Layer Powering Activity and Stability of Electrocatalyst for High-Energy Lithium-Sulfur Pouch Cell." *Nature Communications* 16 (1): 1649. doi:10.1038/s41467-025-56606-2.

Kim, Soyoung, Chibeom Park, Minkyung Lee, Intek Song, Jungah Kim, Minhui Lee, Jaehoon Jung, Yousoo Kim, Hyunseob Lim, and Hee Cheul Choi. 2017. "Rapid Photochemical Synthesis of Sea-Urchin-Shaped Hierarchical Porous COF-5 and Its Lithography-Free Patterned Growth." *Advanced Functional Materials* 27 (32): 1700925. doi:10.1002/adfm.201700925.

Koenig, Steven P., Luda Wang, John Pellegrino, and J. Scott Bunch. 2012. "Selective Molecular Sieving through Porous Graphene." *Nature Nanotechnology* 7 (11): 728–32. doi:10.1038/nnano.2012.162.

Kong, Deying, Tao Bao, and Zilin Chen. 2017. "In Situ Synthesis of the Imine-Based Covalent Organic Framework LZU1 on the Inner Walls of Capillaries for Electrochromatographic Separation of Nonsteroidal Drugs and Amino Acids." *Microchimica Acta* 184 (4): 1169–76. doi:10.1007/s00604-017-2095-5.

Koo, Brian T., William R. Dichtel, and Paulette Clancy. 2012. "A Classification Scheme for the Stacking of Two-Dimensional Boronate Ester-Linked Covalent Organic Frameworks." *Journal of Materials Chemistry* 22 (34): 17460. doi:10.1039/c2jm32009f.

Kuhn, Pierre, Markus Antonietti, and Arne Thomas. 2008. "Porous, Covalent Triazine-Based Frameworks Prepared by Ionothermal Synthesis." *Angewandte Chemie International Edition* 47 (18): 3450–53. doi:10.1002/anie.200705710.

Lan, Youshi, Xianghao Han, Minman Tong, Hongliang Huang, Qingyuan Yang, Dahuan Liu, Xin Zhao, and Chongli Zhong. 2018. "Materials Genomics Methods for High-Throughput Construction of COFs and Targeted Synthesis." *Nature Communications* 9 (1): 5274. doi:10.1038/s41467-018-07720-x.

Lanni, Laura M., R. William Tilford, Muktha Bharathy, and John J. Lavigne. 2011. "Enhanced Hydrolytic Stability of Self-Assembling Alkylated Two-Dimensional Covalent Organic Frameworks." *Journal of the American Chemical Society* 133 (35): 13975–83. doi:10.1021/ja203807h.

Li, He, Xiao Feng, Pengpeng Shao, Jian Chen, Chunzhi Li, Sanjeevi Jayakumar, and Qihua Yang. 2019. "Synthesis of Covalent Organic Frameworks *via in Situ* Salen Skeleton Formation for Catalytic Applications." *Journal of Materials Chemistry A* 7 (10): 5482–92. doi:10.1039/C8TA11058A.

Li, Jian, Xiang Zhou, Jing Wang, and Xiufen Li. 2019. "Two-Dimensional Covalent Organic Frameworks (COFs) for Membrane Separation: A Mini Review." *Industrial & Engineering Chemistry Research* 58 (34): 15394–406. doi:10.1021/acs.iecr.9b02708.

Li, Shu-Ying, Wen-Hao Li, Xing-Long Wu, Yuyang Tian, Jieyu Yue, and Guangshan Zhu. 2019. "Pore-Size Dominated Electrochemical Properties of Covalent Triazine Frameworks as Anode Materials for K-Ion Batteries." *Chemical Science* 10 (33): 7695–7701. doi:10.1039/C9SC02340B.

Li, Xing, and Kian Ping Loh. 2019. "Recent Progress in Covalent Organic Frameworks as Solid-State Ion Conductors." *ACS Materials Letters* 1 (3): 327–35. doi:10.1021/acsmaterialslett.9b00185.

Li, Xing, Hui Wang, Zhongxin Chen, Hai-Sen Xu, Wei Yu, Cuibo Liu, Xiaowei Wang, Kun Zhang, Keyu Xie, and Kian Ping Loh. 2019. "Covalent-Organic-Framework-Based Li–CO_2 Batteries." *Advanced Materials* 31 (48): e1905879. doi:10.1002/adma.201905879.

Li, Xinle, Changlin Zhang, Songliang Cai, Xiaohe Lei, Virginia Altoe, Fang Hong, Jeffrey J. Urban, Jim Ciston, Emory M. Chan, and Yi Liu. 2018. "Facile Transformation of Imine Covalent Organic Frameworks into Ultrastable Crystalline Porous Aromatic Frameworks." *Nature Communications* 9 (1): 2998. doi:10.1038/s41467-018-05462-4.

Li, Zhentao, Qiaoyan Li, Zhuang Hu, Changjun Hu, Xinyue Cui, Yuanyuan Fu, and Zilin Chen. 2022. "Determination of Water in Organic Solvents and Raw Food Products by Fluorescence Quenching of a Crystalline Vinyl-Functionalized COF." *Microchimica Acta* 189 (9): 361. doi:10.1007/s00604-022-05432-0.

Li, Zhongping, Xiao Feng, Yongcun Zou, Yuwei Zhang, Hong Xia, Xiaoming Liu, and Ying Mu. 2014. "A 2D Azine-Linked Covalent Organic Framework for Gas Storage Applications." *Chemical Communications* 50 (89): 13825–28. doi:10.1039/C4CC05665E.

Li, Zonglong, Hui Li, Xinyu Guan, Junjie Tang, Yusran Yusran, Zhan Li, Ming Xue, et al. 2017. "Three-Dimensional Ionic Covalent Organic Frameworks for Rapid, Reversible, and Selective Ion Exchange." *Journal of the American Chemical Society* 139 (49): 17771–74. doi:10.1021/jacs.7b11283.

Liang, Cuiyuan, Huiming Lin, Qian Wang, Erbin Shi, Shenghai Zhou, Feng Zhang, Fengyu Qu, and Guangshan Zhu. 2020. "A Redox-Active Covalent Organic Framework for the Efficient Detection and Removal of Hydrazine." *Journal of Hazardous Materials* 381 (January): 120983. doi:10.1016/j.jhazmat.2019.120983.

Liao, Huaping, Hongmin Wang, Huimin Ding, Xiangshi Meng, Hai Xu, Baoshan Wang, Xinping Ai, and Cheng Wang. 2016. "A 2D Porous Porphyrin-Based Covalent Organic Framework for Sulfur Storage in Lithium–Sulfur Batteries." *Journal of Materials Chemistry A* 4 (19): 7416–21. doi:10.1039/C6TA00483K.

Lin, Chun-Yu, Detao Zhang, Zhenghang Zhao, and Zhenhai Xia. 2018. "Covalent Organic Framework Electrocatalysts for Clean Energy Conversion." *Advanced Materials* 30 (5). doi:10.1002/adma.201703646.

Lin, Guiqing, Huimin Ding, Rufan Chen, Zhengkang Peng, Baoshan Wang, and Cheng Wang. 2017. "3D Porphyrin-Based Covalent Organic Frameworks." *Journal of the American Chemical Society* 139 (25): 8705–9. doi:10.1021/jacs.7b04141.

Liu, Chuanyao, Yunzhe Jiang, Anjaiah Nalaparaju, Jianwen Jiang, and Aisheng Huang. 2019. "Post-Synthesis of a Covalent Organic Framework Nanofiltration Membrane for Highly Efficient Water Treatment." *Journal of Materials Chemistry A* 7 (42): 24205–10. doi:10.1039/C9TA06325K.

Liu, Shengjun, Yanling Xu, Tao Fang, Yan Ma, Bingrui Liu, and Kui Zhang. 2025. "Facile and Rapid Synthesis of Ultrafine RuCo Bimetallic Anchored in N-Doped Porous Carbon for Superior Overall Water-Splitting Performance in Alkaline Media." *International Journal of Hydrogen Energy* 98 (January): 362–69. doi:10.1016/j.ijhydene.2024.12.101.

Liu, Shuai, Lu Yao, Yi Lu, Xiaolin Hua, Jiaqiang Liu, Zhi Yang, Hao Wei, and Yiyong Mai. 2019. "All-Organic Covalent Organic Framework/Polyaniline Composites as Stable Electrode for High-Performance Supercapacitors." *Materials Letters* 236 (February): 354–57. doi:10.1016/j.matlet.2018.10.131.

Liu, Wanting, Qing Su, Pengyao Ju, Bixuan Guo, Hui Zhou, Guanghua Li, and Qiaolin Wu. 2017. "A Hydrazone-Based Covalent Organic Framework as an Efficient and Reusable Photocatalyst for the Cross-Dehydrogenative Coupling Reaction of *N*-Aryltetrahydroisoquinolines." *ChemSusChem* 10 (4): 664–69. doi:10.1002/cssc.201601702.

Lohse, Maria S., Timothée Stassin, Guillaume Naudin, Stefan Wuttke, Rob Ameloot, Dirk De Vos, Dana D. Medina, and Thomas Bein. 2016. "Sequential Pore Wall Modification in a Covalent Organic Framework for Application in Lactic Acid Adsorption." *Chemistry of Materials* 28 (2): 626–31. doi:10.1021/acs.chemmater.5b04388.

Lu, Meng, Mi Zhang, Jiang Liu, Yifa Chen, Jia-Peng Liao, Ming-Yi Yang, Yue-Peng Cai, Shun-Li Li, and Ya-Qian Lan. 2022. "Covalent Organic Framework Based Functional Materials: Important Catalysts for Efficient CO_2 Utilization." *Angewandte Chemie International Edition* 61 (15): e202200003. doi:10.1002/anie.202200003.

Lu, Xing, Songnan Li, Wei Guo, Feng Zhang, and Fengyu Qu. 2021. "A Covalent Organic Polymer–TiO2/Ti3C2 Heterostructure as Nonenzymatic Biosensor for Voltammetric Detection of Dopamine and Uric Acid." *Microchimica Acta* 188 (3): 95. doi:10.1007/s00604-021-04755-8.

Luo, Zhiqiang, Luojia Liu, Jiaxin Ning, Kaixiang Lei, Yong Lu, Fujun Li, and Jun Chen. 2018. "A Microporous Covalent–Organic Framework with Abundant Accessible Carbonyl Groups for Lithium-Ion Batteries." *Angewandte Chemie International Edition* 57 (30): 9443–46. doi:10.1002/anie.201805540.

Ma, Heping, Bailing Liu, Bin Li, Liming Zhang, Yang-Guang Li, Hua-Qiao Tan, Hong-Ying Zang, and Guangshan Zhu. 2016. "Cationic Covalent Organic Frameworks: A Simple Platform of Anionic Exchange for Porosity Tuning and Proton Conduction." *Journal of the American Chemical Society* 138 (18): 5897–5903. doi:10.1021/jacs.5b13490.

Ma, Tianqiong, Eugene A. Kapustin, Shawn X. Yin, Lin Liang, Zhengyang Zhou, Jing Niu, Li-Hua Li, et al. 2018. "Single-Crystal x-Ray Diffraction Structures of Covalent Organic Frameworks." *Science* 361 (6397): 48–52. doi:10.1126/science.aat7679.

Ma, Wenyue, Shan Jiang, Wei Zhang, Bin Xu, and Wenjing Tian. 2020. "Covalent Organic Frameworks with Electron-Rich and Electron-Deficient Structures as Water Sensing Scaffolds." *Macromolecular Rapid Communications* 41 (24): e2000003. doi:10.1002/marc.202000003.

Ma, Xingjian, and Timothy F. Scott. 2018. "Approaches and Challenges in the Synthesis of Three-Dimensional Covalent-Organic Frameworks." *Communications Chemistry* 1 (1): 98. doi:10.1038/s42004-018-0098-8.

Matsumoto, Michio, Lauren Valentino, Gregory M. Stiehl, Halleh B. Balch, Amanda R. Corcos, Feng Wang, Daniel C. Ralph, Benito J. Mariñas, and William R. Dichtel. 2018. "Lewis-Acid-Catalyzed Interfacial Polymerization of Covalent Organic Framework Films." *Chem* 4 (2): 308–17. doi:10.1016/j.chempr.2017.12.011.

Mellah, Abdelkarim, Soraia P. S. Fernandes, Ramón Rodríguez, José Otero, Jairo Paz, Jacobo Cruces, Dana D. Medina, Harik Djamila, Begoña Espiña, and Laura M. Salonen. 2018. "Adsorption of Pharmaceutical Pollutants from Water Using Covalent Organic Frameworks." *Chemistry – A European Journal* 24 (42): 10601–5. doi:10.1002/chem.201801649.

Meng, Yi, Guiqing Lin, Huimin Ding, Huaping Liao, and Cheng Wang. 2018. "Impregnation of Sulfur into a 2D Pyrene-Based Covalent Organic Framework for High-Rate Lithium–Sulfur Batteries." *Journal of Materials Chemistry A* 6 (35): 17186–91. doi:10.1039/C8TA05508D.

Meng, Zheng, Aylin Aykanat, and Katherine A. Mirica. 2019. "Proton Conduction in 2D Aza-Fused Covalent Organic Frameworks." *Chemistry of Materials* 31 (3): 819–25. doi:10.1021/acs.chemmater.8b03897.

Messner, Christoph B., Munazza R. Mirza, Matthias Rainer, Oliver M. D. Lutz, Yüksel Güzel, Thomas S. Hofer, Christian W. Huck, Bernd M. Rode, and Günther K. Bonn. 2013. "Selective Enrichment of Phosphopeptides by a Metal–Organic Framework." *Analytical Methods* 5 (9): 2379. doi:10.1039/c3ay40308d.

Mishra, Soumya Ranjan, Vishal Gadore, Saptarshi Roy, Vishwajit Chavda, and Md. Ahmaruzzaman. 2024. "Advancements in Ni-MOFs: Unravelling the Photocatalytic Tapestry for Organic Contaminant Degradation." *International Journal of Environmental Analytical Chemistry*, 1–19. doi:10.1080/03067319.2024.2343106.

Mishra, Soumya Ranjan, Vishwajit Chavda, Saptarshi Roy, Nityananda Sarkar, Vishal Gadore, Md. Ahmaruzzaman, and Bhari Mallanna Nagaraja. 2025. "Covalent Organic Framework Nanocomposites for Superior Lithium-Ion Storage: Bridging Molecular Design and Battery Technology." *Journal of Energy Storage* 108 (February): 115026. doi:10.1016/j.est.2024.115026.

Mitra, Shouvik, Himadri Sekhar Sasmal, Tanay Kundu, Sharath Kandambeth, Kavya Illath, David Díaz Díaz, and Rahul Banerjee. 2017. "Targeted Drug Delivery in Covalent Organic Nanosheets (CONs) via Sequential Postsynthetic Modification." *Journal of the American Chemical Society* 139 (12): 4513–20. doi:10.1021/jacs.7b00925.

Nagai, Atsushi, Xiong Chen, Xiao Feng, Xuesong Ding, Zhaoqi Guo, and Donglin Jiang. 2013. "A Squaraine-Linked Mesoporous Covalent Organic Framework." *Angewandte Chemie International Edition* 52 (13): 3770–74. doi:10.1002/anie.201300256.

Nandi, Shyamapada, Santosh Kumar Singh, Dinesh Mullangi, Rajith Illathvalappil, Leena George, Chathakudath P. Vinod, Sreekumar Kurungot, and Ramanathan Vaidhyanathan. 2016. "Low Band Gap Benzimidazole COF Supported Ni_3N as Highly Active OER Catalyst." *Advanced Energy Materials* 6 (24): 1601189. doi:10.1002/aenm.201601189.

Nath, Bhaskar, Wen-Hua Li, Jia-Hong Huang, Guan-E. Wang, Zhi-hua Fu, Ming-Shui Yao, and Gang Xu. 2016. "A New Azodioxy-Linked Porphyrin-Based Semiconductive Covalent Organic Framework with I_2 Doping-Enhanced Photoconductivity." *CrystEngComm* 18 (23): 4259–63. doi:10.1039/C6CE00168H.

Ni, Tianhong, Dianwei Zhang, Jing Wang, Shuo Wang, Huilin Liu, and Baoguo Sun. 2018. "Grafting of Quantum Dots on Covalent Organic Frameworks via a Reverse Microemulsion for Highly Selective and Sensitive Protein Optosensing." *Sensors and Actuators B: Chemical* 269 (September): 340–45. doi:10.1016/j.snb.2018.04.172.

Niu, Xiaoying, Sanyuan Ding, Weifeng Wang, Yali Xu, Yinyin Xu, Hongli Chen, and Xingguo Chen. 2016. "Separation of Small Organic Molecules Using Covalent Organic Frameworks-LZU1 as Stationary Phase by Open-Tubular Capillary Electrochromatography." *Journal of Chromatography A* 1436 (March): 109–17. doi:10.1016/j.chroma.2016.01.066.

Pachfule, Pradip, Manas K. Panda, Sharath Kandambeth, S. M. Shivaprasad, David Díaz Díaz, and Rahul Banerjee. 2014. "Multifunctional and Robust Covalent Organic Framework–Nanoparticle Hybrids." *Journal of Materials Chemistry A* 2 (21): 7944–52. doi:10.1039/C4TA00284A.

Palkovits, Regina, Markus Antonietti, Pierre Kuhn, Arne Thomas, and Ferdi Schüth. 2009. "Solid Catalysts for the Selective Low-Temperature Oxidation of Methane to Methanol." *Angewandte Chemie International Edition* 48 (37): 6909–12. doi:10.1002/anie.200902009.

Pandey, Prativa, Alexandros P. Katsoulidis, Ibrahim Eryazici, Yuyang Wu, Mercouri G. Kanatzidis, and SonBinh T. Nguyen. 2010. "Imine-Linked Microporous Polymer Organic Frameworks." *Chemistry of Materials* 22 (17): 4974–79. doi:10.1021/cm101157w.

Pang, Zhong-Fu, Shun-Qi Xu, Tian-You Zhou, Rong-Ran Liang, Tian-Guang Zhan, and Xin Zhao. 2016. "Construction of Covalent Organic Frameworks Bearing Three Different Kinds of Pores through the Heterostructural Mixed Linker Strategy." *Journal of the American Chemical Society* 138 (14): 4710–13. doi:10.1021/jacs.6b01244.

Patra, Bidhan Chandra, Sabuj Kanti Das, Arnab Ghosh, Anish Raj K, Parikshit Moitra, Matthew Addicoat, Sagar Mitra, Asim Bhaumik, Santanu Bhattacharya, and Anirban Pradhan. 2018. "Covalent Organic Framework Based Microspheres as an Anode Material for Rechargeable Sodium Batteries." *Journal of Materials Chemistry A* 6 (34): 16655–63. doi:10.1039/C8TA04611E.

Peng, Yongwu, Guodong Xu, Zhigang Hu, Youdong Cheng, Chenglong Chi, Daqiang Yuan, Hansong Cheng, and Dan Zhao. 2016. "Mechanoassisted Synthesis of Sulfonated Covalent Organic Frameworks with High Intrinsic Proton Conductivity." *ACS Applied Materials & Interfaces* 8 (28): 18505–12. doi:10.1021/acsami.6b06189.

Peng, Yongwu, Ying Huang, Yihan Zhu, Bo Chen, Liying Wang, Zhuangchai Lai, Zhicheng Zhang, et al. 2017. "Ultrathin Two-Dimensional Covalent Organic Framework Nanosheets: Preparation and Application in Highly Sensitive and Selective DNA Detection." *Journal of the American Chemical Society* 139 (25): 8698–8704. doi:10.1021/jacs.7b04096.

Peng, You-Yu, Bilen Akuzum, Narendra Kurra, Meng-Qiang Zhao, Mohamed Alhabeb, Babak Anasori, Emin Caglan Kumbur, Husam N. Alshareef, Ming-Der Ger, and Yury Gogotsi. 2016. "All-MXene (2D Titanium Carbide) Solid-State Microsupercapacitors for on-Chip Energy Storage." *Energy & Environmental Science* 9 (9): 2847–54. doi:10.1039/C6EE01717G.

Plas, Jan, Oleksandr Ivasenko, Natalia Martsinovich, Markus Lackinger, and Steven De Feyter. 2016. "Nanopatterning of a Covalent Organic Framework Host–Guest System." *Chemical Communications* 52 (1): 68–71. doi:10.1039/C5CC07557B.

Puthiaraj, Pillaiyar, Yu-Ri Lee, Siqian Zhang, and Wha-Seung Ahn. 2016. "Triazine-Based Covalent Organic Polymers: Design, Synthesis and Applications in Heterogeneous Catalysis." *Journal of Materials Chemistry A* 4 (42): 16288–311. doi:10.1039/C6TA06089G.

Pyles, David A., Jonathan W. Crowe, Luke A. Baldwin, and Psaras L. McGrier. 2016. "Synthesis of Benzobisoxazole-Linked Two-Dimensional Covalent Organic Frameworks and Their Carbon Dioxide Capture Properties." *ACS Macro Letters* 5 (9): 1055–58. doi:10.1021/acsmacrolett.6b00486.

Qian, Hai-Long, Cheng-Xiong Yang, and Xiu-Ping Yan. 2016. "Bottom-up Synthesis of Chiral Covalent Organic Frameworks and Their Bound Capillaries for Chiral Separation." *Nature Communications* 7 (1): 12104. doi:10.1038/ncomms12104.

Qian, Hai-Long, Cheng-Xiong Yang, Wen-Long Wang, Cheng Yang, and Xiu-Ping Yan. 2018. "Advances in Covalent Organic Frameworks in Separation Science." *Journal of Chromatography A* 1542 (March): 1–18. doi:10.1016/j.chroma.2018.02.023.

Ranjeesh, Kayaramkodath Chandran, Rajith Illathvalappil, Sairam Dnyaneshwar Veer, Joseph Peter, Vivek Chandrakant Wakchaure, Goudappagouda, K. Vipin Raj, Sreekumar Kurungot, and Sukumaran Santhosh Babu. 2019. "Imidazole-Linked Crystalline Two-Dimensional Polymer with Ultrahigh Proton-Conductivity." *Journal of the American Chemical Society* 141 (38): 14950–54. doi:10.1021/jacs.9b06080.

Rao, M. Rajeswara, Yuan Fang, Steven De Feyter, and Dmitrii F. Perepichka. 2017. "Conjugated Covalent Organic Frameworks via Michael Addition–Elimination." *Journal of the American Chemical Society* 139 (6): 2421–27. doi:10.1021/jacs.6b12005.

Rodenas, Tania, Ignacio Luz, Gonzalo Prieto, Beatriz Seoane, Hozanna Miro, Avelino Corma, Freek Kapteijn, Francesc X. Llabrés i Xamena, and Jorge Gascon. 2015. "Metal–Organic Framework Nanosheets in Polymer Composite Materials for Gas Separation." *Nature Materials* 14 (1): 48–55. doi:10.1038/nmat4113.

Rowan, Stuart J., Stuart J. Cantrill, Graham R. L. Cousins, Jeremy K. M. Sanders, and J. Fraser Stoddart. 2002. "Dynamic Covalent Chemistry." *Angewandte Chemie International Edition* 41 (6): 898–952. doi:10.1002/1521-3773(20020315)41:6<898::aid-anie898>3.0.co;2-e.

Royuela, Sergio, Joaquín Almarza, María J. Mancheño, Juan C. Pérez-Flores, Enrique G. Michel, María M. Ramos, Félix Zamora, Pilar Ocón, and José L. Segura. 2019. "Synergistic Effect of Covalent Bonding and Physical Encapsulation of Sulfur in the Pores of a Microporous COF to Improve Cycling Performance in Li-S Batteries." *Chemistry – A European Journal* 25 (53): 12394–404. doi:10.1002/chem.201902052.

Sasmal, Himadri Sekhar, Harshitha Barike Aiyappa, Siddheshwar N. Bhange, Suvendu Karak, Arjun Halder, Sreekumar Kurungot, and Rahul Banerjee. 2018. "Superprotonic Conductivity in Flexible Porous Covalent Organic Framework Membranes." *Angewandte Chemie International Edition* 57 (34): 10894–98. doi:10.1002/anie.201804753.

Schwab, Matthias Georg, Birgit Fassbender, Hans Wolfgang Spiess, Arne Thomas, Xinliang Feng, and Klaus Müllen. 2009. "Catalyst-Free Preparation of Melamine-Based Microporous Polymer Networks through Schiff Base Chemistry." *Journal of the American Chemical Society* 131 (21): 7216–17. doi:10.1021/ja902116f.

Scicluna, Marie Christine, and Liana Vella-Zarb. 2020. "Evolution of Nanocarrier Drug-Delivery Systems and Recent Advancements in Covalent Organic Framework–Drug Systems." *ACS Applied Nano Materials* 3 (4): 3097–3115. doi:10.1021/acsanm.9b02603.

Shah, Priyanka A., Vishwajit Chavda, Darshna Hirpara, Rushikesh Pokar, Sanjeev Kumar, and Pranav S. Shrivastav. 2024. "Advances in Deep Eutectic Solvent-Based Synthesis of Nanomaterials for Environmental Remediation." *Journal of Molecular Liquids* 416 (December): 126482. doi:10.1016/j.molliq.2024.126482.

Sheberla, Dennis, John C. Bachman, Joseph S. Elias, Cheng-Jun Sun, Yang Shao-Horn, and Mircea Dincă. 2017. "Conductive MOF Electrodes for Stable Supercapacitors with High Areal Capacitance." *Nature Materials* 16 (2): 220–24. doi:10.1038/nmat4766.

Shinde, Digambar Balaji, Harshitha Barike Aiyappa, Mohitosh Bhadra, Bishnu P. Biswal, Pritish Wadge, Sharath Kandambeth, Bikash Garai, Tanay Kundu, Sreekumar Kurungot, and Rahul Banerjee. 2016. "A Mechanochemically Synthesized Covalent Organic Framework as a Proton-Conducting Solid Electrolyte." *Journal of Materials Chemistry A* 4 (7): 2682–90. doi:10.1039/C5TA10521H.

Sick, Torben, Alexander G. Hufnagel, Jonathan Kampmann, Ilina Kondofersky, Mona Calik, Julian M. Rotter, Austin Evans, et al. 2018. "Oriented Films of Conjugated 2D Covalent Organic Frameworks as Photocathodes for Water Splitting." *Journal of the American Chemical Society* 140 (6): 2085–92. doi:10.1021/jacs.7b06081.

Smith, Brian J., and William R. Dichtel. 2014. "Mechanistic Studies of Two-Dimensional Covalent Organic Frameworks Rapidly Polymerized from Initially Homogenous Conditions." *Journal of the American Chemical Society* 136 (24): 8783–89. doi:10.1021/ja5037868.

Song, Yanpei, Qi Sun, Briana Aguila, and Shengqian Ma. 2019. "Opportunities of Covalent Organic Frameworks for Advanced Applications." *Advanced Science* 6 (2): 1801410. doi:10.1002/advs.201801410.

Spitler, Eric L., and William R. Dichtel. 2010. "Lewis Acid-Catalysed Formation of Two-Dimensional Phthalocyanine Covalent Organic Frameworks." *Nature Chemistry* 2 (8): 672–77. doi:10.1038/nchem.695.

Spitler, Eric L., Brian T. Koo, Jennifer L. Novotney, John W. Colson, Fernando J. Uribe-Romo, Gregory D. Gutierrez, Paulette Clancy, and William R. Dichtel. 2011. "A 2D Covalent Organic Framework with 4.7-Nm Pores and Insight into Its Interlayer Stacking." *Journal of the American Chemical Society* 133 (48): 19416–21. doi:10.1021/ja206242v.

Stegbauer, Linus, Katharina Schwinghammer, and Bettina V. Lotsch. 2014. "A Hydrazone-Based Covalent Organic Framework for Photocatalytic Hydrogen Production." *Chemical Science* 5 (7): 2789–93. doi:10.1039/C4SC00016A.

Sun, Tao, Jian Xie, Wei Guo, Dong-Sheng Li, and Qichun Zhang. 2020. "Covalent–Organic Frameworks: Advanced Organic Electrode Materials for Rechargeable Batteries." *Advanced Energy Materials* 10 (19). doi:10.1002/aenm.201904199.

Tilford, R. William, William R. Gemmill, Hans-Conrad zur Loye, and John J. Lavigne. 2006. "Facile Synthesis of a Highly Crystalline, Covalently Linked Porous Boronate Network." *Chemistry of Materials* 18 (22): 5296–5301. doi:10.1021/cm061177g.

Trewin, Abbie, and Andrew I. Cooper. 2009. "Predicting Microporous Crystalline Polyimides." *CrystEngComm* 11 (9): 1819. doi:10.1039/b906280g.

Tylianakis, Emmanuel, Emmanouel Klontzas, and George E. Froudakis. 2011. "Multi-Scale Theoretical Investigation of Hydrogen Storage in Covalent Organic Frameworks." *Nanoscale* 3 (3): 856. doi:10.1039/c0nr00505c.

Uribe-Romo, Fernando J., Christian J. Doonan, Hiroyasu Furukawa, Kounosuke Oisaki, and Omar M. Yaghi. 2011. "Crystalline Covalent Organic Frameworks with Hydrazone Linkages." *Journal of the American Chemical Society* 133 (30): 11478–81. doi:10.1021/ja204728y.

Uribe-Romo, Fernando J., Joseph R. Hunt, Hiroyasu Furukawa, Cornelius Klöck, Michael O'Keeffe, and Omar M. Yaghi. 2009. "A Crystalline Imine-Linked 3-D Porous Covalent Organic Framework." *Journal of the American Chemical Society* 131 (13): 4570–71. doi:10.1021/ja8096256.

Valentino, Lauren, Michio Matsumoto, William R. Dichtel, and Benito J. Mariñas. 2017. "Development and Performance Characterization of a Polyimine Covalent Organic Framework Thin-Film Composite Nanofiltration Membrane." *Environmental Science & Technology* 51 (24): 14352–59. doi:10.1021/acs.est.7b04056.

Varghese, Seba, Saino Varghese, Sundaram Swaminathan, Krishna Singh, and Vikas Mittal. 2015. "Two-Dimensional Materials for Sensing: Graphene and Beyond." *Electronics* 4 (3): 651–87. doi:10.3390/electronics4030651.

Vishnoi, Pratap, Dhananjayan Kaleeswaran, and Ramaswamy Murugavel. 2018. "1,3,5-Triphenylbenzene: A Versatile Photoluminescent Chemo-Sensor Platform and Supramolecular Building Block." *RSC Advances* 8 (31): 17535–50. doi:10.1039/C8RA02658K.

Wan, Shun, Felipe Gándara, Atsushi Asano, Hiroyasu Furukawa, Akinori Saeki, Sanjeev K. Dey, Lei Liao, et al. 2011. "Covalent Organic Frameworks with High Charge Carrier Mobility." *Chemistry of Materials* 23 (18): 4094–97. doi:10.1021/cm201140r.

Wan, Shun, Jia Guo, Jangbae Kim, Hyotcherl Ihee, and Donglin Jiang. 2008. "A Belt-Shaped, Blue Luminescent, and Semiconducting Covalent Organic Framework." *Angewandte Chemie International Edition* 47 (46): 8826–30. doi:10.1002/anie.200803826.

Wan, Shun, Jia Guo, Jangbae Kim, Hyotcherl Ihee, and Donglin Jiang. 2009. "A Photoconductive Covalent Organic Framework: Self-Condensed Arene Cubes Composed of Eclipsed 2D Polypyrene Sheets for Photocurrent Generation." *Angewandte Chemie International Edition* 48 (30): 5439–42. doi:10.1002/anie.200900881.

Wang, Mingyue, Hao Guo, Rui Xue, Qi Li, Hui Liu, Ning Wu, Wenqin Yao, and Wu Yang. 2019. "Covalent Organic Frameworks: A New Class of Porous Organic Frameworks for Supercapacitor Electrodes." *ChemElectroChem* 6 (12): 2984–97. doi:10.1002/celc.201900298.

Wang, Peiyuan, Mengmeng Kang, Shumin Sun, Qiang Liu, Zhihong Zhang, and Shaoming Fang. 2014. "Imine-Linked Covalent Organic Framework on Surface for Biosensor." *Chinese Journal of Chemistry* 32 (9): 838–43. doi:10.1002/cjoc.201400260.

Wang, Qinzhi, Rui Li, Yijian Zhao, Taotao Zhe, Tong Bu, Yingnan Liu, Xinyu Sun, et al. 2020. "Surface Morphology-Controllable Magnetic Covalent Organic Frameworks: A Novel Electrocatalyst for Simultaneously High-Performance Detection of p-Nitrophenol and o-Nitrophenol." *Talanta* 219 (November): 121255. doi:10.1016/j.talanta.2020.121255.

Wang, Shan, Qianyou Wang, Pengpeng Shao, Yuzhen Han, Xing Gao, Li Ma, Shuai Yuan, et al. 2017. "Exfoliation of Covalent Organic Frameworks into Few-Layer Redox-Active Nanosheets as Cathode Materials for Lithium-Ion Batteries." *Journal of the American Chemical Society* 139 (12): 4258–61. doi:10.1021/jacs.7b02648.

Wang, Xiaoyan, Linjiang Chen, Samantha Y. Chong, Marc A. Little, Yongzhen Wu, Wei-Hong Zhu, Rob Clowes, et al. 2018. "Sulfone-Containing Covalent Organic Frameworks for Photocatalytic Hydrogen Evolution from Water." *Nature Chemistry* 10 (12): 1180–89. doi:10.1038/s41557-018-0141-5.

Wei, Pi-Feng, Ming-Zhu Qi, Zhi-Peng Wang, San-Yuan Ding, Wei Yu, Qiang Liu, Li-Ke Wang, Huai-Zhen Wang, Wan-Kai An, and Wei Wang. 2018. "Benzoxazole-Linked Ultrastable Covalent Organic Frameworks for Photocatalysis." *Journal of the American Chemical Society* 140 (13): 4623–31. doi:10.1021/jacs.8b00571.

Werber, Jay R., Chinedum O. Osuji, and Menachem Elimelech. 2016. "Materials for Next-Generation Desalination and Water Purification Membranes." *Nature Reviews Materials* 1 (5): 16018. doi:10.1038/natrevmats.2016.18.

Wu, Chenyu, Yamei Liu, Hui Liu, Chenghao Duan, Qingyan Pan, Jian Zhu, Fan Hu, et al. 2018. "Highly Conjugated Three-Dimensional Covalent Organic Frameworks Based on Spirobifluorene for Perovskite Solar Cell Enhancement." *Journal of the American Chemical Society* 140 (31): 10016–24. doi:10.1021/jacs.8b06291.

Wu, Ming-Xue, and Ying-Wei Yang. 2017. "Applications of Covalent Organic Frameworks (COFs): From Gas Storage and Separation to Drug Delivery." *Chinese Chemical Letters* 28 (6): 1135–43. doi:10.1016/j.cclet.2017.03.026.

Wu, Xiaowei, Xing Han, Qisong Xu, Yuhao Liu, Chen Yuan, Shi Yang, Yan Liu, Jianwen Jiang, and Yong Cui. 2019. "Chiral BINOL-Based Covalent Organic Frameworks for Enantioselective Sensing." *Journal of the American Chemical Society* 141 (17): 7081–89. doi:10.1021/jacs.9b02153.

Xie, Yi, Mengli Xu, Li Wang, Huihui Liang, Linyu Wang, and Yonghai Song. 2020. "Iron-Porphyrin-Based Covalent-Organic Frameworks for Electrochemical Sensing H2O2 and PH." *Materials Science and Engineering: C* 112 (July): 110864. doi:10.1016/j.msec.2020.110864.

Xiong, Yifeng, Qiaobo Liao, Zhengping Huang, Xin Huang, Can Ke, Hengtian Zhu, Chenyu Dong, et al. 2020. "Ultrahigh Responsivity Photodetectors of 2D Covalent Organic Frameworks Integrated on Graphene." *Advanced Materials* 32 (9): e1907242. doi:10.1002/adma.201907242.

Xu, Junsong, Can Yang, Shuai Bi, Wenyan Wang, Yafei He, Dongqing Wu, Qifeng Liang, Xinchen Wang, and Fan Zhang. 2020. "Vinylene-Linked Covalent Organic Frameworks (COFs) with Symmetry-Tuned Polarity and Photocatalytic Activity." *Angewandte Chemie International Edition* 59 (52): 23845–53. doi:10.1002/anie.202011852.

Xu, Junsong, Yafei He, Shuai Bi, Mao Wang, Peng Yang, Dongqing Wu, Jianjian Wang, and Fan Zhang. 2019. "An Olefin-Linked Covalent Organic Framework as a Flexible Thin-Film Electrode for a High-Performance Micro-Supercapacitor." *Angewandte Chemie* 131 (35): 12193–97. doi:10.1002/ange.201905713.

Xu, Lirong, Xin Zhou, Yanxia Yu, Wei Quan Tian, Jun Ma, and Shengbin Lei. 2013. "Surface-Confined Crystalline Two-Dimensional Covalent Organic Frameworks *via* on-Surface Schiff-Base Coupling." *ACS Nano* 7 (9): 8066–73. doi:10.1021/nn403328h.

Xu, Qing, Shanshan Tao, Qiuhong Jiang, and Donglin Jiang. 2018. "Ion Conduction in Polyelectrolyte Covalent Organic Frameworks." *Journal of the American Chemical Society* 140 (24): 7429–32. doi:10.1021/jacs.8b03814.

Raghav, Sapna, Jyoti Raghav, Ma'aruf Abdulmumin Muhammad, Madhur Babu Singh, Pallavi Jain, Dinesh Kumar. 2025. Recent advancements in synthesis and applications of covalent organic frameworks. *Applied Materials Today*, 44(June): 102694. doi:10.1016/j.apmt.2025.102694

Yaghi, Omar M., Markus J. Kalmutzki, and Christian S. Diercks. 2019. *Introduction to Reticular Chemistry*. Wiley. doi:10.1002/9783527821099.

Yahiaoui, Oussama, Andrew N. Fitch, Frank Hoffmann, Michael Fröba, Arne Thomas, and Jérôme Roeser. 2018. "3D Anionic Silicate Covalent Organic Framework with Srs Topology." *Journal of the American Chemical Society* 140 (16): 5330–33. doi:10.1021/jacs.8b01774.

Yan, Xu, Yingpan Song, Jiameng Liu, Nan Zhou, Chunlin Zhang, Linghao He, Zhihong Zhang, and Zhongyi Liu. 2019. "Two-Dimensional Porphyrin-Based Covalent Organic Framework: A Novel Platform for Sensitive Epidermal Growth Factor Receptor and Living Cancer Cell Detection." *Biosensors and Bioelectronics* 126 (February): 734–42. doi:10.1016/j.bios.2018.11.047.

Yang, Cheng-Xiong, Chang Liu, Yi-Meng Cao, and Xiu-Ping Yan. 2015. "Facile Room-Temperature Solution-Phase Synthesis of a Spherical Covalent Organic Framework for High-Resolution Chromatographic Separation." *Chemical Communications* 51 (61): 12254–57. doi:10.1039/C5CC03413B.

Yang, Jin, Amitava Acharjya, Meng-Yang Ye, Jabor Rabeah, Shuang Li, Zdravko Kochovski, Sol Youk, et al. 2021. "Protonated Imine-Linked Covalent Organic Frameworks for Photocatalytic Hydrogen Evolution." *Angewandte Chemie International Edition* 60 (36): 19797–803. doi:10.1002/anie.202104870.

Yao, Linli, Yuexing Zhang, Hang-Xing Wang, Yun Guo, Zi-Min Zhuang, Wei Wen, Xiuhua Zhang, and Shengfu Wang. 2020. "An Energy and Charge Transfer Synergetic Donor–Acceptor Heterostructure 2D-COF in Photovoltaics." *Journal of Materials Chemistry A* 8 (17): 8518–26. doi:10.1039/D0TA00818D.

Ying, Yunpan, Dahuan Liu, Jing Ma, Minman Tong, Weixin Zhang, Hongliang Huang, Qingyuan Yang, and Chongli Zhong. 2016. "A GO-Assisted Method for the Preparation of Ultrathin Covalent Organic Framework Membranes for Gas Separation." *Journal of Materials Chemistry A* 4 (35): 13444–49. doi:10.1039/C6TA04579K.

Yuan, Hongye, Nanxi Li, Jiajun Linghu, Jinqiao Dong, Yuxiang Wang, Avishek Karmakar, Jiaren Yuan, et al. 2020. "Chip-Level Integration of Covalent Organic Frameworks for Trace Benzene Sensing." *ACS Sensors* 5 (5): 1474–81. doi:10.1021/acssensors.0c00495.

Yuan, Shushan, Xin Li, Junyong Zhu, Gang Zhang, Peter Van Puyvelde, and Bart Van der Bruggen. 2019. "Covalent Organic Frameworks for Membrane Separation." *Chemical Society Reviews* 48 (10): 2665–81. doi:10.1039/C8CS00919H.

Yue, Jie-Yu, Xuan-He Liu, Bing Sun, and Dong Wang. 2015. "The On-Surface Synthesis of Imine-Based Covalent Organic Frameworks with Non-Aromatic Linkage." *Chemical Communications* 51 (76): 14318–21. doi:10.1039/C5CC05689F.

Zeng, Mengqi, Yao Xiao, Jinxin Liu, Kena Yang, and Lei Fu. 2018. "Exploring Two-Dimensional Materials toward the Next-Generation Circuits: From Monomer Design to Assembly Control." *Chemical Reviews* 118 (13): 6236–96. doi:10.1021/acs.chemrev.7b00633.

Zeng, Shuai, Hui Wang, Xiangyang Li, Hailin Guo, Linfeng Dong, Chuanhang Guo, Zhenghong Chen, et al. 2025. "Composition Design of Fullerene-Based Hybrid Electron Transport Layer for Efficient and Stable Wide-Bandgap Perovskite Solar Cells." *Journal of Energy Chemistry* 102 (March): 172–78. doi:10.1016/j.jechem.2024.10.046.

Zeng, Yongfei, Ruqiang Zou, and Yanli Zhao. 2016. "Covalent Organic Frameworks for CO_2 Capture." *Advanced Materials* 28 (15): 2855–73. doi:10.1002/adma.201505004.

Zhan, Xuejun, Zhong Chen, and Qichun Zhang. 2017. "Recent Progress in Two-Dimensional COFs for Energy-Related Applications." *Journal of Materials Chemistry A* 5 (28): 14463–79. doi:10.1039/C7TA02105D.

Zhang, Cong, Min Cui, Jujie Ren, Yifei Xing, Na Li, Haiyan Zhao, Peng Liu, Xueping Ji, and Meng Li. 2020. "Facile Synthesis of Novel Spherical Covalent Organic Frameworks Integrated with Pt Nanoparticles and Multiwalled Carbon Nanotubes as Electrochemical Probe for Tanshinol Drug Detection." *Chemical Engineering Journal* 401 (December): 126025. doi:10.1016/j.cej.2020.126025.

Zhang, Gen, Masahiko Tsujimoto, Daniel Packwood, Nghia Tuan Duong, Yusuke Nishiyama, Kentaro Kadota, Susumu Kitagawa, and Satoshi Horike. 2018. "Construction of a Hierarchical Architecture of Covalent Organic Frameworks via a Postsynthetic Approach." *Journal of the American Chemical Society* 140 (7): 2602–9. doi:10.1021/jacs.7b12350.

Zhang, Kai, Song-Liang Cai, Yi-Lun Yan, Zi-Hao He, Hui-Mei Lin, Xiao-Ling Huang, Sheng-Run Zheng, Jun Fan, and Wei-Guang Zhang. 2017. "Construction of a Hydrazone-Linked Chiral Covalent Organic Framework–Silica Composite as the Stationary Phase for High Performance Liquid Chromatography." *Journal of Chromatography A* 1519 (October): 100–109. doi:10.1016/j.chroma.2017.09.007.

Zhang, Kang, Zhongjin He, Krishna M. Gupta, and Jianwen Jiang. 2017. "Computational Design of 2D Functional Covalent–Organic Framework Membranes for Water Desalination." *Environmental Science: Water Research & Technology* 3 (4): 735–43. doi:10.1039/C7EW00074J.

Zhang, Panpan, Faxing Wang, Minghao Yu, Xiaodong Zhuang, and Xinliang Feng. 2018. "Two-Dimensional Materials for Miniaturized Energy Storage Devices: From Individual Devices to Smart Integrated Systems." *Chemical Society Reviews* 47 (19): 7426–51. doi:10.1039/C8CS00561C.

Zhang, Shuaihua, Qian Yang, Xingtao Xu, Xiaohong Liu, Qian Li, Jingru Guo, Nagy L. Torad, et al. 2020. "Assembling Well-Arranged Covalent Organic Frameworks on MOF-Derived Graphitic Carbon for Remarkable Formaldehyde Sensing." *Nanoscale* 12 (29): 15611–19. doi:10.1039/D0NR03041D.

Zhang, Xiaozhi, Qi Lu, Yaojie Yan, Tingting Zhang, Shujuan Liu, Meirong Cai, Qian Ye, Feng Zhou, and Weimin Liu. 2023. "Tribochemical Synthesis of Functionalized Covalent Organic Frameworks for Anti-Wear and Friction Reduction." *Friction* 11 (10): 1804–14. doi:10.1007/s40544-022-0696-4.

Zhang, Yuanyuan, Xiao Feng, Haiwei Li, Yifa Chen, Jingshu Zhao, Shan Wang, Lu Wang, and Bo Wang. 2015. "Photoinduced Postsynthetic Polymerization of a Metal–Organic Framework toward a Flexible Stand-Alone Membrane." *Angewandte Chemie International Edition* 54 (14): 4259–63. doi:10.1002/anie.201500207.

Zhang, Yuwei, Xiaochen Shen, Xiao Feng, Hong Xia, Ying Mu, and Xiaoming Liu. 2016. "Covalent Organic Frameworks as PH Responsive Signaling Scaffolds." *Chemical Communications* 52 (74): 11088–91. doi:10.1039/C6CC05748A.

Zhao, Xiaojia, Pradip Pachfule, and Arne Thomas. 2021. "Covalent Organic Frameworks (COFs) for Electrochemical Applications." *Chemical Society Reviews* 50 (12): 6871–6913. doi:10.1039/D0CS01569E.

Zhi, Yongfeng, Ziping Li, Xiao Feng, Hong Xia, Yumin Zhang, Zhan Shi, Ying Mu, and Xiaoming Liu. 2017. "Covalent Organic Frameworks as Metal-Free Heterogeneous Photocatalysts for Organic Transformations." *Journal of Materials Chemistry A* 5 (44): 22933–38. doi:10.1039/C7TA07691F.

Zhou, Tian-You, Shun-Qi Xu, Qiang Wen, Zhong-Fu Pang, and Xin Zhao. 2014. "One-Step Construction of Two Different Kinds of Pores in a 2D Covalent Organic Framework." *Journal of the American Chemical Society* 136 (45): 15885–88. doi:10.1021/ja5092936.

Zhu, Youlong, Shun Wan, Yinghua Jin, and Wei Zhang. 2015. "Desymmetrized Vertex Design for the Synthesis of Covalent Organic Frameworks with Periodically Heterogeneous Pore Structures." *Journal of the American Chemical Society* 137 (43): 13772–75. doi:10.1021/jacs.5b09487.

Zhu, Yuanzhi, Man Qiao, Wenchao Peng, Yang Li, Guoliang Zhang, Fengbao Zhang, Yafei Li, and Xiaobin Fan. 2017. “Rapid Exfoliation of Layered Covalent Triazine-Based Frameworks into N-Doped Quantum Dots for the Selective Detection of Hg^{2+} Ions.” *Journal of Materials Chemistry A* 5 (19): 9272–78. doi:10.1039/C7TA01438D.

Zhuang, Xiaodong, Wuxue Zhao, Fan Zhang, Yu Cao, Feng Liu, Shuai Bi, and Xinliang Feng. 2016. “A Two-Dimensional Conjugated Polymer Framework with Fully Sp^2-Bonded Carbon Skeleton.” *Polymer Chemistry* 7 (25): 4176–81. doi:10.1039/C6PY00561F.

Zwaneveld, Nikolas A. A., Rémy Pawlak, Mathieu Abel, Daniel Catalin, Didier Gigmes, Denis Bertin, and Louis Porte. 2008. “Organized Formation of 2D Extended Covalent Organic Frameworks at Surfaces.” *Journal of the American Chemical Society* 130 (21): 6678–79. doi:10.1021/ja800906f.

2 Advanced Synthesis Techniques for Covalent Organic Frameworks

Vishnu Kumar Khandelwal and Bhawana Jangir

2.1 INTRODUCTION

By encouraging the regulated development and creation of regular, porous, and crystalline structures, reticular chemistry, a relatively young field in material science and organic chemistry, focuses on the logical and strategic synthesis of novel organic molecules (Yaghi, 2016; Yaghi et al., 2019). Solids with a regular, predetermined structure can be produced by selecting suitable building blocks with well-defined symmetry. These solids' secondary building units are repeating elements with certain geometrical properties that might be intriguing from a technological or industrial perspective (Yaghi et al., 2003).

The creation of a new class of materials known as metal-organic frameworks (MOFs) was the first noteworthy outcome of using ideas from reticular chemistry. As a subclass of coordination polymers, these new structures consist of ions from metals (or clusters) and multifunctional organic binding agents that may work together to form two- (2D) or three-dimensional (3D) subunits (Yaghi et al., 2019).

The chemical ligands that make up the backbone of an MOF's reticulated 3D structure are typically stiff and geometrically well-defined. However, the metal ions function as coordination points or vertices. Li and colleagues created the first synthetic MOF structure (MOF-5) by coordinating Zn ions with dicarboxylate ligands (Li et al., 1999). The $Zn_4(O)O_{12}C_6$ clusters that provide the linking points in this framework are made up of four Zn^{2+} atoms around an oxygen atom. The six 1,4-benzene dicarboxylate ligands' carboxylic groups are subsequently bound to these atoms. A regular, permanent cubic-shaped structure is produced by the ligands' bifunctionality and symmetry, which further coordinates with two additional adjacent zinc ions from neighboring clusters (Li et al., 1999).

MOFs are defined as materials that are crystalline, stiff, highly porous, and stabile (Safaei et al., 2019). Energy storage, gas and pollutant adsorption, chemical sensing, and heterogeneous catalysis are some selective applications that can be achieved by controlling the crystalline framework and the shape of the pores, depending on the ligands and corresponding tridimensional structures (Li et al., 1999; Czaja et al., 2009; Yu et al., 2019, 2021).

The scientific community has recently become interested in COFs due to their special properties and the potential they have shown for the synthesis of highly

 DOI: 10.1201/9781003605386-2

precise, targeted materials with spatial accuracy at the atomic scale (Lohse et al., 2018). Covalent bonds' directionality and binding strength enable the creation of previously unheard-of organic structures in the field of polymer chemistry, leading to materials with low density, stiffness, stability, and resistance as well as well-defined, permanent porosity (Raghav et al., 2025).

The creation of MOFs (metal-organic hybrid class) and a number of other (usually less crystalline) organic polymer subclasses such as porous aromatic frameworks, conjugated microporous polymers, and polymers with intrinsic microporosity hyper cross-linked polymers (Waller et al., 2015; Lohse et al., 2018).

Different pore sizes, forms, and distributions can be found throughout the material as a result of condensed polymers with varying shapes and point group symmetries, which will produce structures with varying topologies and degrees of reversibility (Lyle et al., 2019; Geng et al., 2020). The dimensional structure of the theoretical unit cell used to depict the crystalline framework determines whether COFs are referred to as 2D or 3D. Pyromellitic dianhydride and tris(4-aminophenyl) amine (Fang et al., 2014) form the imide-type structure known as PI-COF, and it is a 2D COF example (Figure 2.1). The dense packing of those planes via dipolar interactions between molecules or, in the case of aromatic-rich structures, π–π orbital stacking, constitutes the 3D shape of the actual material that is formed.

The resulting COF will be 3D if one monomer is 3D. A 3D COF is COF-320, which has the structure of an imine-type constructed from 4,4′-biphenyldicarbaldehyde and tetracids(4-aminophenyl) methane (TAPM) (Zhang et al., 2013) (Figure 2.1). Côté et al. created the first COFs documented in the literature in 2005 by reacting derivatives of boronic acid (Côté et al., 2005). One of the materials, COF-1, was created by self-condensing diboronic acid, which produced a substance with a 2D structure

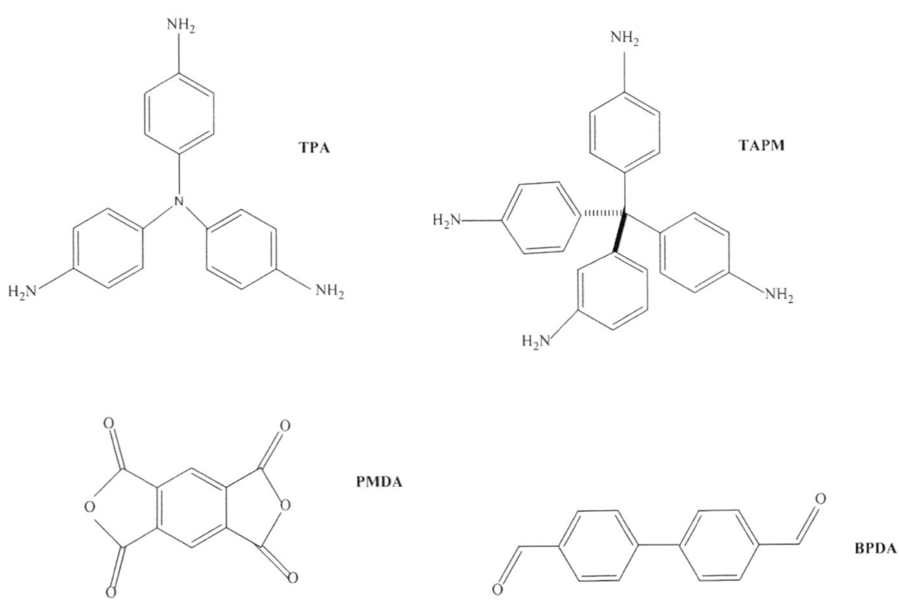

FIGURE 2.1 (a) PI-COF and (b) COF-320 constructing monomer structures.

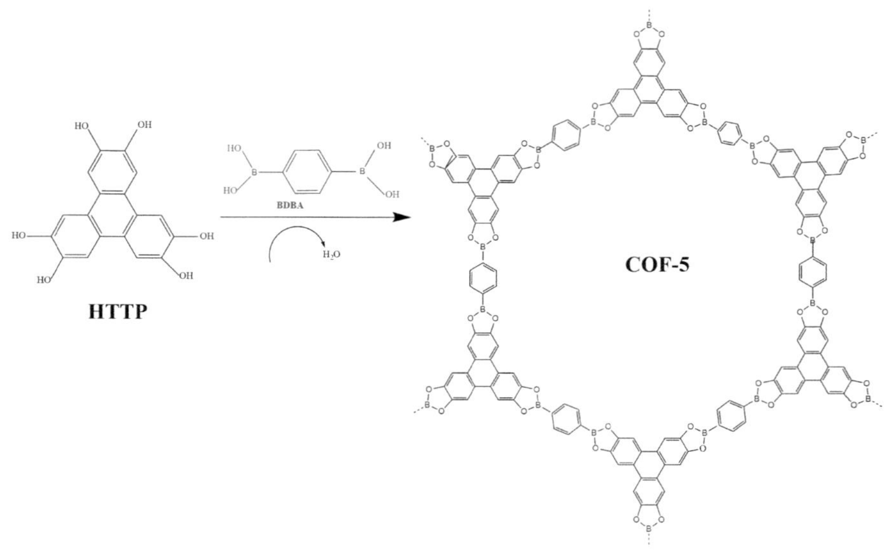

SCHEME 2.1 COF-5 synthesis.

made of boroxine-type linkages that had a diameter of about 1.5 nm (Côté et al., 2005). The second chemical, COF-5, which has a reticulated structure of boronic ester linkages, was produced by the reaction of BDBA with 2,3,6,7,10,11-hexahydro xytriphenylene (Scheme 2.1).

However, the most thermodynamically likely result in practice should be the creation of an amorphous substance. Therefore, one of the primary problems in effective COF synthesis is to induce order to the structure throughout the synthetic process (Côté et al., 2005; Waller et al., 2015; Lohse et al., 2018). Controlling the reversibility of the condensation reaction is one way to do this (Song et al., 2019). Retaining water in the reactant mixture is essential in the case of COF-5 because it permits the reaction to smoothly return to the initial monomers, allowing for the "correction" of conformational "errors" in the polymer structure and the emergence of a more ordered framework.

2.2 TYPES OF COF LINKAGES: STRUCTURE AND CHARACTERISTICS

Based on various kinds of chemical bonds and linkages, a broad variety of monomers have been studied to create new COFs with a range of characteristics and uses. Particularly intriguing among them are COFs containing imine, hydrazone, azine, imide, and β-ketoenamine (Lohse et al., 2018). Polyaldehydes and polyamines condense to produce imine-type COFs. Because they are generally more resistant to hydrolysis, they are more stable than COFs that include boronic ester bonds (Lohse et al., 2018). Uribe-Romo and colleagues produced the first imine-type COF, known as COF-300, by condensation between TAPM and BDA (Uribe-Romo et al., 2009) (Figure 2.2).

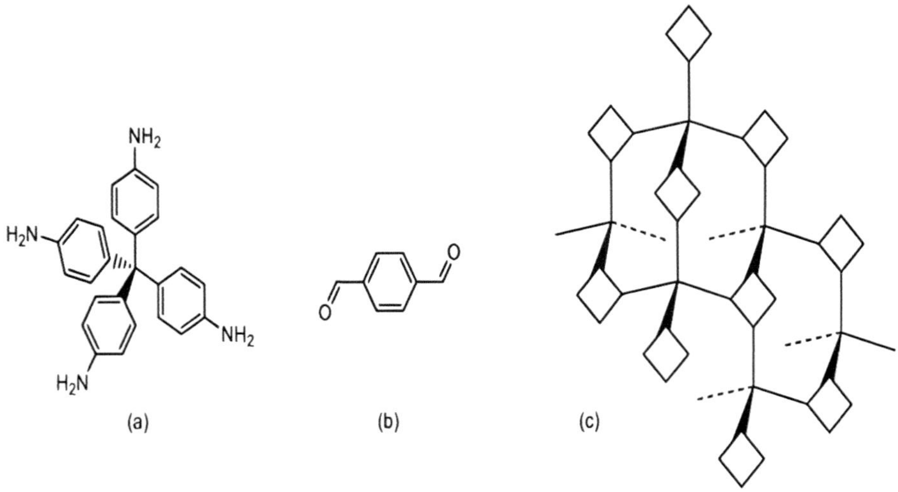

FIGURE 2.2 The schematic 3D structure of COF-300, as well as the structures of (a) TAPM and (b) BDA monomers.

Ding et al. developed the first imine-type COF with a 2D unit cell, COF-LZU1, utilizing 4-phenylenediamine and 1,3,5-triformylbenzene (Ding et al., 2011). Nearby layers formed by nitrogen atoms only 3.7 Å apart in its eclipsed state, the material is very nitrogen-rich and facilitates beneficial metal ion coordination. The diameter of its pores is 1.8 nm. With exceptional conversion rates, COF-LZU 1 was the first COF structure to be assessed using this property in heterogeneous catalysis (Ding et al., 2011).

According to the current study, COFs with imine groups grow through distinct, separate, isolable fragments that eventually join to form supramolecular structures, as opposed to boronic ester-type COFs, which progressively extend from the initially produced crystal cores (Smith et al., 2016). Symmetric aldehydes and hydrazides or hydrazine, respectively, go through a reversible reaction to produce COFs with hydrazone and azine linkages, just as imine-type COFs. By mixing 2,5-diethoxy with hydrazone (DETH) with trigonal tri-aldehydes of varying dimensions, Uribe-Romo and colleagues reported synthesizing the first two COFs having hydrazone activity (Uribe-Romo et al., 2011). One of the materials made from DETH and (1,3,5-Triformylbenzene) (TFB) COF-42, has thermal stability up to 280°C, as shown in Figure 2.3.

Dynamic synthetic conditions can be used to modify the crystallization process since the hydrazone bond reversibility is especially pH sensitive (Yaghi et al., 2019). COF-42 has 2.8 nm porosity and a 720 $m^2 g^{-1}$ surface area because of the larger size of the building monomers (Uribe-Romo et al., 2011).

Polyaldehydes and hydrazine undergo a condensation process to create azoine-type COFs. Dalapati et al. synthesized the first structure of its sort, known as Py-Azine COF, using 1,3,6,8-tetrakis (4-formyl phenyl) pyrene as a building component. A rhombic shape is induced in the crystalline structure by the planar C2 metaldehyde

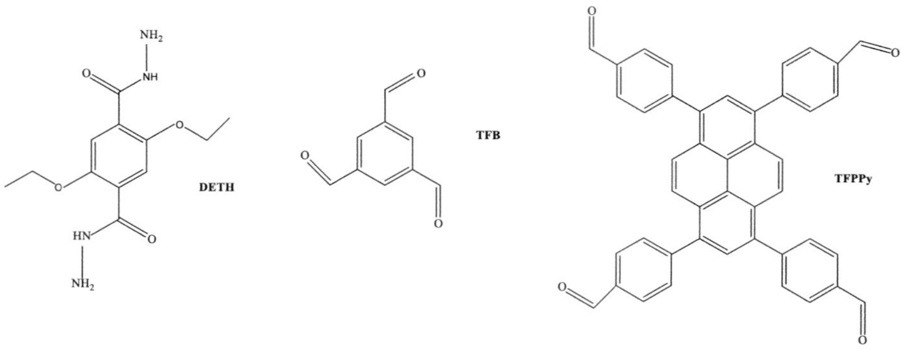

FIGURE 2.3 The building components for COF-42 (a) and Py-Azine COF (b).

Topy (Dalapati et al., 2013) (Figure 2.3). Because hydrazine is a much smaller molecule than the analogous Topy monomer, it stacks favorably across 2D layers to form a crystalline and rigid structure.

Because they develop and crystallize in two distinct phases, COFs of the β-ketenimine type are a specific example of this family of materials. The standard C=N reversible reaction between an amine and an aldehyde makes up the first step. The second, irreversible process is the tautomerization of keto enol, which results in an enamine bond (Lohse et al., 2018).

When the aldehyde employed as a building block has a hydroxyl group next to the formyl group that is reacting (bonded to the β-carbon), tautomerization takes place. One common monomer for this kind of reaction is 1,3,5-triformylphloroglucinol (TFP). Protons from the OH group go to the nearby N atoms during the tautomerization process, creating an enamine group (Scheme 2.2). β-ketenimine COFs have unique chemical resistance due to their irreversible reactivity, maintaining their structural integrity even under extreme pH settings (Yaghi et al., 2019). However, because conformational error correction is impossible after tautomerization, they are usually less crystalline materials (Lohse et al., 2018)

Tappe-COF (Kandambeth et al., 2012) is the first β-ketoenamine COF, called, which was made from TFP and 4-phenylenediamine (Scheme 2.2). In addition to its regular, crystalline form, the material in question demonstrated exceptional chemical resilience, retaining its structural integrity in both an aqueous solution at 100°C and a solution of 9 mol•dm^{-3} HCl (Annabeth et al., 2012).

2.3 GROUP REPLACEMENTS AND MODIFICATIONS

Since external substances must interact with the walls, internal surfaces, and intrapore regions of typical COF applications, it is crucial to identify and regulate key COF properties. The chemical environment is determined by the surface atoms' intrinsic chemical properties and any interactions they may have with the guest molecules. Adding substitution groups to the building blocks is a useful method of altering the chemical properties and customizing them for specific, focused applications. COF changes can be made in situ, after, or before (Yaghi et al., 2019) (Scheme 2.3).

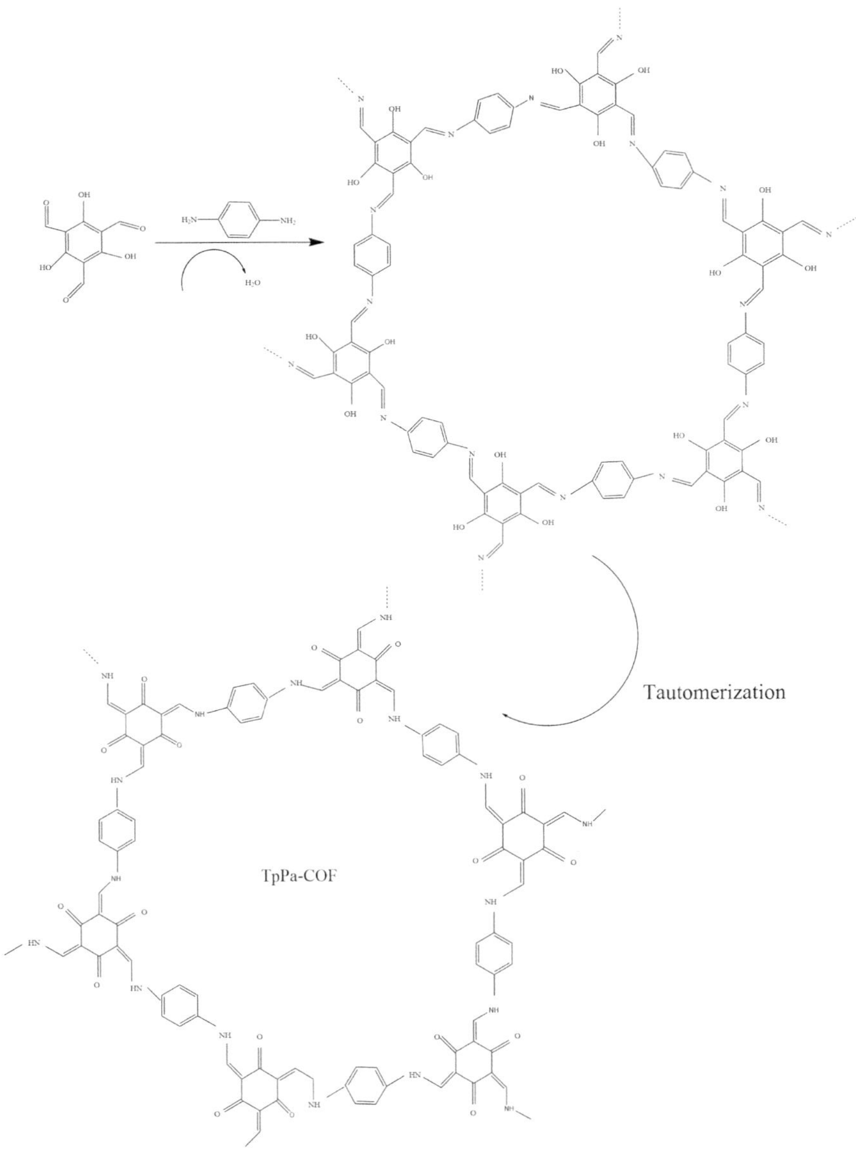

SCHEME 2.2 Tappe-COF synthesis and the ensuing keto-enol tautomerization.

Presynthetic modifications entail giving the building block structures enough functional groups prior to the material's synthesis. A possible drawback of this approach is the need to make sure the functional groups do not interfere with the COF crystallization process and stay stable in reacting conditions. Depending on the functional groups and conditions, this could be a difficult process.

According to Scheme 2.3, a covalent postsynthetic functionalization requires the addition of reactive sites to the original monomers. Ensuring that the COF has

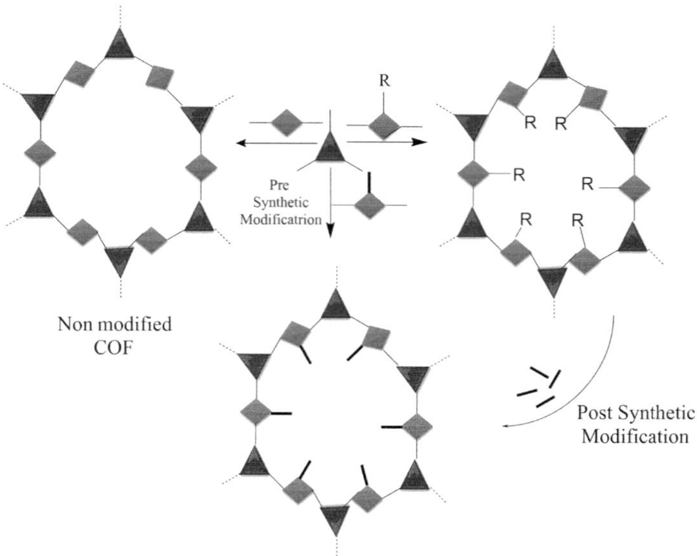

SCHEME 2.3 Illustration of the paths for COF construction. (Modified from Dogru, 2012.)

sufficient chemical stability under the novel postsynthetic reaction conditions may ultimately prove to be a difficult compromise.

After COF is ready, a heterogeneous reaction takes place during the functionalization process. The reaction should produce the fewest secondary products, be favorable, and have a high yield (Lyle et al., 2019). Furthermore, one special characteristic of COFs that has led to this class being referred to as both products and starting reagents for organic synthesis is postsynthetic modification (Lyle et al., 2019; Song et al., 2019).

When COF is treated with an organic azide in the presence of Cu(I), a cycloaddition process takes place, forming a heterocyclic 5-atom ring with the selected functional group. A different functional group, including ethyl, acetate, hydroxyl, carboxyl, or amine groups, may be introduced using click reactions, which are typically high yielding and can be carried out under mild reacting conditions (Yaghi et al., 2019).

For the introduction of carboxylic groups, for instance, ring-opening reactions using succinic anhydride offer an alternate postsynthetic method (Lyle et al., 2019; Yaghi et al., 2019). Acidic group's postsynthetic functionalization is much more crucial since their presence might potentially interfere with the regular crystallization process of COF. Various uses have also been investigated for adding amine groups to COF pore walls. This summarizes a number of additional postsynthetic functionalization techniques for adding various groups, including ether, thiol, amide, amidoxime, or quaternary amines.

To change the chemical characteristics inside the porous areas, an alternative strategy involves making structural changes to the COF walls. These methods can be useful when COF syntheses with certain linkage types are not desirable or practical; the problem can be handled by generating an analogous COF and then chemically changing its links. For example, postsynthetic oxidation processes involving sodium chlorite and 2-methylbut-2-ene in a dioxane/acetic acid mixture can convert imidine-type COFs into amide-type structures; sulfur treatment can convert them into thiazole-type COFs; and a reaction with 2,5-diaminehydroquinone in the presence of water, DMF, and molecular oxygen can change them into benzoxazole-type COFs (Fernandes et al., 2019; Lyle et al., 2019; Yaghi et al., 2019).

2.4 NONCONVENTIONAL METHODS (SOLVOTHERMAL SYNTHESIS)

Microwave irradiation offers significant benefits over conventional techniques like electric heating, sand baths, and oil baths. First, since the heat is generated directly in the material that is required, the heating system does not need to use as much energy, making it less expensive to run. Second, there is a shorter heating period and some convectional heat transfer from the surface due to microwave radiation. Consequently, Microwaves (MWs) reduce the distinctive temperature profile from outside to inside created by conventional heating techniques by producing a better temperature distribution and rigorously and precisely controlling the experimental settings.

Given the aforementioned benefits, various functional materials involving metal oxides, MOFs, and carbon quantum dots (CDs) have been efficiently, quickly, and neatly produced by microwave irradiation (Y. Liang et al., 2021; Y. Liu et al., 2021). Researchers will undoubtedly be very interested in this simple and ecofriendly microwave-assisted synthesis. The first example of COFs (donated as COF-5 and COF-102) made using microwave-aided synthesis was shown by Cooper et al. in 2009. Hexahydroxytriphenylene (HHTP) and 1,4-Benzenediboronic acid (BDBA) (a molar ratio of 2:3) were condensed for around 20 minutes in a sealed glass microwave tube shielded by N_2 using 1,4-dioxane (20 mL) as solvents and mesitylene to create COF-5.

A technique for the development of industrial scales is provided by the 200-fold acceleration of the synthesis speed compared to traditional solvothermal synthesis, which significantly lowers the time and energy costs. The COF-5 produced by MWs possesses superior material characteristics, including a greater yield (up to 95%) and size (less than 500 nm).

At first, it was only employed in the production of borate COFs. However, because it readily breaks down in the presence of water or other organic solvents, this type of COF's application range is restricted to anhydrous or inert conditions. The Schiff base reaction of 1,3,5-trimethylbenzene/1,4-dioxane/3 mol L^{-1} acetic acid (HAc) (7 mL, 3:3:1) as the solvent under microwave heating for 60 minutes (Figure 2.2) (Vazquez-Molina et al., 2018) was used by Zhang et al. (2015) to get TpPa–COF–MW.

In 2018, Uribe-Romo's group created a series of COFs with different side chains by condensation between Tp and the side-chain derivatives of 4,4-diamino-p-terphenyl (TP-R) using a mixture of 1,2-dichlorethane (DCE), 1,4-dioxane as the solvent and 8 mol L^{-1} HAc as the catalyst (v/v/v 17:17:1) at 170°C and 250 W for 20 minutes (L. Chen et al., 2020). By adding -F groups, Zhang et al. (2010) created extremely hydrophobic COFs, which they then effectively employed for oil–water separation.

Microwave-assisted approach is also used to build covalent triazine frameworks (CTFs) because of its various benefits. Lewis's acid zinc chloride ($ZnCl_2$) catalyzes the trimerization of aromatic nitriles at high temperatures to produce CTF. The development of microwave-assisted synthesis offers a more straightforward, efficient, and cost-effective way to create CTF materials. More significantly, $ZnCl_2$ is a microwave reaction medium that works well and helps the process heat up quickly.

Using $ZnCl_2$ as a catalyst and as a DCB reaction monomer for 10–60 minutes at 280 W, Zhang et al. (E. Vitaku et al., 2017) produced CTF materials with varying specific surface areas for the first time in 2010. This resulted in a considerable reduction in reaction time. Subsequently, the microwave-assisted approach was used to successfully manufacture more COFs (Yang et al., 2012; Das et al., 2017; Wang et al., 2020). Thus, based on the experimental findings, we may conclude that, within a specific range, microwave-assisted solve-thermal synthesis can emerge as a potent substitute for the solvothermal approach. We should not overlook their drawbacks while appreciating their benefits. Clearly, its general applicability is less than that of solvothermal techniques. Therefore, investigating various techniques to synthesize COFs remains a significant problem going forward.

2.5 OTHER ENERGY METHODS

Inspired by microwave-assisted synthesis, other cutting-edge energy sources, including sonochemistry synthesis (Kim et al., 2017), photochemical processes synthesis (Zhang et al., 2020), electron beam radiation-induced synthesis (Jiang et al., 2021), and plasma synthesis techniques (Zhao et al., 2022), have also been used gradually to create COFs. In 2012, COF-1 and COF-5 were produced in ambient conditions for one to two hours utilizing the soon-chemistry technique (Kim et al., 2017). The resulting COF-5 has an average crystal size of around 250 nm, which is about 100 times smaller than its traditional solvothermal cousin. Additionally, the effects of ultrasonication time and monomer concentration on the experimental data are investigated, demonstrating that ultrasound influences the reaction's temperature and improves the mixture's monomer homogeneity. Stated differently, ultrasonography influences the reaction's temperature and helps ensure that the monomers in the mixture are homogeneous. As a result, ultrasonography causes a very high local temperature and pressure by hastening the formation and dissolution of bubbles in the solution.

This accelerates the pace of heating and cooling as well as the rate of crystallization. Sonochemistry has recently advanced significantly in the synthesis of COFs once more. Cooper et al. prepared various imine-linked COFs in aqueous acetic acid solution in less than 60 minutes (Gan et al., 2022). Additionally, they found that COFs made using this technique have similar or superior crystallinity, porosity, and photocatalytic activity to those made using the solvothermal approach.

2.6 SYNTHESIS AT ROOM TEMPERATURE

Compared to the previous methods, synthesis at room temperature is, in some ways, a more useful and energy-efficient method of material formation. This approach offers several outstanding additional features, such as the following, in addition to the benefits of easy operation and no heating. High-temperature and high-pressure experimental conditions are dangerous from the standpoint of product synthesis and safety, for example, most commonly used non-aqueous solvents, including dimethyl sulfoxide and mesitylene, easily volatilize at high temperatures, causing environmental contamination.

However, many monomers become unstable at high temperatures, leading to various property changes that may produce by-products that affect the form of COFs and reduce their stability. On the other hand, room temperature synthesis in appropriate solvents can produce COFs with improved characteristics and controlled shape, making up for the drawbacks of the previously listed techniques. Preparing COFs with consistent and manageable morphology and improving their processing capabilities in application design are the biggest application challenges. The development of this approach offers further approaches to the aforementioned problems.

Banerjee's group verified the Ostwald ripening hypothesis in 2016, which they feel may adequately detail the creation mechanism of the HFH-COFs (Mahmoud et al., 2013; Annabeth et al., 2015; Wang et al., 2022). For the synthesis of COFs with regulated morphology, we think this is quite valuable. Simultaneously, our team used the same approach to create flower-like COFs (De la Peña Ruigómez et al., 2015).

The straightforward and gentle room-temperature synthesis process was also used to carry out a number of intriguing investigations pertaining to real-world applications. Using a straightforward one-pot reaction technique at room temperature, Zamora and colleagues created COFs with a hexagonal structure (donated as RT–COF–1) (Mow et al., 2020). They also showed how soft lithography and inkjet printing methods might be used to design RT-COF-1 on certain solid or soft-supported surfaces.

The vast majority of COFs cannot be processed, which further limits the variety of possible future applications for COFs. Braunecker's group was able to get around the issues by creating stable colloidal suspended particles in the 200–600 nm range using nitrile-based solvent suspension COFs. This work supports postprocessing and is crucial for increasing the useful applications of COFs (Dey et al., 2017).

The production of COF films at room temperature has, surprisingly, been the subject of several papers, which is crucial for real-world applications (Zhu et al., 2021). Then, a machinable COF micellar aqueous solution was also created utilizing the room temperature emulsion system consisting of water and surfactant.

Researchers have recently become interested in the quick synthesis of COFs using different catalysts. For the first time, Zhu et al. used nitrate of transition metal as a catalyst to create a series of imine COFs at ambient temperature. $Fe(NO_3)_3{\bullet}9H_2O$ has the highest catalytic performance, synthesizing a range of 2D imine COFs, 3D COFs, and azine-linked COFs in less than 10 minutes. It also exhibits superior universality. Their approach offers a straightforward and affordable method of creating COFs.

Simultaneously, the synthesis of COFs was catalyzed by various metal oxides, and the experimental procedure was greatly simplified by the one-step synthesis of metal oxide/COF composite materials (Smith et al., 2016). When compared to the solvothermal method, room temperature synthesis is typically more cost-effective, convenient, and environmentally friendly than synthesis. More precisely, because of volatilization, it can lessen the risk that organic solvents pose to researchers and the environment. In the meantime, this approach can prepare COFs with various morphologies by better controlling the reaction process. Furthermore, the produced COFs typically demonstrated regulated shape, outstanding crystallinity, and other characteristics. However, lower reaction temperatures are detrimental to the building blocks' ability to dissolve in the solvent, which could restrict the use of amines and aldehydes containing many benzene rings in the creation of COFs.

2.7 NOVEL REACTION MEDIUM FOR COF SYNTHESIS: SYNTHESIS MEDIATED BY THE AQUEOUS PHASE

Since water is the most abundant and environmentally benign solvent found in nature, it has attracted a lot of interest as a reaction medium in the synthesis of COFs. The link between amorphous imine polymers and crystalline 2D COFs, for instance, as well as the primary factors influencing the transition from the amorphous state to the crystalline form in water were studied by Dichter's group using the imine condensation reaction of TPB and PDA as a model (Stewart et al., 2017).

By preserving the reversibility of reaction equilibrium throughout the imine condensation process, water has shown the capacity to enhance the crystallinity of COFs within a specific range. To produce COFs, scientists have therefore proposed using water as the solvent. But since most reaction monomers are organic amines or aldehydes with several aromatic rings, they are little soluble in water, if at all. This poses a number of challenges for the formation of crystal structures and the occurrence of condensation between aldehyde and amine. A few high-temperature and high-pressure hydrothermal methods using water as the only solvent as well as an aqueous phase synthesis approach using water and additional solvents or gases as the reaction media have been proposed in light of the aforementioned. The ultimate objective of green chemistry has always been to use water as the sole reaction solvent. To produce hexagonal layered 2D and diamondoid 3D amide-linked COFs, respectively, Rosinsky's group (Lu et al., 2019) employed water as the sole reaction solvent in a condensation reaction between 1,3,5-benzenetricarboxylic acid or trans-1,4-diamino cyclohexane and 4,4′,4″,4‴-methane tetra benzoic acid in 2017. For the first time, hydrothermal synthesis in hot water was used to create amide COFs that were extremely stable and well-organized.

The clean synthesis of HCOF-1 was then carried out in 2019 using this hydrothermal technique Martín-Illán et al., 2020). First, top and hydrazine hydrate were dissolved in water using a standard process. After 10 minutes of complete decontamination in a liquid nitrogen bath, the mixture was sealed and allowed to sit at 120°C for several hours. This resulted in a red solid that was moderately crystalline and had a high BET surface area (617 $m^2 g^{-1}$). Compared to earlier studies (Lu et al., 2019), this method showed a number of advantages. High pressure, for instance, is not necessary and is better suited for straightforward processes and large-scale manufacturing.

In 2020, 1,3,5-benzenetricarbaldehyde (BTCA) and 1,3,5-tris-(4-aminophenyl) benzene (TAPB) were condensed in a mixed solution of H_2O and hack at 80°C for five days, resulting in a highly crystalline COF known as TAPB-BTCA-COF (Ma et al., 2022). Highly crystalline COFs are simple, rapid, and environmentally safe, and they can be easily produced at the gram level. Since they produce COFs with consistent morphology and provide the previously mentioned benefits, many water-soluble additives, including surfactants, can be thought of as morphological regulators or template-directing agents (Franco et al., 2020).

As a result, more ecologically friendly methods for creating premium COFs with controlled morphology are developed. For a succinct evaluation of the current corpus of linked literature, the following will select the representative literature (Zhang et al., 2018; Martín-Illán et al., 2020). Because monomer solubility in solvent is important, COF-LZU1 and Cu-COF-LZU1 were synthesized at room temperature using the low-energy, environmentally safe green CO_2-assisted protocol, which involved dissolving CO_2 in water under high pressure to create a CO_2–H_2O system as the reaction system (Welton et al., 1999).

This method offers the following advantages over other protocols in the design stage:

1. By altering CO_2 pressure, one can change the structure and performance of COFs, which also alters the solvent's performance.
2. Carbon dioxide can be used directly as a catalyst in the reaction with H_2O to form carbonic acid.
3. The product's post-treatment is straightforward, and CO_2 can be extracted straight through decompression.
4. The solvent is safe, nontoxic, and environmentally friendly. In terms of material properties, the COFs produced with this method have excellent performance as metal catalyst carriers, good crystallinity, nanoscale size, and superior surface area.

The production of COFs with a consistent shape and adjustable size has gradually become a research hotspot in recent years, along with the production of new COFs. In 2020, using micelles as the reaction chamber, crystalline COF colloids were created at room temperature and ambient pressure, inspired by the biological system's minuscule volume (Zhang et al., 2018). The best option for a reaction solvent is H_2O because it is inexpensive, readily available, and environmentally friendly. However, the monomer's weak water solubility makes it difficult to create COFs, let alone maintain their regular shape, which restricts their usefulness in industry. Therefore, more research is needed to determine if water can be used generally as a reaction medium for the preparation of COFs.

2.8 IONIC LIQUID-MEDIATED SYNTHESIS

Apart from the green solvents that were previously considered, such as water and supercritical carbon dioxide, ionic liquids (ILs) have emerged as a novel alternative (Rantwijk et al., 2007). In 1914, Walden reported the first ethyl ammonium nitrate IL ([$EtNH_3$] [NO_3]) with a melting point of 12°C (Zheng et al., 2019). Later, he described ILs using halogen salt as the anion and imidazolium salt as the cation. ILs have garnered a lot of attention since the concept of green environmental protection emerged and the necessity for chemicals and solvents in industry increased. Commonly referred to as room-temperature molten salts or ILs, organic ILs are low-temperature molten salts composed only of anions and cations that are liquids between −30°C and −50°C (Pei et al., 2022).

Strong solubility, high cohesion, a broad liquid range, nonvolatilization, a designable structure, a strong molecular recognition ability, and good stability are only a few of the notable properties of these liquid materials, which represent a distinct type of green solvent (Zhang et al., 2021). The ability to alter the type and ratio of anions and cations or carry out particular functional modifications of anions and cations to alter their inherent properties for particular applications, such as viscosity and solvation ability, is the main way that structural designability among these is demonstrated. These days, electrochemistry (Rodrigues et al., 2020), extraction (Liu et al., 2016; Flieger et al., 2020; Jiang et al., 2021), separation (Kore et al., 2017; Zuo et al., 2020; McNeice et al., 2021), catalysis (Ma et al., 2014; Jon et al., 2019), and nanomaterial fabrication (Abd Razik et al., 2014; Kinki et al., 2017 ; Guan et al., 2018; Yang et al., 2019) are all fields that make extensive use of ILs.

When Qiu's group studied ILs as green catalysts and reaction media in an open system at room temperature and ambient pressure, they reported the first example of the rapid synthesis of 3D COFs in 2018 (Dong et al., 2019). In particular, TFP forms crystalline solid materials in acidic 1-(4-sulfobutyl)-3-methyl imidazole hydrogen sulfate ([Besmir] HSO_4) when it combines with Pa, 5-dimethyl-1,4-phenylenediamine (MPA), or other substances (Gao et al., 2021). Almost simultaneously, reactive monomers Pa and 1,3,6,8-tetramethyl (4-methylphenyl) pyrene (Topy) were used to successfully construct 2D COFs in ILs (Abd Razik et al., 2014). The crystallinity of COFs in 1-butyl-3-methyl imidazolium salt (Bim+) with various counter anions has also been examined by the scientists, as previously noted.

It should be noted explicitly that the reaction in ILs takes place in the ionic medium and that the cationic part's ILs have the ability to cause interaction between the cation Π-Π units. This ultimately results in high-quality microcrystalline products by supporting the aromatic structure's ordered arrangement (Mele et al., 2003; Yang et al., 2019). In addition to the often-utilized building blocks, IL-mediated synthesis of a functionalized COF, COF-NHC (N-Heterocyclic carbenes) can also be performed directly rather than postmodified (Francisco et al., 2013). Nonetheless, COF–NHC may have a moderate surface area ($45.6\,cm^2g^{-1}$) due to the functional groups. It is interesting that there was significant coordination between the metal ions and the NHC functional group. Under mild circumstances, functional group–NHC has shown an outstanding activity in the Suzuki–Miyaura coupling process, indicating that COF–NHC can serve as a carrier for immobilized metal ions.

2.9 DEEP EUTECTIC SOLVENT-MEDIATED SYNTHESIS

Another innovative class of ecofriendly liquid compounds is thought to be deep eutectic solvents (Abbott et al., 2003). Interestingly, it has a lower melting point than any of the mixture's constituents (El Achkar et al., 2021). As of right now, deep eutectic solvents (DESs) may be broadly classified into four categories using the formula Cat^{+} X– zee, where X is a Lewis base, often a halide anion, and Cat+ (cation) typically stands for ammonium, sulfonium, or phosphine (Taghizadeh et al., 2021).

On the other hand, z indicates how much of the Y molecule interacts with the corresponding anion, and Y stands for a Brønsted or Lewis acid (Abbott et al., 2007; Rubin et al., 2021). On the one hand, DESs share many of the same physical and chemical properties as conventional ILs, including good thermal stability, a wide range of liquids, and low vapor pressure (Chang et al., 2021). They do, however, also have special qualities that set them apart from ILs, including greater biodegradability, reduced toxicity, cheaper raw materials, easier preparation techniques, and environmental friendliness (Maia et al., 2021).

Due to these significant advantages, DESs have generated a lot of attention in various fields, including polymer synthesis (Zhang et al., 2012; Tomé et al., 2018; Chen et al., 2020; Lan et al., 2021), organic processes (Alonso et al., 2016), separation science (Cai et al., 2019; Hu et al., 2020), electrochemistry (Wang et al., 2022), enzyme reactions (Mamtani et al., 2021), and biodiesel treatment (Hansen et al., 2020; Qiu et al., 2020).

Wang et al. [99] created the first example for the synthesis of numerous 2D and 3D COFs at room temperature in an open reaction system in 2–24 hours in 2020 in an effort to completely eliminate or reduce the use of hazardous organic reagents. In contrast, Y stands for a Brønsted or Lewis acid and z indicates the amount of Y molecule that interacts with the corresponding anion (Abbott et al., 2007; Rubin et al., 2021).

On the one hand, DESs share many of the same physical and chemical properties as conventional ILs, including good thermal stability, a wide range of liquids, and low vapor pressure (Chang et al., 2021). More significantly, compared to alternative approaches, the COFs produced using this process provide a number of exceptional benefits. For example, Tappe-Gly's process time is reduced from 72 to 2 hours when compared to traditional methods, which is most likely due to the monomers' full dissolution in solvents.

The successful creation of several 2D COFs so far has prompted research into 3D COFs with large areas of particular surface and lots of open channel locations. Actually, there are not many 3D COFs that have been released yet. In addition to the previously stated Chill eutectic solvents, Dong et al. recently synthesized new carboxyl functionalized COFs (Tappe-COOH) using Bu4NBr/I'm, which is made up of Bu4NBr and imidazole at a molar ratio of 3:7, as the reaction solvent in an open container. According to PXRD analysis, the absence of dense peaks corresponding to the (100), (110), and (001) planes indicates that the degree of crystallinity in this material was moderate and significantly lower than the previously reported Tappe without carboxyl groups (Gao et al., 2021). Here, they make the audacious assumption that monomer structures are generally thought to play a significant part in the creation of crystalline COFs. Furthermore, through the optimization of COF synthesis employing DESs as solvents, Gao et al. produced a series of COFs with high specific surface area and high crystallinity. They next employed these COFs to adsorb flavonoids, indicating the remarkable adsorption performance of the COF-DES-based solid-phase extraction columns on real complex materials (Wagare et al., 2021).

In this regard, we may say that they have contributed by offering unique perspectives on green-synthesized crystalline COFs. DESs as the reaction medium used for the synthesis of COFs are still rare, because DESs' numerous benefits cannot outweigh their drawbacks, particularly the influence of viscosity. Given the existing circumstances, further DES types with superior qualities may be discovered for the future production of COFs. Besides DESs, sustainable solvents have also caught on in a number of applications and entered the ranks of COFs (Kumar et al., 2021) via 12 biobased green solvents as reaction media. Szekely et al. created 43 COFs via a solvothermal technique (Xiao et al., 2022). In addition to developing a green approach for mild-condition COF synthesis, our study may offer fresh insights into the control of COF morphology (Biswal et al., 2013). As a result, we think there is greater potential for sustainable solvents to advance in the production of COFs.

As was previously said, compared to solvothermal methods, the DESs approach's reaction conditions for building the COFs are incredibly straightforward, thermally compatible, and ecofriendly. Compared to the preceding three days, the reaction time can drop to a few hours. Furthermore, DESs may be recycled and used as a structural guide to create new COFs with hierarchical pore structures that are not possible with conventional solvothermal techniques without suffering a major loss of activity. The resultant COFs had a large surface area and good crystallinity in terms of material qualities; some of the values were even greater than or equivalent to those obtained using the solvothermal approach. The disadvantage is that this synthetic method is currently only applied to particular COFs and has very little predictability.

2.10 SYNTHESIS WITHOUT THE USE OF SOLVENTS

An ecologically friendly substitute for traditional solution synthesis is solvent-free synthesis. Banerjee's group was the first to synthesize TpPa-1 and TpPa-2 using the method, which contrasts with many of the COFs that have been prepared. The compounds' persistent existence in boiling water, 9 mol L^{-1} HCl, and 3 mol L^{-1} NaOH solutions demonstrated their exceptional chemical stability (Das et al., 2014). More significantly, the produced COFs exhibited layered morphology (removed layers) resembling graphene, which could result in new eliminated COF designs.

In contrast to solvothermal techniques, the crystallinity is quite low. To increase their specific surface area and crystallinity, the same innovation team produced a range of COFs by including tiny amounts of liquid aggregation reaction monomers (Karak et al., 2017). The approach showed better generality in terms of COF kinds and crystallinity. With the quick synthesis of 12 highly crystalline COFs in 2017 using a salt-mediated crystallization approach, the researchers achieved yet another remarkable achievement (Spitzer et al., 2017). By protonating the nitrogen atom on the amine, the sulfonic acid (PTSA-H_2O) created the PTSA-ammonium salt structure, playing a crucial role.

The surface area can exceed 3,000 $cm^2 g^{-1}$, which is noteworthy. Additionally, by simulating the organic terracotta process, the technology may be processed in situ to create various sculptures and shapes, which is highly advantageous for industrial use. In addition to the solvent-free technique for creating COF powder, some intriguing research on COF films is progressively coming to light (Kuhn et al., 2008; Cheung et al., 2019). The solvent-free synthesis of COFs is yet in its early stages, although the application in the manufacturing and synthesis of CTF is well known. In contrast to conventional synthetic methods, solvent addition as a reaction medium is not necessary throughout the CTF preparation process. Thomas' group used molten $ZnCl_2$ salt as the catalyst and reaction media in 2008 to create CTF-1 with hexagonal pores and improved crystallinity at high temperatures (Zhang et al., 2022). However, the usage of hazardous organic solvents was circumvented by using molten salts as reaction media. Triazine ring's reversibility and self-healing were facilitated by the greater reaction temperature and duration, but they also caused partial thermal breakdown or carbonization of the monomer, which somewhat restricted the use. Although solvent-free reactions are somewhat less useful than solution-based reactions, they may be thought of as an additional step in the solvothermal material production process.

2.11 CONCLUSION

In conclusion, a significant bottleneck in the field has been broken by the active exploration of different new synthesis methods based on ecofriendly routes. This is because common synthesis routes need high energy consumption and large organic solvents. The goal of these strategies is to produce COFs on a large scale, in an efficient and green manner. They also provided an overview of the features, benefits, drawbacks, and formation mechanism of each COF preparation technique. We also anticipate that there will be many obstacles to overcome in the future, such as the novel synthesis approach and other typical issues in the field of covalent organic frameworks.

1. The creation of growth mechanisms utilizing reaction medium and non-traditional heating methods is still in its infancy as well. For example, IL-mediated synthesis, microwave synthesis, and other techniques offer a relatively quick, easy, and ecofriendly synthetic pathway for COFs. Nevertheless, little is known about how new reaction media or unconventional energy sources affect the development, growth, and crystallization of COFs. As a result, expanding the use of atypical synthetic techniques is fairly difficult.
2. There is an issue with "crystallinity" and "stability" together. While stability dictates the range of applications, crystallinity has a direct correlation with COF performance. Consequently, both "stability" and "crystallinity" are significant elements influencing the future evolution of COFs.
3. While little study has been done for 3D COFs, nontraditional synthesis techniques are mostly used for 2D COFs. It should be noted that 3D COFs could have a greater specific surface area and a lower density. Thus, 3D COFs should be the main emphasis.
4. Attention is also being paid to the ordered regular morphology, which is very important for its uses in chromatography, optical sensing, and other areas.

The preparation process is understood by altering the energy, temperature, and structure, simulating the creation, growth, and nucleation of COFs using theoretical modeling. Synthetic approaches and theoretical computations are used to create COFs using unusual heating techniques and reaction fluids. Then, rather than selecting the ideal experimental circumstances by a great deal of "trial and error," we select the reaction solvent and energy source based on the real needs. Regarding the compatibility of stability and crystallinity, Zhu et al. (2022) group recently used frame reconstruction to prepare and produce crystalline COFs on a wide scale with good stability under standard synthesis circumstances. It is crucial to remember that theoretical calculations are used to explain the reconstruction synthesis. The synthesis approach may be enhanced more successfully with the use of theoretical models, resulting in a more ecologically friendly and green synthesis strategy.

The following elements mostly represent this:

1. Look for novel reactive solvent systems, such as supercritical fluids, green solvents, and a combination of water and other organic solvents.
2. Within certain limits, to improve atomic utilization and crystallinity, scientists can develop new chemical methods, which can minimize or completely do away with the requirement for organic solvent.
3. Look for novel catalysts to reduce reaction time and attain energy-efficient green preparation. Furthermore, it is important to remember that different catalysts have the ability to regulate the shape of COFs.

In conclusion, further research and analysis are still needed to develop COFs. It is widely accepted that when people do in-depth research on the mechanism of COFs, the current issues will be enhanced and expanded, and there will be ample room for further advancement. Last but not least, it is certain that simpler or greener processes for creating COFs will be created, and it will be possible to manufacture them in large quantities to meet industrial needs.

REFERENCES

Abbott, Andrew P., Glen Capper, David L. Davies, Raymond K. Rasheed, and Vasuki Tambyrajah. "Novel solvent properties of choline chloride/urea mixtures." *Chemical Communications* 1 (2003): 70–71.

Abbott, Andrew P., John C. Barron, Karl S. Ryder, and David Wilson. "Eutectic-based ionic liquids with metal-containing anions and cations." *Chemistry–A European Journal* 13, no. 22 (2007): 6495–6501.

Abd Razik, Basma M., Hasnah Osman, Alireza Basiri, Abdussalam Salhin, Yalda Kia, Mohammed Oday Ezzat, and Vikneswaran Murugaiyah. "Ionic liquid mediated synthesis and molecular docking study of novel aromatic embedded Schiff bases as potent cholinesterase inhibitors." *Bioorganic Chemistry* 57 (2014): 162–168.

Alonso, Diego A., Alejandro Baeza, Rafael Chinchilla, Gabriela Guillena, Isidro M. Pastor, and Diego J. Ramón. "Deep eutectic solvents: the organic reaction medium of the century." *European Journal of Organic Chemistry* 2016, no. 4 (2016): 612–632.

Annabeth L., Alejandro Baeza, and Rafael Chinchilla. "Deep eutectic solvents: syntheses, properties and applications." *Chemical Society Reviews* 41, no. 21 (2012): 2108–2116.

Biswal, Bishnu P., Suman Chandra, Sharath Kandambeth, Binit Lukose, Thomas Heine, and Rahul Banerjee. "Mechanochemical synthesis of chemically stable isoreticular covalent organic frameworks." *Journal of the American Chemical Society* 135, no. 14 (2013): 5328–5331.

Cai, Tianpei, and Hongdeng Qiu. "Application of deep eutectic solvents in chromatography: a review." *TrAC Trends in Analytical Chemistry* 120 (2019): 115623.

Chang, Xin Xiong, Nabisab Mujawar Mubarak, Shaukat Ali Mazari, Abdul Sattar Jatoi, Awais Ahmad, Mohammad Khalid, Rashmi Walvekar et al. "A review on the properties and applications of chitosan, cellulose and deep eutectic solvent in green chemistry." *Journal of Industrial and Engineering Chemistry* 104 (2021): 362–380.

Chen, Jia, Mohammad Chand Ali, Ruirui Liu, Jean Claude Munyemana, Zhan Li, Honglin Zhai, and Hongdeng Qiu. "Basic deep eutectic solvents as reactant, template and solvents for ultra-fast preparation of transition metal oxide nanomaterials." *Chinese Chemical Letters* 31, no. 6 (2020): 1584–1587.

Chen, Li, Jingcheng Du, Wei Zhou, Huizhen Shen, Luxi Tan, Cailong Zhou, and Lichun Dong. "Microwave-assisted solvothermal synthesis of covalent organic frameworks (COFs) with stable superhydrophobicity for oil/water separation." *Chemistry–An Asian Journal* 15, no. 21 (2020): 3421–3427.

Cheung, Po Ling, Sze Koon Lee, and Clifford P. Kubiak. "Facile solvent-free synthesis of thin iron porphyrin COFs on carbon cloth electrodes for CO_2 reduction." *Chemistry of Materials* 31, no. 6 (2019): 1908–1919.

Cote, Adrien P., Annabelle I. Benin, Nathan W. Ockwig, Michael O'Keeffe, Adam J. Matzger, and Omar M. Yaghi. "Porous, crystalline, covalent organic frameworks." *Science* 310, no. 5751 (2005): 1166–1170.

Czaja, Alexander U., Natalia Trukhan, and Ulrich Müller. "Industrial applications of metal–organic frameworks." *Chemical Society Reviews* 38, no. 5 (2009): 1284–1293.

Dalapati, Sasanka, Shangbin Jin, Jia Gao, Yanhong Xu, Atsushi Nagai, and Donglin Jiang. "An azine-linked covalent organic framework." *Journal of the American Chemical Society* 135, no. 46 (2013): 17310–17313.

Das, Gobinda, Digambar Balaji Shinde, Sharath Kandambeth, Bishnu P. Biswal, and Rahul Banerjee. "Mechanosynthesis of imine, β-ketoenamine, and hydrogen-bonded imine-linked covalent organic frameworks using liquid-assisted grinding." *Chemical Communications* 50, no. 84 (2014): 12615–12618.

Das, Gobinda, Tina Skorjanc, Sudhir Kumar Sharma, Felipe Gándara, Matteo Lusi, D. S. Shankar Rao, Sridurai Vimala et al. "Viologen-based conjugated covalent organic networks via Zincke reaction." *Journal of the American Chemical Society* 139, no. 28 (2017): 9558–9565.

De la Peña Ruigómez, Alejandro, David Rodríguez-San-Miguel, Kyriakos C. Stylianou, Massimiliano Cavallini, Denis Gentili, Fabiola Liscio, Silvia Milita et al. "Direct on-surface patterning of a crystalline laminar covalent organic framework synthesized at room temperature." *Chemistry–A European Journal* 21, no. 30 (2015): 10666–10670.

Dey, Kaushik, Manas Pal, Kanhu Charan Rout, Shebeeb Kunjattu H, Anuja Das, Rabibrata Mukherjee, Ulhas K. Kharul, and Rahul Banerjee. "Selective molecular separation by interfacially crystallized covalent organic framework thin films." *Journal of the American Chemical Society* 139, no. 37 (2017): 13083–13091.

Ding, San-Yuan, Jia Gao, Qiong Wang, Yuan Zhang, Wei-Guo Song, Cheng-Yong Su, and Wei Wang. "Construction of covalent organic framework for catalysis: Pd/COF-LZU1 in Suzuki–Miyaura coupling reaction." *Journal of the American Chemical Society* 133, no. 49 (2011): 19816–19822.

Dogru, Mirjam. Functionalization of Covalent Organic Frameworks. Ph.D. Thesis, University of Munich Ludwig-Maximilians, Munich, Germany, 2012.

Dong, Bin, Wen-Jing Wang, Sun-Chang Xi, Dong-Yue Wang, and Ren Wang. "A carboxyl-functionalized covalent organic framework synthesized in a deep eutectic solvent for dye adsorption." *Chemistry–A European Journal* 27, no. 8 (2021): 2692–2698.

Dong, Bin, Wen-Jing Wang, Wei Pan, and Guo-Jun Kang. "Ionic liquid as a green solvent for ionothermal synthesis of 2D keto-enamine-linked covalent organic frameworks." *Materials Chemistry and Physics* 226 (2019): 244–249.

El Achkar, Tracy, Hélène Greige-Gerges, and Sophie Fourmentin. "Basics and properties of deep eutectic solvents: a review." *Environmental Chemistry Letters* 19 (2021): 3397–3408.

Fang, Qianrong, Zhongbin Zhuang, Shuang Gu, Robert B. Kaspar, Jie Zheng, Junhua Wang, Shilun Qiu, and Yushan Yan. "Designed synthesis of large-pore crystalline polyimide covalent organic frameworks." *Nature Communications* 5, no. 1 (2014): 4503.

Fernandes, Soraia P. S., Vanesa Romero, Begoña Espiña, and Laura M. Salonen. "Tailoring covalent organic frameworks to capture water contaminants." *Chemistry–A European Journal* 2019, 25, 6461–6473.

Flieger, Jolanta, Joanna Feder-Kubis, and Małgorzata Tatarczak-Michalewska. "Chiral ionic liquids: structural diversity, properties and applications in selected separation techniques." *International Journal of Molecular Sciences* 21, no. 12 (2020): 4253.

Francisco, María, Adriaan van den Bruinhorst, and Maaike C. Kroon. "Low- transition- temperature mixtures (LTTMs): a new generation of designer solvents." *Angewandte Chemie International Edition* 52, no. 11 (2013): 3074–3085.

Franco, Carlos, David Rodríguez-San-Miguel, Alessandro Sorrenti, Semih Sevim, Ramon Pons, Ana E. Platero-Prats, Marko Pavlovic et al. "Biomimetic synthesis of sub-20 nm covalent organic frameworks in water." *Journal of the American Chemical Society* 142, no. 7 (2020): 3540–3547.

Gan, Shi-Xian, Chao Jia, Qiao-Yan Qi, and Xin Zhao. "A facile and scalable synthetic method for covalent organic nanosheets: ultrasonic polycondensation and photocatalytic degradation of organic pollutants." *Chemical Science*, 13 (2022): 1009–1015.

Gao, Manjie, Dandan Wang, Linlin Deng, Shaochi Liu, Kailian Zhang, Tian Quan, Lijuan Yang, Xun Kang, Zhining Xia, and Die Gao. "High-crystallinity covalent organic framework synthesized in deep eutectic solvent: potentially effective adsorbents alternative to macroporous resin for flavonoids." *Chemistry of Materials* 33, no. 20 (2021): 8036–8051.

Geng, Keyu, Ting He, Ruoyang Liu, Sasanka Dalapati, Ke Tian Tan, Zhongping Li, Shanshan Tao, Yifan Gong, Qiuhong Jiang, and Donglin Jiang. "Covalent organic frameworks: design, synthesis, and functions." *Chemical Reviews* 120, no. 16 (2020): 8814–8933.

Guan, Xinyu, Yunchao Ma, Hui Li, Yusran Yusran, Ming Xue, Qianrong Fang, Yushan Yan, Valentin Valtchev, and Shilun Qiu. "Fast, ambient temperature and pressure ionothermal synthesis of three-dimensional covalent organic frameworks." *Journal of the American Chemical Society* 140, no. 13 (2018): 4494–4498.

Hansen, Benworth B., Stephanie Spittle, Brian Chen, Derrick Poe, Yong Zhang, Jeffrey M. Klein, Alexandre Horton et al. "Deep eutectic solvents: a review of fundamentals and applications." *Chemical Reviews* 121, no. 3 (2020): 1232–1285.

Hu, Yongxing, Tianpei Cai, Haijuan Zhang, Jia Chen, Zhan Li, Liang Zhao, Zuguang Li, and Hongdeng Qiu. "Two copolymer-grafted silica stationary phases prepared by surface thiol-ene click reaction in deep eutectic solvents for hydrophilic interaction chromatography." *Journal of Chromatography A* 1609 (2020): 460446.

Jiang, Danni, Jia Chen, Ming Guan, and Hongdeng Qiu. "Octadecylimidazolium ionic liquids-functionalized carbon dots and their precursor co-immobilized silica as hydrophobic chromatographic stationary phase with enhanced shape selectivity." *Talanta* 233 (2021): 122513.

Jon, Chol-San, Long-Yue Meng, and Donghao Li. "Recent review on carbon nanomaterials functionalized with ionic liquids in sample pretreatment application." *TrAC Trends in Analytical Chemistry* 120 (2019): 115641.

Kandambeth, Sharath, Arijit Mallick, Binit Lukose, Manoj V. Mane, Thomas Heine, and Rahul Banerjee. "Construction of crystalline 2D covalent organic frameworks with remarkable chemical (acid/base) stability via a combined reversible and irreversible route." *Journal of the American Chemical Society* 134, no. 48 (2012): 19524–19527.

Karak, Suvendu, Sharath Kandambeth, Bishnu P. Biswal, Himadri Sekhar Sasmal, Sushil Kumar, Pradip Pachfule, and Rahul Banerjee. "Constructing ultraporous covalent organic frameworks in seconds via an organic terracotta process." *Journal of the American Chemical Society* 139, no. 5 (2017): 1856–1862.

Kim, Soyoung, Chibeom Park, Minkyung Lee, Intek Song, Jungah Kim, Minhui Lee, Jaehoon Jung, Yousoo Kim, Hyunseob Lim, and Hee Cheul Choi. "Rapid photochemical synthesis of sea-urchin-shaped hierarchical porous COF-5 and its lithography-free patterned growth." *Advanced Functional Materials* 27, no. 32 (2017): 1700925.

Kinik, Fatma Pelin, Alper Uzun, and Seda Keskin. "Ionic Liquid/Metal–Organic Framework Composites: From Synthesis to Applications." *ChemSusChem* 10, no. 14 (2017): 2842–2863.

Kore, Rajkumar, Paula Berton, Steven P. Kelley, Pavankumar Aduri, Sanjeev S. Katti, and Robin D. Rogers. "Group IIIA halometallate ionic liquids: speciation and applications in catalysis." *ACS Catalysis* 7, no. 10 (2017): 7014–7028.

Kuhn, Pierre, Markus Antonietti, and Arne Thomas. "Porous, covalent triazine-based frameworks prepared by ionothermal synthesis." *Angewandte Chemie International Edition* 47, no. 18 (2008): 3450–3453.

Kumar, Sushil, Gergo Ignacz, and Gyorgy Szekely. "Synthesis of covalent organic frameworks using sustainable solvents and machine learning." *Green Chemistry* 23, no. 22 (2021): 8932–8939.

Lan, Shanshan, Yaofen Li, Song Xue, Tao Zhu, and Hongyuan Yan. "Application of deep eutectic solvents modified oxidized hydrogen-substituted graphyne in adsorption and electrochemistry." *Journal of Molecular Liquids* 335 (2021): 116532.

Li, Hailian, Mohamed Eddaoudi, Michael O'Keeffe, and Omar M. Yaghi. "Design and synthesis of an exceptionally stable and highly porous metal-organic framework." *Nature* 402, no. 6759 (1999): 276–279.

Liang, Yu, Chenhui Li, Lanjun Chen, Jia Huo, Mohammed Loubidi, Yangyang Zhou, and Yanbo Liu. "Microwave-assisted acid-induced formation of linker vacancies within Zr-based metal organic frameworks with enhanced heterogeneous catalysis." *Chinese Chemical Letters* 32, no. 2 (2021): 787–790.

Liu, Yinghui, Dandan Guo, Ying Gao, Bihai Tong, Yu Li, and Yuezhao Zhu. "Non-thermal effect of microwave on the chemical structure and luminescence properties of biomass-derived carbon dots via hydrothermal method." *Applied Surface Science* 552 (2021): 149503.

Liu, Zhongling, Wei Yu, Hanqi Zhang, Fanbin Gu, and Xiangqun Jin. "Salting-out homogenous extraction followed by ionic liquid/ionic liquid liquid–liquid micro-extraction for determination of sulfonamides in blood by high performance liquid chromatography." *Talanta* 161 (2016): 748–754.

Lohse, Maria S., and Thomas Bein. "Covalent organic frameworks: structures, synthesis, and applications." *Advanced Functional Materials* 28, no. 33 (2018): 1705553.

Lu, Jian, Feng Lin, Qiang Wen, Qiao-Yan Qi, Jia-Qiang Xu, and Xin Zhao. "Large-scale synthesis of azine-linked covalent organic frameworks in water and promoted by water." *New Journal of Chemistry* 43, no. 16 (2019): 6116–6120.

Lyle, Steven J., Peter J. Waller, and Omar M. Yaghi. "Covalent organic frameworks: organic chemistry extended into two and three dimensions." *Trends in Chemistry* 1, no. 2 (2019): 172–184.

Ma, Ming-Guo, Jie-Fang Zhu, Ying-Jie Zhu, and Run-Cang Sun. "The microwave-assisted ionic-liquid method: a promising methodology in nanomaterials." *Chemistry–An Asian Journal* 9, no. 9 (2014): 2378–2391.

Ma, Xueji, Wan Xu, Xiaoqian Liang, and Jikuan Qiu. "Low-temperature and gram-scale synthesis of chemically stable covalent organic frameworks in an aqueous medium." *New Journal of Chemistry* 46, no. 10 (2022): 4558–4561.

Mahmoud, Mahmoud A., Radha Narayanan, and Mostafa A. El-Sayed. "Enhancing colloidal metallic nanocatalysis: sharp edges and corners for solid nanoparticles and cage effect for hollow ones." *Accounts of Chemical Research* 46, no. 8 (2013): 1795–1805.

Maia, Renata A., Benoit Louis, and Stéphane A. Baudron. "Deep eutectic solvents for the preparation and post-synthetic modification of metal-and covalent organic frameworks." *CrystEngComm* 23, no. 29 (2021): 5016–5032.

Mamtani, Kapil, Kaveh Shahbaz, and Mohammed M. Farid. "Deep eutectic solvents–Versatile chemicals in biodiesel production." *Fuel* 295 (2021): 120604.

Martín-Illán, Jesús Á., David Rodríguez-San-Miguel, Carlos Franco, Inhar Imaz, Daniel Maspoch, Josep Puigmartí- Luis, and Félix Zamora. "Green synthesis of imine-based covalent organic frameworks in water." *Chemical Communications* 56, no. 49 (2020): 6704–6707.

McNeice, Peter, Patricia C. Marr, and Andrew C. Marr. "Basic ionic liquids for catalysis: the road to greater stability." *Catalysis Science & Technology* 11, no. 3 (2021): 726–741.

Mele, Andrea, Chieu D. Tran, and Silvia H. De Paoli Lacerda. "The structure of a room-temperature ionic liquid with and without trace amounts of water: the role of C—H···O and C—H···F interactions in 1-n-butyl-3-methylimidazolium tetrafluoroborate." *Angewandte Chemie International Edition* 42, no. 36 (2003): 4364–4366.

Mow, Rachel E., Andrew S. Lipton, Sarah Shulda, E. Ashley Gaulding, Thomas Gennett, and Wade A. Braunecker. "Colloidal three-dimensional covalent organic frameworks and their application as porous liquids." *Journal of Materials Chemistry A* 8, no. 44 (2020): 23455–23462.

Pei, Yuanchao, Yaxin Zhang, Jie Ma, Maohong Fan, Suojiang Zhang, and Jianji Wang. "Ionic liquids for advanced materials." *Materials Today Nano* 17 (2022): 100159.

Qiu, Jikuan, Pengxin Guan, Yuling Zhao, Zhiyong Li, Huiyong Wang, and Jianji Wang. "Syntheses of two- and three-dimensional covalent organic frameworks in deep eutectic solvents." *Green Chemistry* 22, no. 21 (2020): 7537–7542.

Raghav, Sapna, Jyoti Raghav, Ma'aruf Abdulmumin Muhammad, Madhur Babu Singh, Pallavi Jain, Dinesh Kumar. 2025. "Recent advancements in synthesis and applications of covalent organic frameworks." *Applied Materials Today* 44 (June): 102694.

Rantwijk. Van, Fred, and Roger A.Sheldon. "Biocatalysis in ionic liquids." *Chemical Reviews* 107, no. 6 (2007): 2757–2785.

Rodrigues, Renata Débora Pinto, Amanda Sousa e Silva, Thalita Adrielly Viana Carlos, Ana Karine Pessoa Bastos, Rilvia Saraiva de Santiago-Aguiar, and Maria Valderez Ponte Rocha. "Application of protic ionic liquids in the microwave-assisted extraction of phycobiliproteins from Arthrospira platensis with antioxidant activity." *Separation and Purification Technology* 252 (2020): 117448.

Rubin Pedrazzo, Alberto, Claudio Cecone, Francesco Trotta, and Marco Zanetti. "Mechanosynthesis of β-cyclodextrin polymers based on natural deep eutectic solvents." *ACS Sustainable Chemistry & Engineering* 9, no. 44 (2021): 14881–14889.

Safaei, Mohadeseh, Mohammad Mehdi Foroughi, Nasser Ebrahimpoor, Shohreh Jahani, Ali Omidi, and Mehrdad Khatami. "A review on metal-organic frameworks: Synthesis and applications." *TrAC Trends in Analytical Chemistry* 118 (2019): 401–425.

Smith, Brian J., Anna C. Overholts, Nicky Hwang, and William R. Dichtel. "Insight into the crystallization of amorphous imine-linked polymer networks to 2D covalent organic frameworks." *Chemical Communications* 52, no. 18 (2016): 3690–3693.

Song, Yanpei, Qi Sun, Briana Aguila, and Shengqian Ma. "Opportunities of covalent organic frameworks for advanced applications." *Advanced Science* 6, no. 2 (2019): 1801410.

Spitzer, Saskia, Atena Rastgoo-Lahrood, Klaus Macknapp, Viktoria Ritter, Stefan Sotier, Wolfgang M. Heckl, and Markus Lackinger. "Solvent-free on-surface synthesis of boroxine COF monolayers." *Chemical Communications* 53, no. 37 (2017): 5147–5150.

Stewart, David, Dmytro Antypov, Matthew S. Dyer, Michael J. Pitcher, Alexandros P. Katsoulidis, Philip A. Chater, Frédéric Blanc, and Matthew J. Rosseinsky. "Stable and ordered amide frameworks synthesised under reversible conditions which facilitate error checking." *Nature Communications* 8, no. 1 (2017): 1102.

Taghizadeh, Mohsen, Ali Taghizadeh, Vahid Vatanpour, Mohammad Reza Ganjali, and Mohammad Reza Saeb. "Deep eutectic solvents in membrane science and technology: fundamental, preparation, application, and future perspective." *Separation and Purification Technology* 258 (2021): 118015.

Tomé, Luciana IN, Vanessa Baião, Wanderson da Silva, and Christopher MA Brett. "Deep eutectic solvents for the production and application of new materials." *Applied Materials Today* 10 (2018): 30–50.

Uribe-Romo, Fernando J., Christian J. Doonan, Hiroyasu Furukawa, Kounosuke Oisaki, and Omar M. Yaghi. "Crystalline covalent organic frameworks with hydrazone linkages." *Journal of the American Chemical Society* 133, no. 30 (2011): 11478–11481.

Uribe-Romo, Fernando J., Joseph R. Hunt, Hiroyasu Furukawa, Cornelius Klock, Michael O'Keeffe, and Omar M. Yaghi. "A crystalline imine-linked 3-D porous covalent organic framework." *Journal of the American Chemical Society* 131, no. 13 (2009): 4570–4571.

Vazquez-Molina, Demetrius A., Giovanna M. Pope, Andrew A. Ezazi, Jose L. Mendoza-Cortes, James K. Harper, and Fernando J. Uribe-Romo. "Framework vs. side-chain amphidynamic behaviour in oligo-(ethylene oxide) functionalised covalent-organic frameworks." *Chemical Communications* 54, no. 50 (2018): 6947–6950.

Vitaku, Edon, and William R. Dichtel. "Synthesis of 2D imine-linked covalent organic frameworks through formal transimination reactions." *Journal of the American Chemical Society* 139, no. 37 (2017): 12911–12914.

Wagare, Devendra S., Sagar E. Shirsath, Mujahed Shaikh, and Prashant Netankar. "Sustainable solvents in chemical synthesis: a review." *Environmental Chemistry Letters* 19, no. 4 (2021): 3263–3282.

Waller, Peter J., Felipe Gándara, and Omar M. Yaghi. "Chemistry of covalent organic frameworks." *Accounts of Chemical Research* 48, no. 12 (2015): 3053–3063.

Wang, Yuting, Jia Chen, Ganping Wang, Yongliang Yu, Jianhua Wang, and Hongdeng Qiu. "Fluorescent determination of cysteine and homocysteine via adjustable synthesis of flower-shaped covalent organic frameworks." *Sensors and Actuators B: Chemical* 359 (2022): 131555.

Welton, Thomas. "Room-temperature ionic liquids. Solvents for synthesis and catalysis." *Chemical Reviews* 99, no. 8 (1999): 2071–2084.

Xiao, Jing, Jia Chen, and Hongdeng Qiu. "Morphological control of covalent organic frameworks in a PEG–H_2O system." *Green Chemistry* 24, no. 5 (2022): 2193–2202.

Yaghi, Omar M. "Reticular chemistry construction, properties, and precision reactions of frameworks." *Journal of the American Chemical Society* 138, no. 48 (2016): 15507–15509.

Yaghi, Omar M., Markus J. Kalmutzki, and Christian S. Diercks. *Introduction to Reticular Chemistry: Metal-Organic Frameworks and Covalent Organic Frameworks*. John Wiley & Sons, 2019.

Yaghi, Omar M., Michael O'Keeffe, Nathan W. Ockwig, Hee K. Chae, Mohamed Eddaoudi, and Jaheon Kim. "Reticular synthesis and the design of new materials." *Nature* 423, no. 6941 (2003): 705–714.

Yang, Guang, Hongye Huang, Junyu Chen, Defu Gan, Fengjie Deng, Qiang Huang, Yuanqing Wen, Meiying Liu, Xiaoyong Zhang, and Yen Wei. "Preparation of ionic liquids functionalized nanodiamonds-based composites through the Michael addition reaction for efficient removal of environmental pollutants." *Journal of Molecular Liquids* 296 (2019): 111874.

Yang, Junjuan, Yuanyuan Wu, Xiaowei Wu, Wenjing Liu, Yaofang Wang, and Junwen Wang. "An N-heterocyclic carbene-functionalised covalent organic framework with atomically dispersed palladium for coupling reactions under mild conditions." *Green Chemistry* 21, no. 19 (2019): 5267–5273.

Yang, Seung-Tae, Jun Kim, Hye-Young Cho, Sangho Kim, and Wha-Seung Ahn. "Facile synthesis of covalent organic frameworks COF-1 and COF-5 by sonochemical method." *RSC Advances* 2, no. 27 (2012): 10179–10181.

Yu, Deyou, Liping Wang, Taoyu Yang, Guangpeng Yang, Dong Wang, Huagang Ni, and Minghua Wu. "Tuning Lewis acidity of iron-based metal-organic frameworks for enhanced catalytic ozonation." *Chemical Engineering Journal* 404 (2021): 127075.

Yu, Deyou, Minghua Wu, Qian Hu, Lili Wang, Chencheng Lv, and Lu Zhang. "Iron-based metal-organic frameworks as novel platforms for catalytic ozonation of organic pollutant: efficiency and mechanism." *Journal of Hazardous Materials* 367 (2019): 456–464.

Zhang, Fanyu, Jianling Zhang, Bingxing Zhang, Xiuniang Tan, Dan Shao, Jinbiao Shi, Dongxing Tan et al. "Room-temperature synthesis of covalent organic framework (COF-LZU1) nanobars in CO_2/water solvent." *ChemSusChem* 11, no. 20 (2018): 3576–3580.

Zhang, Mingxing, Junchang Chen, Shitong Zhang, Xiaoqi Zhou, Linwei He, Matthew V. Sheridan, Mengjia Yuan, et al. "Electron beam irradiation as a general approach for the rapid synthesis of covalent organic frameworks under ambient conditions." *Journal of the American Chemical Society*142 (2020): 9169–9174.

Zhang, Qinghua, Karine De Oliveira Vigier, Sébastien Royer, and François Jérôme. "Deep eutectic solvents: syntheses, properties and applications." *Chemical Society Reviews* 41, no. 21 (2012): 7108–7146.

Zhang, Wang, Cun Li, Yu-Peng Yuan, Ling-Guang Qiu, An-Jian Xie, Yu-Hua Shen, and Jun-Fa Zhu. "Highly energy- and time-efficient synthesis of porous triazine-based framework: microwave-enhanced ionothermal polymerization and hydrogen uptake." *Journal of Materials Chemistry* 20, no. 31 (2010): 6413–6415.

Zhang, Weiwei, Linjiang Chen, Sheng Dai, Chengxi Zhao, Cheng Ma, Lei Wei, Minghui Zhu et al. "Reconstructed covalent organic frameworks." *Nature* 604, no. 7904 (2022): 72–79.

Zhang, Wu, and Batric Pesic. "Electrochemistry of $PdCl_2$ at glassy carbon electrode in 1-ethyl-3-methylimidazolium chloride ionic liquid." *Electrochimica Acta* 370 (2021): 137818.

Zhang, Yue-Biao, Jie Su, Hiroyasu Furukawa, Yifeng Yun, Felipe Gándara, Adam Duong, Xiaodong Zou, and Omar M. Yaghi. "Single-crystal structure of a covalent organic framework." *Journal of the American Chemical Society* 135, no. 44 (2013): 16336–16339.

Zhao, Wei, Peiyao Yan, Haofan Yang, Mounib Bahri, Alex M. James, Hongmei Chen, Lunjie Liu et al. "Using sound to synthesize covalent organic frameworks in water." *Nature Synthesis* 1, no. 1 (2022): 87–95.

Zheng, Anqi, Tingting Guo, Fulin Guan, Xuwei Chen, Yang Shu, and Jianhua Wang. "Ionic liquid mediated carbon dots: preparations, properties and applications." *TrAC Trends in Analytical Chemistry* 119 (2019): 115638.

Zhu, Dongyang, Zhuqing Zhang, Lawrence B. Alemany, Yilin Li, Njideka Nnorom, Morgan Barnes, Safiya Khalil, Muhammad M. Rahman, Pulickel M. Ajayan, and Rafael Verduzco. "Rapid, ambient temperature synthesis of imine covalent organic frameworks catalyzed by transition-metal nitrates." *Chemistry of Materials* 33, no. 9 (2021): 3394–3400.

Zhu, Yifan, Dongyang Zhu, Qianqian Yan, Guanhui Gao, Jianan Xu, Yifeng Liu, Sampath B. Alahakoon et al. "Metal oxide catalysts for the synthesis of covalent organic frameworks and one-step preparation of covalent organic framework-based composites." *Chemistry of Materials* 33, no. 15 (2021): 6158–6165.

Zuo, Lin, Xianquan Ao, and Yu Guo. "Study on the synthesis of dual-chain ionic liquids and their application in the extraction of flavonoids." *Journal of Chromatography A* 1628 (2020): 461446.

Zhu, Mu, Z., Y., Li, B., Dong, A., Wang, B. and Feng, X., "Covalent organic frameworks with record pore apertures." *Journal of the American Chemical Society* 144, no. 11 (2022): 5145–5154.

3 Characterization Methods for Covalent Organic Frameworks

Ramadan M. Ramadan, Mohamed M. Hammam, Pallavi Jain, and Amani F.H. Noureldeen

3.1 INTRODUCTION

Covalent organic frameworks (COFs) are a category of crystalline porous organic polymers that synthesized from different organic units consisting of nonmetal atoms, such as C, H, N, O and B, and supramolecularly arrayed into two-dimensional (sheets) or three-dimensional (frameworks) networks connected via covalent bonds (Figure 3.1) (Tan et al. 2023; Y. Liu et al. 2024). On the other hand, combining covalent and noncovalent bond interactions is another method to perform a composition of complex structures that can be used to define functions such as mimicking macromolecular biological systems like nucleic acids and proteins. To determine the structure of the COF polymers, material investigations and in silico computational simulation are deeply desired to identify the purity and features of COFs (Sharma et al. 2020; Abuzeid et al. 2021).

Notably, COFs have better advantages over metal-organic frameworks (MOFs), where the densities of COFs are generally low. In addition, they exhibit good stability in common organic solvents in acidic, basic, oxidative or reductive conditions, retaining their ordered structure, crystallinity and large surface area (Raghav et al. 2025). This could be attributed to the strong covalent bonds in COFs rather than coordinate bonds in MOFs. On the one hand, the possible existence of different types of hydrogen bonding as well as C-H···π and π···π stacking interactions in COFs further strengthens the skeletal structure and prevents them from hydrolysis (Figure 3.2) (Elsheemy et al. 2024). Therefore, COFs are examined and surveyed in numerous applications due to such advantages. These fascinating preferences have found many applications in fields such as gas and energy storage, sensing, adsorption and extraction of metal ions, catalysis, optoelectronics and drug delivery.

The features of COFs stem from their chemical and physical stability, highly porous structure and structural tunability. Type of bonding, crystallization methods, powder X-ray diffraction (PXRD), single-crystal X-ray diffraction (SXRD), porosity, thermal stability and surface morphology are thus leading factors that are performed with the different instrumental techniques. Furthermore, more insight into the lattice spacing of COF nanocrystals can be explored by running scanning electron microscopy (SEM) and transmission electron microscopy (TEM) to analyze their crystal

DOI: 10.1201/9781003605386-3

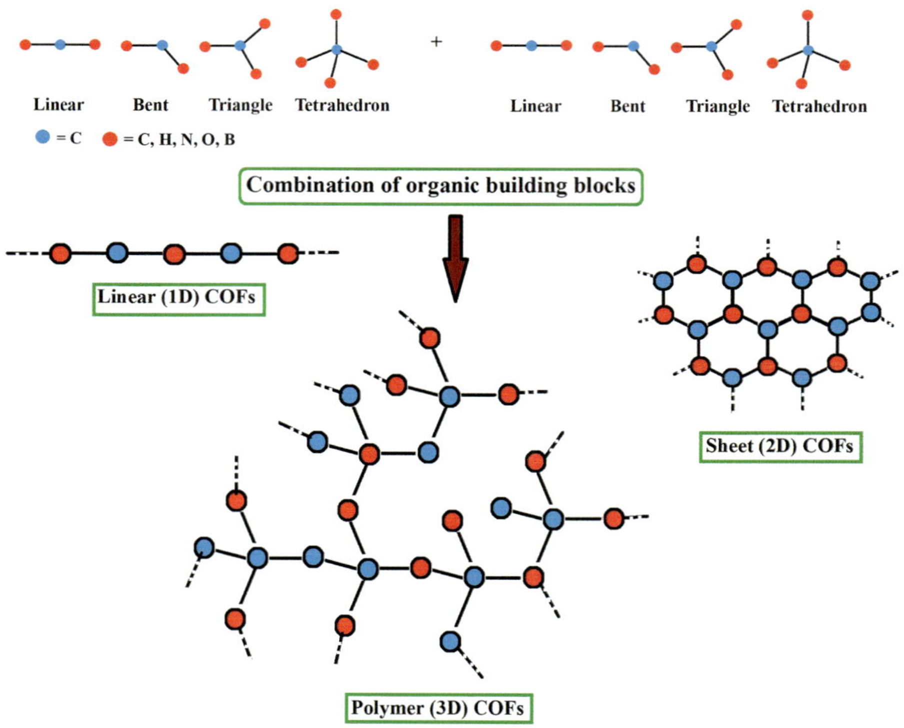

FIGURE 3.1 Representative examples of the different COFs originated from organic building blocks.

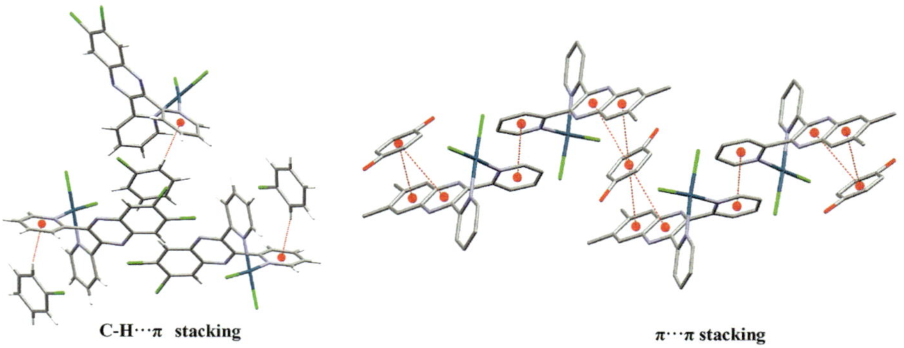

FIGURE 3.2 C-H···π and π···π stacking interactions in COFs.

structure and surface. Moreover, computational simulation is required to study the specific interactions between COF frameworks and guest species. In the following sections, detailed descriptions of the important techniques commonly used for the elucidation of the molecular structures and properties of the COFs are given and discussed.

3.2 MOLECULAR STRUCTURE OF COVALENT ORGANIC FRAMEWORKS

3.2.1 Infrared Spectroscopy

Infrared spectroscopic technique is an important method that was employed to analyze the vibrational properties of COFs as well as to determine the effective functional groups. Both Fourier Transform infrared (FTIR) spectroscopy and diffuse reflectance infrared fourier transform spectroscopy (DRIFTS) are IR-based techniques that are used to analyze molecular vibrations, functional groups and chemical bonds (Figure 3.3). These methods work on the principle of infrared absorption, but they differ in how the sample interacts with infrared light. In FTIR spectroscopy, Infrared (IR) light passes directly through a thin solid or liquid sample by allowing for transmission-based measurements. DRIFTS works by scattering IR light off the rough or powdered surface of a sample with the diffusely reflected light then being analyzed. Because of their unique methodologies, FTIR and DRIFTS serve as complementary techniques in IR spectroscopy each tuned for specific sample types and analytical purposes. FTIR is best for homogeneous samples that can be prepared in

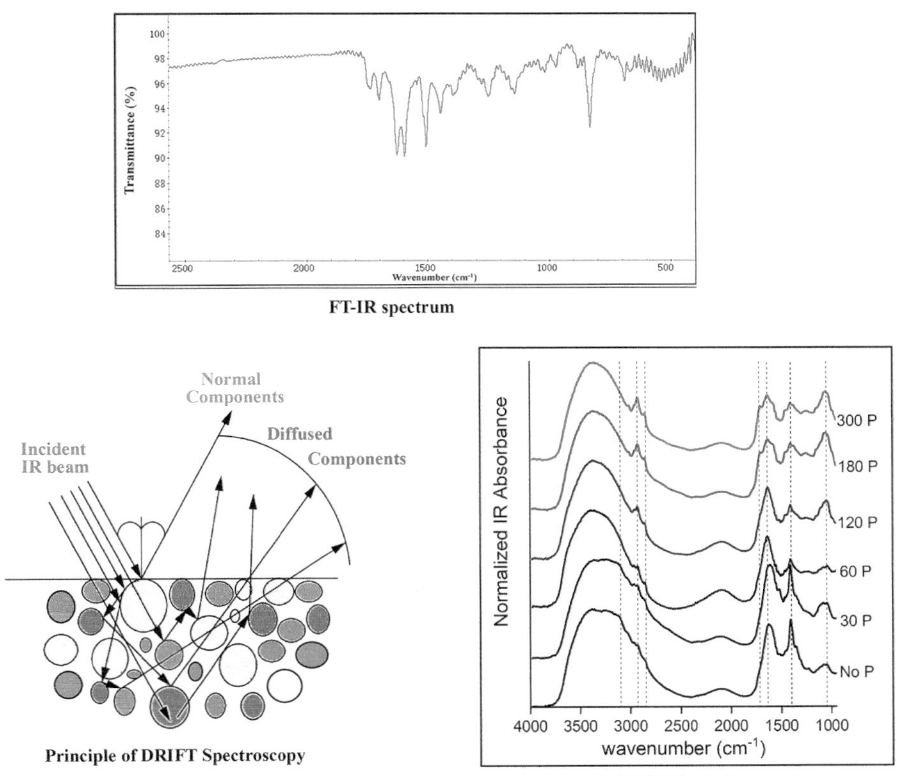

FIGURE 3.3 Representations of FTIR and DRIFT spectra.

solid discs, liquids or solutions, whereas DRIFTS is mainly advantageous for studying powders, rough surfaces and surface-sensitive materials. Under in situ conditions, IR spectroscopy can monitor the formation of products and surface chemical adsorption states of COFs as well as track processes like photocatalytic CO_2 conversion. This ability enables further investigation into the reaction mechanisms of such transformations.

A sensitive and reversible fluorescent probe that depends on a β-ketoenamine COF (TD-COF) was made through a Schiff base condensation reaction between 1,3,5-triformylphloroglucinol (TTD) and 2,6-diaminoanthraquinone (DA). This TD-COF was mainly engineered for the detection and efficient removal of p-phenylenediamine (PPD). The FTIR spectrum of TD-COF has a characteristic C-N stretching vibration at 1,258 cm^{-1} along with a partially diminished N-H stretching band at 3,424 cm^{-1} that confirms the successful condensation between TTD and DA. Additionally, the C=O stretching band of DA observed at 1,659 cm^{-1} shifted to a lesser wavenumber aligning closely with the C=O frequency of β-ketoenamine at 1,618 cm^{-1}. The presence of distinct C=O and C=C bands at 1,672 cm^{-1} and 1,566 cm^{-1} further verified the existence of β-ketoenamine in both its enol and keto tautomeric forms within the solid phase (L. Li et al. 2024).

The TD-COF has a π-conjugated framework with π-deficient electron moieties such as β-ketone carbonyl groups. These structural characteristics provide interactions with π electron-rich sites having compounds containing nitrogen or oxygen donor atoms via H-bonding and π-π interaction. Depending on this, the capacity of TD-COF for fluorescence dependence detection of amines was explored. Upon the introduction of PPD, the fluorescence of TD-COF was entirely quenched. As the PPD concentration increased, the amount of the dual fluorescence emission at 406 nm and 431 nm reduced (L. Li et al. 2024).

COFs are prominent classes of organic porous materials that offer promising hope for the photocatalytic reduction of CO_2 because of their tunable properties, high surface areas and photochemical stability in the photocatalytic CO_2 reduction process. Conversion of CO_2 into different useful products like methane, ethylene, methanol, carbon monoxide, formaldehyde and formic acid with the use of renewable energy production processes represents a great challenge for recent investigations. It has been indicated that the use of FTIR and DRIFT spectroscopies is crucial to follow up on the formation of these interesting products. Such products have various applications in the chemical industries and are used as building blocks to synthesize vast interesting organic materials. Many publications focus on the use of COFs in the photocatalytic reduction of CO_2 (Yeo et al. 2024; Hussain et al. 2023; Wang et al. 2023; Yousaf 2020).

3.2.2 Nuclear Magnetic Resonance Spectroscopy

NMR spectroscopy, such as 1H-NMR and 13C-NMR, is an important tool in analyzing the chemical environment and structure of COFs. Solid-state NMR gives complete details about the local environment of specific atoms within COFs and helps point out the covalent bonds formed and evaluate the material's structural composition. NMR is employed in verifying the chemical structure and connectivity

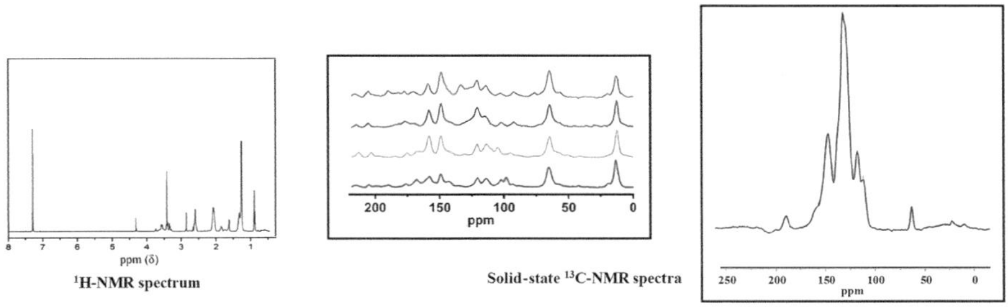

FIGURE 3.4 ^{1}H-NMR and solid-state ^{13}C-NMR spectra.

of monomers, confirming the formation of the intended covalent bonds. It is also utilized to analyze the reaction kinetics of COFs during photocatalytic processes. Changes in the NMR spectra, such as changes in signal intensity and chemical shifts of reactant and product, can be used to determine reaction rates, identify intermediate products and analyze the overall reaction mechanism (Dey et al. 2023; Zhang et al. 2024; Alhashmi et al. 2024; Mitra et al. 2017)] (Figure 3.4).

3.3 CRYSTALLINITY OF COVALENT ORGANIC FRAMEWORKS

COFs have a crystalline structure due to their highly ordered covalently linked porous networks. COFs crystallinity is a deciding factor that affects their stability, porosity and functional properties, which makes them appropriate for applications in gas storage, catalysis and drug delivery. The degree of crystallinity in COFs is examined by the nature of the building blocks, reaction conditions and postsynthetic modifications. Highly crystalline COFs give well-defined pore structures and enhanced mechanical stability, whereas lower crystallinity can lead to structural defects that affect performance. To analyze the structure of COFs, PXRD and SXRD are commonly used.

3.3.1 Powder X-Ray Diffraction

PXRD is employed to find the crystalline phases available in a compound and also used to collect the detail about its crystallographic structure. PXRD is rapidly employed for the characterization of COFs due to their crystallinity and porosity that result from organic molecules joined together by strong covalent bonds. PXRD analysis has many important features that can be summarized as: (a) Structure confirmation: COFs are designed and synthesized with specified geometric arrangements. Therefore, PXRD is carried out to determine whether the resulting COFs match the desired crystalline structure by evaluating with the experimental diffraction pattern with simulated patterns. (b) The phase purity: The COF products can sometimes have multiple crystalline phases or impurities. However, the PXRD assists in determining the purity of the final COFs product by recognizing all the phases available in the sample. (c) Lattice parameters definition: The lattice parameters of the COFs can be calculated from the obtained PXRD diffraction data. These parameters provide

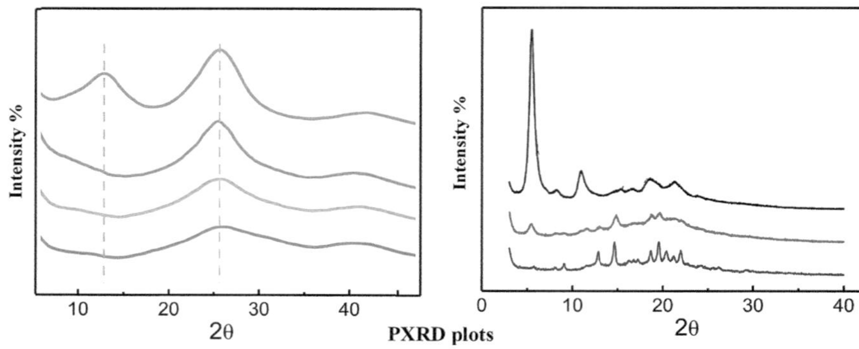

FIGURE 3.5 Representation of PXRD plots for COFs.

details about the size and shape of the unit cell and consequently valuable parameters including the pore size and surface area of compound can be determined. (d) The size and imperfection of the crystals: The broadening of the PXRD peaks can refer to the availability of small crystals or defects in the COF crystal structures. The imperfection of the crystals can affect their mechanical strength, chemical stability, and catalytic activity. (e) Modifications of post-preparation: PXRD can be employed to check the structural changes in COFs after any possible postmodification like functionalization or doping in order to ensure that the applied modifications do not affect or damage the crystalline structure (Ding et al. 2023; Gong et al. 2023; Tang et al. 2023; Akinnawo 2024)

3.3.2 Single-Crystal X-Ray Diffraction

SXRD is a powerful technique employed for the analysis of detailed structure of crystalline solid materials. SXRD provides detailed information about the internal crystal lattice and structure of crystalline compounds. Crystallization operations are prominent processes in determining the chemistry of COFs. Some time ago, the structural characterization of COFs was limited to molecular modeling, powder X-ray, or electron diffraction data. Recently, single-crystalline COFs have gained more interest due to their unique advantages, like well-defined structures, bond lengths and bond angles data, and porosity. Single-crystalline COFs are required to understand their pore chemistry and functions. However, the development of the preparation of suitable large single crystals has been restricted due to the lack of comprehensive synthetic techniques and successful characterization procedures. SXRD of COFs requires high-quality crystals that allow the collection of the diffraction data (Figure 3.5) (Yu et al. 2024; Qian et al. 2023).

3.4 MORPHOLOGY OF COVALENT ORGANIC FRAMEWORKS

Designing COFs for use in different applications, such as photocatalysis and CO_2 reduction, requires recognition of their crystal morphology. This is because the crystal surface, shape, and size of COFs usually affect the catalytic efficiency, light absorption, charge displacement and better access to their active sites (Figure 3.6).

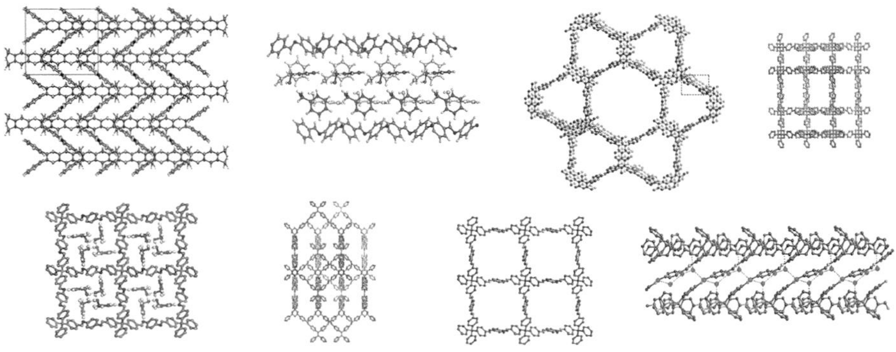

FIGURE 3.6 Representative single-crystal X-ray structures for COFs.

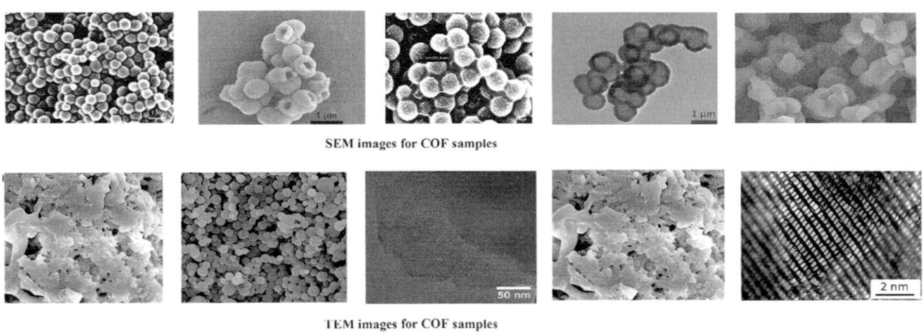

FIGURE 3.7 Representative SEM and TEM images for COFs.

3.4.1 Scanning Electron Microscopy and Transmission Electron Microscopy

SEM is an electron microscope that provides images of a crystal by scanning the surface with a beam of electrons. Electrons' interaction with atoms of the sample generates various signals and provides a clear picture of surface topography and composition of the sample. On the other hand, in TEM microscopy, a beam of electrons is transmitted via a crystal sample to produce an image. The sample should be a very thin section of about 100 nm thick. Interaction of the electrons with the sample generates an image of the crystal surface. Therefore, SEM and TEM are very useful techniques to characterize and analyze the morphology of COFs (Figure 3.7). The SEM method analyzes the surface morphology, the particle shape, and the size distribution on microscale. It yields images for the surface texture and structural characteristics of the COFs. For the use of COFs in photocatalysis, SEM can help in identifying the best morphology, which facilitates the light absorption and the effective diffusion of both reactants and products. In addition, the SEM technique can determine whether the crystal surface consists of a uniform and well-defined species or nanosheets that are essential to increase photocatalytic activity. The TEM method, on the other hand, is used for the investigation of the internal structure, morphology, and size of the crystal on nanoscale. This advantage can indicate the presence of the nanoscale pores, channels or layered structure arrangements, which could affect the light absorption and charge transport.

The distinctive crystalline structures of COFs can provide different adaptable pore sizes, low density, very large surface area and craftily designed functionalities. These excellent characteristics make a great attraction in many applications such as photocatalysis, electrocatalysis and gas adsorption. SEM and TEM are used for evaluating the microstructural properties of COFs, mainly in applications such as photocatalytic carbon dioxide conversion. These methods give detailed information about morphology, porosity and structural integrity that guide to optimize performance and refine the designs for increased catalytic efficiency. SEM and TEM also make a good contribution in COF applications such as water treatment. They guide us in understanding how COFs interact with contaminants, which makes them important for extracting heavy metals, isolating dyes and identifying pollutants that can also contribute to seawater desalination. By giving high-resolution visualization, these methods have significantly contributed to advancing COF-based environmental and energy solutions (X. Liu et al. 2023).

3.4.2 High-Resolution Transmission Electron Microscopy

The HRTEM technique provides high-resolution visualization of atomic structures. It is employed to determine the properties of different materials at the atomic scale such as semiconductors, metals, nanoparticles, organic crystals, graphene and carbon nanotubes. HRTEM has an important contribution in characterizing the crystallinity and structural properties of COFs. HRTEM helps in determining the crystalline size of COFs and distinguishing between different crystalline phases available in a sample. Furthermore, it also gives valuable information about interlayer stacking arrangements and orientations, which contributes to a good understanding of the structural characteristics of COFs (Zhan et al. 2024).

3.4.3 Atomic Force Microscopy

AFM is employed for analyzing crystalline surfaces with nanoscale accuracy. It gives high-resolution images that make it a necessary instrument for the evaluation of surface topography, roughness and structural properties at the atomic level. Advantages of AFM can be used in both air and liquid media that allow for real-time analysis of materials under various conditions. AFM is useful for characterizing nanoparticles, thin films and biological samples by giving valuable information regarding their morphology and mechanical properties. Qualitative and quantitative details can be obtained for many physical properties such as the size, morphology, surface texture and roughness of the tested material. AFM gives high-resolution and good-resolution 3D images. AFM is also used to analyze the morphology of COF materials, especially for the 2D COFs having thin layers (Akinnawo 2024; Ishida 2024).

3.5 POROSITY OF COFs

COFs are crystalline porous materials characterized by their stable porosity that results from the connection of organic building blocks through reversible covalent bonds. Consequently, the design and synthesis of COFs primarily focused on porosity, structural precision and functional properties. Despite these considerations, getting optimal porosity remains a main objective in COF development. Therefore, a well-planned design approach is necessary to effectively regulate and tailor their porosity. Thus, the terrific feature of high porosity of COFs along with the large surface area makes these materials valuable for a range of applications. Characterization of the porosity of COFs is, therefore, critical for understanding their possible use in many applications like electro- and photocatalysis (Altaf et al. 2021).

3.5.1 Gas Adsorption Isotherms

Gas adsorption is a procedure where a gaseous constituent is separated from a gas stream by the use of a solid material. Gas adsorption process is highly important for the characterization of many porous materials. Gas adsorption isotherm is a relation between pressure and adsorption amount at a constant temperature. Gas adsorption on solid surface and in pore wall is a complicated situation that uses mass and energy interactions as well as phase changes. When a solid is exposed to a gas stream, the gas molecules hit and adhere to the solid surface for a limited time. The N_2 adsorption isotherms are used for characterizing the porosity of COFs. In 2D COFs, organic building blocks are connected to form atomic layers or sheets that are stacked together through π-π interactions within their crystalline structure. The way these layers align in the stacked arrangement influences the material properties mainly pore accessibility that has an important role in adsorption capacity. To analyze the surface area of these materials, the Brunauer–Emmett–Teller (BET) method was used to measure nitrogen adsorption isotherms. Meanwhile, the Barrett–Joyner–Halenda (BJH) technique is applied to find the pore size distribution of mesoporous materials (Figure 3.8). This method employed a modified version of the Kelvin equation to relate the amount of adsorbate removed from the

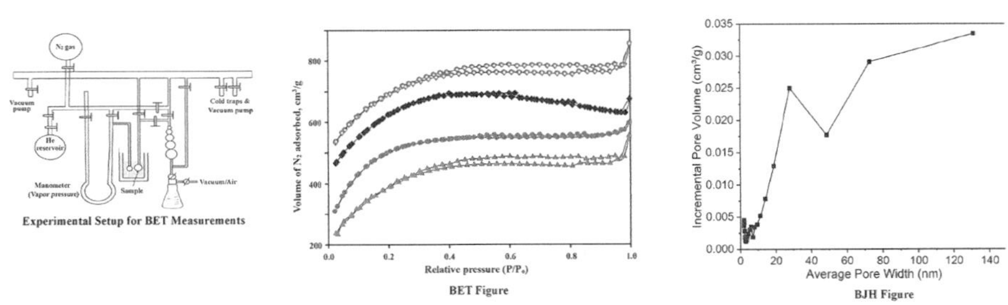

FIGURE 3.8 Experimental setup for BET analysis and representative BET and BJH drawing.

material pores to the reduction in the relative pressure that varies depending on the pore size. Nitrogen is used for adsorption studies and argon, a nonpolar and chemically inert gas, can provide precise data for characterizing the microporosity of COFs. Since argon adsorption at low temperatures reduces the influence of certain interactions that affect pore size distribution, it offers a more accurate assessment compared to nitrogen (Kessler et al. 2021).

The characterization of COFs primarily relies on nitrogen adsorption isotherms, which play an important role in evaluating porosity and distinguishing different pore types. A widely used approach involves nitrogen adsorption at 77 K, offering valuable information on isotherm classification, pore size distribution, and BET surface area. In addition to gas adsorption studies, pore characteristics can also be examined using techniques such as cryogenic transmission electron microscopy, small-angle X-ray scattering, small-angle neutron scattering and mercury intrusion porosimetry. These complementary methods provide a more comprehensive understanding of COF structures aiding in their design for applications in catalysis, gas storage and separation technologies.

3.6 VALENCE STATE

Valence state in COFs has importance in finding electronic properties, stability and functionality. The valence state of the constituent heteroatom elements like nitrogen, oxygen and sulfur can affect the electronic distribution and redox behavior of the COFs. This has importance in various applications such as catalysis, energy storage and sensing, where electron transfer and charge mobility are important factors. By changing the valence states through chemical modifications or external stimuli, the properties of COFs can be altered to increase their efficacy in various advanced material applications. Many techniques can be used to find the properties of COFs like X-ray photoelectron spectroscopy (XPS) and X-ray absorption near edge structure (XANES), among others.

3.6.1 X-Ray Photoelectron Spectroscopy

XPS is a spectroscopic technique employed for the analysis of approximately 200 atoms within a region of 10 nm of a material surface. It is mainly used to find the valence states of metals and doped elements in COFs. XPS works by employing X-ray photons from sources like Al Kα (1486.6 eV), Mg Kα (1253.6 eV) or energy synchrotron radiation to eject core electrons. Each element has different binding energies, and XPS makes accurate elemental identification. Evaluating changes in binding energy by X-ray excitation provides details about the oxidation states and local chemical environments of elements. This technique is necessary for knowing surface chemistry, electronic structure and potential catalytic sites in COFs (Akinnawo 2024; Ma et al. 2024).

XPS is also used for analyzing structural changes in porous COF materials and makes it an important instrument in applications such as catalysis, sensing and gas storage. A set of 2D COF films was made at the air–water interface employing cellulose nanocrystals as templates. XPS analysis revealed distinct peaks at 400.1 eV and

398.4 eV, which correspond to the NH group associated with pyrrole nitrogen and the H-bonding interaction (O-H---N=C) employing imine nitrogen. This confirms the presence of imine nitrogen within the porphyrin unit. Evaluating peaks such as C1s, N1s and O1s XPS gives detailed data on the chemical bonds and valence states of elements that make it a necessary tool in the characterization of COF-based catalysts (Al-dolaimy et al. 2024). It can be done by analyzing peaks such as C1s, N1s and O1s to provide detail about the chemical bonds that refer to the different peaks as well as the valence states of the elements.

3.6.2 X-Ray Absorption Near Edge Structure

XANES is used to study the X-ray absorption spectra of condensed materials. It mainly depends on the photo-absorption cross section that explains electronic transitions from an atom core level to higher energy states within a range of 50–100 eV above its ionization energy. In this energy region, wavelength of the photoelectron is longer than the distance between the absorbing atom and its nearest neighboring atom. X-rays are a form of ionizing radiation having sufficient energy to eject a core electron from an atom with each core shell having a different binding energy. When the X-ray energy is gradually increased and surpasses the binding energy of a specific core shell, a rapid rise in the absorption cross section occurs that is known as an absorption edge. Each absorption edge has the binding energy of a specific core electron and is classified depending on the principal quantum number of the excited electron: the K edge for $n = 1$, the L edge for $n = 2$, the M edge for $n = 3$ and so on (Penner-Hahn 2003,). As shown in Figure 3.9a, the L edge consists of three distinct sub-edges L1, L2 and L3 arranged in a descending energy order. Figure 3.9b presents a X-ray absorption spectrum that shows the energy ranges of XANES and EXAFS. EXAFS gives detailed knowledge about interatomic distances, coordination numbers of nearby atoms and lattice dynamics. The absorption edge is not merely a sharp jump in absorption but has intricate structural variations both near the edge and at significantly higher energies. The features observed close to the edge are classified as XANES while the oscillatory patterns extending beyond the edge can persist for more than 1,000 eV ("Penner-Hahn2003," n.d.).

XANES is also used for determining the electronic structure and valence states of elements in COFs. By analyzing the shape and energy of absorption, researchers

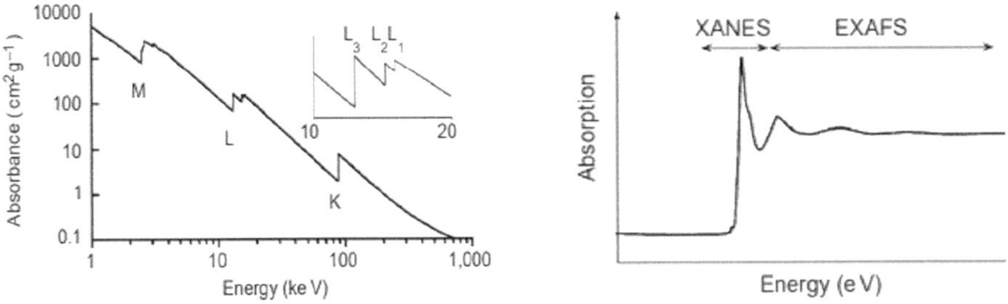

FIGURE 3.9 (a) Low-resolution X-ray absorption of Pb. (b) Illustration of an X-ray absorption spectrum showing XANES and EXAFS energy ranges ("Penner-Hahn2003," n.d.).

can find valuable information about the oxidation states, making XANES useful for analyzing transition metals incorporated into COFs. These methods provide qualitative and quantitative details and make them good for bulk sample analysis (Zheng et al. 2020).

In photocatalytic analysis, XANES has been used for studying CO_2 adsorption on catalyst surfaces and finding the chemical states of reaction intermediates. The spectral variations found in the process of photocatalytic reaction give details about the behaviors of adsorbed species and the chemical environment within the catalyst. XANES was employed to find the local atomic and electronic structures of Cu ions in a synthesized 2,2-bipyridine COF. The study showed that Cu^{2+} ions were monoatomically spread within the COF, which shows the ability of the method to give accurate structural and electronic information (Hosokawa et al. 2021).

3.7 ENERGY BAND STRUCTURE

Energy band structure tells the relationship between energy and momentum of charge carriers in a solid. Solid-state physics shows the range of energy levels in which electrons can be accommodated within a material. The energy band structure of solid COFs can be investigated by various techniques. To begin with, the valence band (VB) energy levels of COFs are measured using X-ray photoelectron valence band spectroscopy (VB-XPS). On the other hand, conduction band (CB) energy levels are measured through Mott–Schottky analysis and electrochemical characterization mainly by examining the cyclic voltammetry behavior of the catalysts. Mott–Schottky plots are used to determine key system properties such as flat band potential, doping density and Helmholtz capacitance.

3.7.1 X-Ray Photoelectron Spectroscopy-Based Valence Band

VB-XPS is employed to find the VB of a photocatalyst. This technique is based on spectral data mainly obtained near the Fermi energy level of the photocatalyst. As a result, it has been found to be highly effective in characterizing COFs with photocatalytic properties. For example, in the case of a reversible photochromic catalyst (DBTB-DETH-COF; DBTB = 4,4-dihydroxybiphenyl-3,3,5,5-tetracarbaldehyde and DETH = 2,5-diethoxyterephthalohydrazide), XPS played a key role in analyzing valence state changes during photochromism (X. T. Li et al. 2024). The C1s and N1s spectra were examined and their characteristic peaks were assigned to the corresponding chemical bonds. On the other hand, the O 1s binding energy peaks were assigned for C-OH/C-O-C and C=O, which have shown a significant shift in their binding energy values before and after photochromism (from 531.3 eV and 533.1 eV to 531.5 eV and 532.2 eV). Both peaks have shifted to higher binding energy, which suggests a reduced electron cloud density of the C-OH/C-O-C and C=O bond. This observation may indicate an electron transfer from the investigated COF to O_2 molecule. Another example demonstrated that the VB values were crucial in evaluating the success of constructing a p-n heterojunction by growing TpPa-2-COF in situ by covalent bonding to Bi_2O_2S nanosheets. Using VB-XPS, the values of the VB were analyzed for both Bi_2O_2S and TpPa-2-COF (M. Q. Li et al. 2020).

3.7.2 The Mott–Schottky Test

The Mott–Schottky method determines the type of semiconductor and provides its CB energy level. The redox potential of a semiconductor is first governed by the position of its CB and VB. Knowing the arrangement of these bands is important for obtaining photocatalytic mechanisms and analyzing the feasibility of redox reactions on the catalyst surface. Mott–Schottky analysis was employed to find the CB energy levels of synthesized FeP-COFs and $ZnFe_2O_4$, confirming their n-type semiconductor behaviors (Ma et al. 2024). The Mott–Schottky method was applied to find the CB place of a synthesized DBTB-DETH-COF before and after photochromism (X. T. Li et al. 2024) as shown in Figure 3.10. The positive slopes of the Mott–Schottky plots show that the material is a n-type semiconductor. The CB energy before photochromism was measured at 0.73 eV vs. SCE, a more negative potential in comparison to its post photochromism value of −0.34 eV vs. NHE. The flat band potential before photochromism was negative to facilitate the reduction of O_2 to$\bullet O_2^-$ (⁻0.33 eV). Furthermore, it was determined that the bandgap of DBTB-DETH-COF decreased after undergoing photochromism.

3.7.3 Cyclic Voltammetry

Electrochemical technique using cyclic voltammetry (CV) gives a complementary approach to Mott–Schottky tests for calculating the CB values of COFs. CV is a mainly applied electrochemical technique for determining electron transfer mechanisms within electroactive biofilms (EABs). This method includes cycling the working electrode potential between two predetermined values while recording the corresponding current response. As a result, CV can be employed to assess the electrode potential of biofilms by identifying the redox active species they secrete and the potential at which these compounds undergo oxidation and reduction whether they are freely diffusing or surface adsorbed. CV curves give insight into the rate-determining steps of electrochemical reactions at the electrode–electrolyte interface (Ma et al. 2024). By analyzing CV curves, the onset redox potential of an organic semiconductor can be determined, which allowed for the calculation of ionization

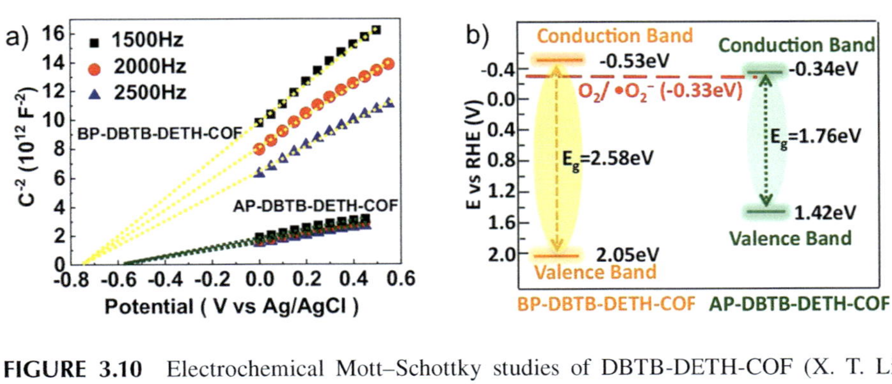

FIGURE 3.10 Electrochemical Mott–Schottky studies of DBTB-DETH-COF (X. T. Li et al. 2024).

potential, electron affinity and the bandgap energy (Eg) that ultimately enables the calculation of the semiconductor CB energy.

The CV of the prepared Cu_3-BTDE-COF (BTDE=(benzo-2,1,3-thiadiazole-4,7-diyl) dianiline) in the light and dark conditions was investigated. The study not only confirmed the improved catalytic performance under light exposure but also find the CB values of the materials using various characterization techniques (Chang et al. 2024). The CV analysis showed that the synthesized COF has a high specific surface area, a well-structured porous framework and numerous active sites. Furthermore, a study investigated the COF-BTT-SO_3H material (BTT= benzotrithiophene) aimed to function as an electrode for Zn^{2+}/H^+ energy storage through a two-electrode cell where a COF-BTT-SO_3H functions as a cathode coupled with a Zn foil anode in an aqueous electrolytic solution of $Zn(CF_3SO_3)_2$. Investigating the redox peaks elucidated the insertion/extraction reversible process of Zn^{2+} and H^+ within the COF electrode. Further information revealed the propensity of the COF to be employed as a cathode in comparison to the reported COF-based electrodes in Zn^{2+} energy storage (Peng et al. 2024).

3.8 CONCLUSION

COFs have emerged as a versatile and tunable porous material having unique structural and physicochemical properties that make them fit for various applications. COF molecular design, covalent bonding and good crystallinity make it suitable for specific functionalities such as catalysis, gas storage, energy conversion and drug delivery. Characterization of COFs remains a challenge for the development and also for making a deep understanding of their molecular structure, morphology, porosity, valence states and electronic properties. Various analytical techniques such as IR and NMR spectroscopy, XRD, electron microscopy and electrochemical methods give information about COFs structure and functional ability. These characterization methods confirm the synthesis of COFs and also help in optimizing their design for targeted applications. Combining computational simulations with experimental characterizations will enhance the ability to make COFs with increased stability, selectivity and performance. COFs have many abilities but challenges such as long-term stability, scalability in synthesis and more industrial applications need to be addressed in future.

REFERENCES

Abuzeid, Hesham R., Ahmed F.M. EL-Mahdy, and Shiao Wei Kuo. 2021. "Covalent Organic Frameworks: Design Principles, Synthetic Strategies, and Diverse Applications." Giant 6 (June): 100054. doi:10.1016/J.GIANT.2021.100054.

Akinnawo, Solomon Oluwaseun. 2024. "Covalent Organic Frameworks: Design, Synthesis, Characterization, and Applications." *ChemPhysMater* 3 (1): 36–63. doi:10.1016/j.chphma.2023.08.003.

Al-dolaimy, Faraas, Shelesh Krishna Saraswat, Baydaa Abed Hussein, Uday Abdul Reda Hussein, Shakir Mahmood Saeed, Ashwaq Talib Kareem, Alzahraa S. Abdulwahid, et al. 2024. "A Review of Recent Advancement in Covalent Organic Framework (COFs) Synthesis and Characterization with a Focus on Their Applications in Antibacterial Activity." *Micron* 179: 103595. doi:10.1016/j.micron.2024.103595.

Alhashmi, Shamma, Shu Yan Jiang, Aristarchos Mavridis, Rahul Raveendran Nair, Xin Zhao, and Carmine D'Agostino. 2024. "Diffusion and Adsorption in Covalent Organic Frameworks (COFs) Probed by Nuclear Magnetic Resonance Methods." *Materials Today Chemistry* 41 (October): 102245. doi:10.1016/J.MTCHEM.2024.102245.

Altaf, Amna, Nadeem Baig, Manzar Sohail, Muhammad Sher, Anwar Ul-Hamid, and Muhammad Altaf. 2021. "Covalent Organic Frameworks: Advances in Synthesis and Applications." *Materials Today Communications* 28 (September): 102612. doi:10.1016/J. MTCOMM.2021.102612.

Chang, Jia Nan, Shan Li, Qi Li, Jian Hui Wang, Can Guo, Yi Rong Wang, Yifa Chen, Shun Li Li, and Ya Qian Lan. 2024. "Redox Molecular Junction Metal-Covalent Organic Frameworks for Light-Assisted CO_2 Energy Storage." *Angewandte Chemie International Edition* 63 (23): e202402458. doi:10.1002/ANIE.202402458.

Dey, Anupam, Faruk Ahamed Rahimi, Soumitra Barman, Arpan Hazra, and Tapas Kumar Maji. 2023. "Metal-Free 3D Donor–Acceptor COF with Low Exciton Binding for Solar Fuel Production Based on CO_2 Reduction." *Journal of Materials Chemistry A* 11 (25): 13615–22. doi:10.1039/D3TA00901G.

Ding, Jiehua, Xinyu Guan, Jia Lv, Xiaohong Chen, Yi Zhang, Hui Li, Daliang Zhang, Shilun Qiu, Hai Long Jiang, and Qianrong Fang. 2023. "Three-Dimensional Covalent Organic Frameworks with Ultra-Large Pores for Highly Efficient Photocatalysis." *Journal of the American Chemical Society* 145 (5): 3248–54. doi:10.1021/JACS.2C13817.

Elsheemy, Walid M., Ayman A. Abdel Aziz, Ramadan M. Ramadan, Anna Kozakiewicz-Piekarz, and Mostafa A. Sayed. 2024. "Crystal Structure, Hirshfeld Surface Analysis, Molecular Modeling, Electrochemical Properties, and Potential Medicinal Activity of a Novel Binuclear Co(II) Complex." *Applied Organometallic Chemistry* 38 (11). doi:10.1002/ AOC.7641.

Gong, Li Juan, Li Yao Liu, Shao Shuai Zhao, Shuai Long Yang, Duan Hui Si, Qiu Jin Wu, Qiao Wu, Yuan Biao Huang, and Rong Cao. 2023. "Rapid Charge Transfer in Covalent Organic Framework via Through-Bond for Enhanced Photocatalytic CO_2 Reduction." *Chemical Engineering Journal* 458 (February): 141360. doi:10.1016/J.CEJ.2023.141360.

Hosokawa, Tomoya, Masaki Tsuji, Kosei Tsuchida, Kazuyuki Iwase, Takashi Harada, Shuji Nakanishi, and Kazuhide Kamiya. 2021. "Metal-Doped Bipyridine Linked Covalent Organic Framework Films as a Platform for Photoelectrocatalysts." *Journal of Materials Chemistry A* 9 (17): 11073–80. doi:10.1039/D1TA00396H.

Hussain, Altaf, Muhammad Nazim Lakhan, Abdul Hanan, Irfan Ali Soomro, Mukhtiar Ahmed, Faiza Bibi, and Irum Zehra. 2023. "Recent Progress on Green Synthesis of Selenium Nanoparticles – A Review." *MTSus* 23 (September): 100420. doi:10.1016/J. MTSUST.2023.100420.

Ishida, Naoyuki. 2024. "Atomic Force Microscopy." *Non-Destructive Material Characterization Methods* (January): 89–125. doi:10.1016/B978-0-323-91150-4.00011-2.

Kessler, Christopher, Johannes Eller, Joachim Gross, and Niels Hansen. 2021. "Adsorption of Light Gases in Covalent Organic Frameworks: Comparison of Classical Density Functional Theory and Grand Canonical Monte Carlo Simulations." *Microporous and Mesoporous Materials* 324 (September): 111263. doi:10.1016/J.MICROMESO.2021.111263.

Li, Lihua, Yinghu Ma, Weizhi Ou, Haohao Yang, and Xiaomei Yang. 2024. "β-Ketoenamine-Based Covalent Organic Framework for Detection and Removal of p-Phenylenediamine." *Microporous and Mesoporous Materials* 375 (July): 113162. doi:10.1016/j. micromeso.2024.113162.

Li, Meng Qiu, Le Yang Dang, Gui Gen Wang, Fei Li, Mao Han, Zhi Peng Wu, Gui Zhong Li, Zheng Liu, and Jie Cai Han. 2020. "Bismuth Oxychalcogenide Nanosheet: Facile Synthesis, Characterization, and Photodetector Application." *Advanced Materials Technologies* 5 (7): 2000180. doi:10.1002/ADMT.202000180.

Li, Xue Tian, Meng Jing Li, Yuan Liang Tian, Shu Lin Han, Lei Cai, Hui Chao Ma, Ying Qiang Zhao, Gong Jun Chen, and Yu Bin Dong. 2024. "A Reversible Photochromic Covalent Organic Framework." *Nature Communications* 15 (1): 1–9. doi:10.1038/s41467-024-52788-3.

Liu, Xiaolu, Yang Li, Zhongshan Chen, Hui Yang, Suhua Wang, Zhenwu Tang, and Xiangke Wang. 2023. "Recent Progress of Covalent Organic Frameworks Membranes: Design, Synthesis, and Application in Water Treatment." *Eco-Environment & Health* 2 (3): 117–30. doi:10.1016/J.EEHL.2023.07.001.

Liu, Yikuan, Xiaona Liu, An Su, Chengtao Gong, Shenwei Chen, Liwei Xia, Chengwei Zhang, et al. 2024. "Revolutionizing the Structural Design and Determination of Covalent–Organic Frameworks: Principles, Methods, and Techniques." *Chemical Society Reviews* 53 (1): 502–44. doi:10.1039/D3CS00287J.

Ma, Jiaqi, Lan Li, Yiming Zhang, Jinjie Qian, and Xusheng Wang. 2024. "Covalent Organic Frameworks: Synthesis, Structures, Characterizations and Progress of Photocatalytic Reduction of CO_2." *Chinese Journal of Structural Chemistry* 43 (12): 100466. doi:10.1016/j.cjsc.2024.100466.

Mitra, Shouvik, Himadri Sekhar Sasmal, Tanay Kundu, Sharath Kandambeth, Kavya Illath, David Díaz Díaz, and Rahul Banerjee. 2017. "Targeted Drug Delivery in Covalent Organic Nanosheets (CONs) via Sequential Postsynthetic Modification." *Journal of the American Chemical Society* 139 (12): 4513–20. doi:10.1021/JACS.7B00925.

Peng, Haijun, Verónica Montes-García, Kaiyue Jiang, Dawid Pakulski, Shunqi Xu, Michał Bielejewski, Fanny Richard, Xiaodong Zhuang, Paolo Samorì, and Artur Ciesielski. 2024. "Benzotrithiophene-sulfonate Covalent-organic Frameworks: Supramolecular Proton Pumps for High-Rate Aqueous Zinc-ion Energy Storage Systems." *SmartMat* 5 (6): e1312. doi:10.1002/SMM2.1312.

Penner-Hahn. 2003. "Characterization of Metalloprotein Metal Sites: High Throughput Approaches to Biological X-ray Absorption Spectroscopy-Invited." *Powder Diffraction* 18(2): 174–174

Qian, Cheng, Hongwei Wu, Wei Liang Teo, Yaozu Liao, and Yanli Zhao. 2023. "Single-Crystalline Covalent Organic Frameworks." *Trends in Chemistry* 5 (11): 853–67. doi:10.1016/J.TRECHM.2023.09.002.

Raghav, Sapna, Jyoti Raghav, Ma'aruf Abdulmumin Muhammad, Madhur Babu Singh, Pallavi Jain, Dinesh Kumar. 2025. "*Recent advancements in synthesis and applications of covalent organic frameworks*." *Applied Materials Today* 44(June): 102694. doi:10.1016/j.apmt.2025.102694

Sharma, Rakesh Kumar, Priya Yadav, Manavi Yadav, Radhika Gupta, Pooja Rana, Anju Srivastava, Radek Zbořil, Rajender S. Varma, Markus Antonietti, and Manoj B. Gawande. 2020. "Recent Development of Covalent Organic Frameworks (COFs): Synthesis and Catalytic (Organic-Electro-Photo) Applications." *Materials Horizons* 7 (2): 411–54. doi:10.1039/C9MH00856J.

Tan, Ke Tian, Samrat Ghosh, Zhiyong Wang, Fuxiang Wen, David Rodríguez-San-Miguel, Jie Feng, Ning Huang, et al. 2023. "Covalent Organic Frameworks." *Nature Reviews Methods Primers* 3: 1. doi:10.1038/S43586-022-00181-Z.

Tang, Qingqing, Ying Ying Gu, Jing Ning, Yingkui Yan, Li Shi, Mengshi Zhou, Hongtao Wei, et al. 2023. "Boosting Photocatalysis of Hydrazone-Linked Covalent Organic Frameworks through Introducing Electron-Rich Conjugated Aldehyde." *ChEnJ* 470 (August): 144106. doi:10.1016/J.CEJ.2023.144106.

Wang, Lu, Lin Wang, Saifei Yuan, Liping Song, Hao Ren, Yuankang Xu, Manman He, et al. 2023. "Covalently-Bonded Single-Site Ru-N2 Knitted into Covalent Triazine Frameworks for Boosting Photocatalytic CO_2 Reduction." *Applied Catalysis B: Environmental* 322 (March): 122097. doi:10.1016/J.APCATB.2022.122097.

Yeo, Chien Ing, Yee Seng Tan, Hafiz Taimoor Ahmed Awan, Abdul Hanan, Weng Pin Wong, Rashmi Walvekar, Bey Hing Goh, and Mohammad Khalid. 2024. "A Review on the Advancements in Covalent Organic Frameworks for Photocatalytic Reduction of Carbon Dioxide." *Coordination Chemistry Reviews* 521 (December): 216167. doi:10.1016/J.CCR.2024.216167.

Yousaf, Afifa. 2020. "A Hydrazone-Based Covalent Organic Framework/Iridium (III) Complex for Photochemical CO2 Reduction with Enhanced Efficiency and Durability." *Journal of Catalysis* 392: 49–55.

Yu, Baoqiu, Yu Tao, Xuan Yao, Yucheng Jin, Shan Liu, Tongtong Xu, Hailong Wang, et al. 2024. "Single-Crystalline 3D Covalent Organic Frameworks with Exceptionally High Specific Surface Areas and Gas Storage Capacities." *Journal of the American Chemical Society* 146 (42): 28932–40. doi:10.1021/JACS.4C09680.

Zhan, Zhen, Yuxin Liu, Weizhen Wang, Guangyu Du, Songhua Cai, and Peng Wang. 2024. "Atomic-Level Imaging of Beam-Sensitive COFs and MOFs by Low-Dose Electron Microscopy." *Nanoscale Horizons* 9 (6): 900–33. doi:10.1039/D3NH00494E.

Zhang, Ying, Linlin Zang, Shuting Zhao, Weipeng Cheng, Long Zhang, and Liguo Sun. 2024. "Brominated Metal Phthalocyanine-Based Covalent Organic Framework for Enhanced Selective Photocatalytic Reduction of CO_2." *Journal of Colloid and Interface Science* 655 (February): 1–11. doi:10.1016/J.JCIS.2023.10.111.

Zheng, Jianyun, Yanhong Lyu, Binbin Wu, and Shuangyin Wang. 2020. "Defect Engineering of the Protection Layer for Photoelectrochemical Devices." *EnergyChem* 2 (4): 100039. doi:10.1016/j.enchem.2020.100039.

4 Structural Properties and Strategies of Covalent Organic Frameworks

Anjali Yadav, Meenu Devi, Bhawana Jangir, Rajesh Kumar Meena, Deepti Chaudhary, and Aruna Sharma

4.1 INTRODUCTION

One of the main goals of materials science is to develop porous structures with basic molecular building blocks. The performance of energy storage devices, electrical devices, catalysis, and chemical sensors could all be significantly enhanced by well-planned porous and chemically resistant nanostructured materials. Strong chemical bonds allow molecular building blocks to be precisely arranged to form extended crystalline networks. Reticular chemistry, which is the foundation for the preparation of metal-organic frameworks (MOFs) and, more recently, covalent organic frameworks (COFs), emerged as a result of this tactic. Building open-framework architectures in both COFs and MOFs is made possible by achieving molecular-level control (Altaf et al., 2021, Hegazy et al., 2023).

COFs are formed by assembling organic building blocks, yielding two- and three-dimensional (2D and 3D) polymeric architectures. Their structural framework is composed of light elements like carbon, oxygen, nitrogen, and boron, which are held together by strong covalent bonds (Abuzeid et al., 2021; Raghav et al., 2025). COFs have a crystalline conjugated network with many pores, functional groups, and interconnected channels. COFs are members of the porous material family, which means they are composed of systematically arranged building blocks that assemble in an orderly manner to form well-defined pores.

COFs are excellent materials for adsorption and separation because of their light element composition and organized frame structures, which frequently result in numerous pore structures, low densities, and high surface areas (Ma et al., 2024, Yeo et al., 2024). More significantly, the covalently bound COFs have a greater range of uses as solvent adsorbents, particularly in adverse conditions, because they are far more stable than the well-known MOFs. Furthermore, due to the wide range of monomers available and the ease with which pore environments can be tailored, COFs can be used as versatile platforms for various applications, including sensors, semiconductors, and heterogeneous catalysis (Liu et al., 2021, Jaid et al., 2024). Boronic acid self-condensation reactions or their condensation with catechol to produce boronate esters were the basis for the earliest successful COFs3 examples

 DOI: 10.1201/9781003605386-4

(Figure 4.1a). Despite their weak chemical stability, boron-based COFs have been extensively researched; they easily hydrolyze in water, which is a substantial barrier to many possible uses.

Recent research has shown that Schiff base chemistry (Figure 4.1a) is an effective method to produce COFs with improved chemical stability, which allows them

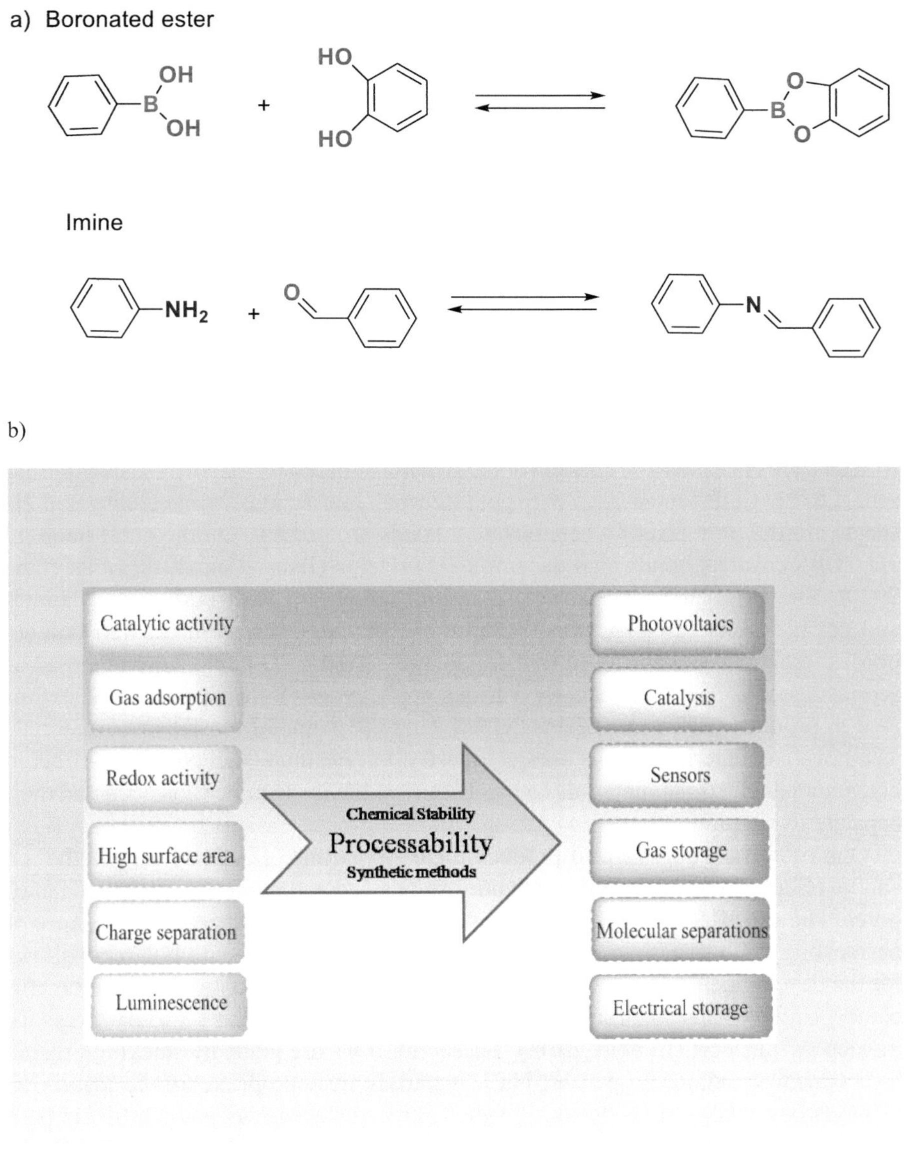

FIGURE 4.1 (a) The formation of boronate esters and imines is one of the most common reversible reactions used in the synthesis of COFs. (b) Sketch indicating some key aspects of material applicability and emphasizing the importance of processability in material applications.

to maintain their porosity and crystallinity in polar and nonpolar organic solvents as well as in basic and acidic inorganic solutions.

Because of its porous structure, the majority of the original uses identified for COFs were storage, separation, and gas adsorption. They have, however, lately been studied as sensors and catalysts for use in electrical device manufacturing or renewable energy applications. COFs must be both extremely stable and have superior properties in order to be used successfully in these applications. Furthermore, their integration into functional devices necessitates effective processing strategies (Figure 4.1b). As a result, structural cohesion-ensuring processing methods are required, as is the ability to shape, position, and align COFs as needed while seamlessly incorporating them into various device components.

In fact, effective processing is critical in determining a material's functionality and optimizing its performance, as evidenced by plastics' widespread success, which stems primarily from their remarkably simple yet efficient processability. The integration of COFs into devices is still difficult, despite their great qualities and promise. This difficulty arises because the majority of synthesis techniques are carried out in solvothermal conditions in sealed glass tubes, which results in the formation of insoluble powders made up of crystallites that have aggregated at random. As a result, most current processing methods, which are designed for materials with varying properties, are not well suited to carbon fibers (Yuan et al., 2019).

COFs are classified as either 2D or 3D structures based on their dimensional properties. In 2D COFs, interlayers only contain weak interactions, while conjugated 2D sheets are the only places where covalent bonds are present. On the other hand, in 3D COFs, covalent bonds bind the entire 3D skeleton (Jiang et al., 2018, Guan et al., 2020). The exfoliated COF structures mimic graphene, with periodical channels and flexible atomic architectures. Because of their large surface area, well-ordered porous structure, and ability to transfer charge carriers, 2D COFs have emerged as promising candidate materials for various applications. Bulk COFs can be exfoliated to produce COF nanosheets (CONs). There is potential for 2D COF materials as innovative materials for the energy industry. For example, the ion shuttle effect in conventional electrode materials causes batteries' life cycle to be shortened and their capacity to degrade.

One promising approach to produce ionic sieves that effectively reduce the ion shuttle effect is to use 2D COFs with uniformly sized and precisely controlled nanopores. The use of 2D materials in separation applications can improve the balance of permeability and selectivity. COFs' remarkable compatibility with polymers allows for the development of perfect membranes, which improves flux and separation efficiency. Unlike graphene, 2D COFs have inherent nanopores that act as pathways for molecular transport (Li et al., 2019). These 2D COFs are made by stacking organic layers and have unique properties. These materials are appropriate for the adsorption of gases like NH- and H- because of their large surface areas and adjustable pore sizes (Hao et al., 2023). Because of their short interlayer spacing and π-conjugated structure, 2D COFs enable electronic interactions between layers, resulting in conductive or semiconductive behavior (Yang et al., 2022).

CONs are a new class of porous, ultrathin 2D nanostructures that can be easily engineered and functionalized. However, there are several obstacles to the scalable production of CON thin films, such as restricted dispersion, layer restacking, and intricate postsynthetic modifications. Furthermore, exfoliated CONs have limited dispersion stability in various solvents, which complicates thin film fabrication techniques such as dip-coating, spin-coating, and drop-casting. Because of their high specific surface area, 3D COFs are also gaining a lot of attention in addition to 2D COFs. However, because of the 3D porous network's propensity to collapse, the synthesis of 3D COFs—particularly novel structures—remains difficult. Imides, boroxine, silicate, and porphyrin linkages have all been used to synthesize reported 3D COFs. Since 3D COFs are mostly made using high-temperature solvothermal techniques, their synthesis is more complicated than that of 2D COFs, which can be produced through self-assembly and simple chemical reactions.

Dynamic covalent chemistry (DCC), which is used in the synthesis of COFs, allows for the self-correction of reaction errors, resulting in the formation of highly ordered and crystalline structures. Furthermore, COFs are easily functionalized and combined with different materials, increasing their adaptability for a range of uses (Feng et al., 2021, Qian et al., 2024).

Several synthetic processes have been used to increase the stability of COFs while also accommodating properties that are incompatible with reversible processes. The processability of COFs has only lately come to light as a major obstacle in their actual use, despite years of research since their inception that has concentrated on the precise molecular design of building blocks to improve COF performance for various possible applications.

This chapter discusses and examines the most promising methods, highlighting their unique advantages and limitations in achieving specific morphologies. Furthermore, we will provide examples of how these processing techniques have been extensively utilized to integrate COFs into functional devices.

4.2 PROPERTIES OF COFs

Using dynamic organic chemistry principles, the error-checking mechanism reduces the occurrence of structural defects while promoting the formation of ordered COF structures. This thermodynamically driven process uses reversible reactions to enable "error checking" and "self-correction." Crystallization occurs during the formation of COFs, which have two major components: linkers and linkages. COF pore sizes and geometries are often linked with linkers, and the type of linkage has a significant impact on their stability and crystallinity (Figure 4.2). The properties of COFs differ depending on the specific linkers and linkages used (Huang et al., 2021; López-Magano et al., 2023). Another widely researched type of Imine-based COFs is developed by condensing aldehydes with amines or hydrazides. These imine-based COFs stand out from other types of COFs due to their exceptional crystallinity and structural uniformity. Additionally, they demonstrate enhanced stability in water and a wide range of organic solvents. Their framework's high nitrogen atom concentration facilitates interactions with various metal species, thereby encouraging metal modification. Furthermore, new COF types like imidazole-linked, aminal-linked, β-ketoenamine-linked, and dioxin-linked COFs expand the potential for multifunctional design and applications (Jiang et al., 2019, Wang et al., 2019) (Table 4.1).

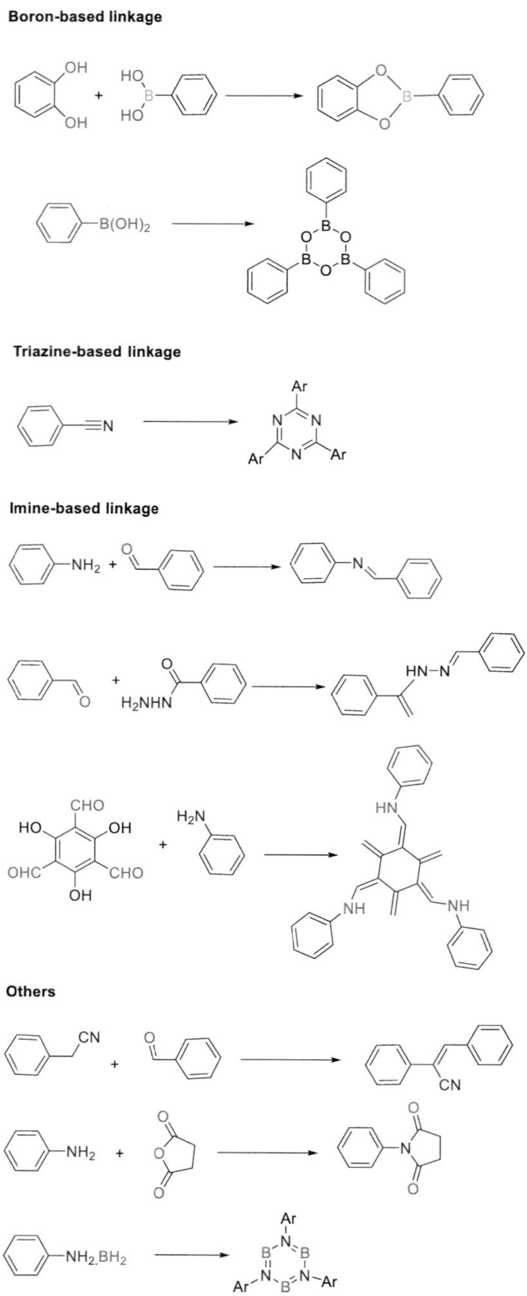

FIGURE 4.2 Common types of dynamic linkages to make COFs.

Although COF design has produced preliminary results, the majority of research is still conducted in laboratories, and the transition to practical applications remains a significant challenge.

TABLE 4.1
The Porosity and Chemical Stability Properties of Typical COFs

Linkage	COF	SBET, $m^2\ g^{-1}$	Pore Shape	Chemical Stability	References
Imine	HPB-COF	965	Triangular	Organic solvents and 1 M NaOH, water	Dalapati et al. (2015)
	DhaTph	1,305	Rhombus	HCl and boiling water	Kandambeth et al. (2013)
	DmaTph	431	Rhombus	Boiling water	Kandambeth et al. (2013)
	TPB-DMTP-COF	2,105	Hexagonal	Water, organic solvents, NaOH, and HCl	Xu et al. (2015)
Triazine	CTF-1	791	Hexagonal	Water, acidic and basic	Kuhn et al. (2008)
	CTF-DCBT	500	Triangular	Boiling water, HCl, and NaOH	Wang et al. (2017)
C=C	sp^2 c-COF	613	Tetragonal	Water, organic solvents, NaOH, and HCl	Jin et al. (2018)
	TpPa-1	535	Hexagonal	Boiling water and HCl	Kandambeth et al., (2012)
Dioxin	COF-316	557	Hexagonal	NaOH and HCl	Zhang et al. (2018)
	COF-318	576	Hexagonal	NaOH and HCl	Zhang et al. (2018)
β-ketoenamine	TpPa-H	632	Hexagonal	Water, organic solvents, and HCl	Singh et al. (2018)
	Dq1Da2Tp	1,400	Hexagonal	Water, organic solvents, and H_2SO_4	Khayum et al. (2018)
Benzoxazole	LZU-190	1,035	Hexagonal	Water, HCl, and NaOH	Wei et al. (2018)
Aminal	Aminal-COF-1	1,169	Tetragonal	Water, organic solvents, and NaOH	Jiang et al. (2019)
Thiazole	COF-921	1,550	Rhombus	HCl, H_2SO_4, H_3PO_4, and NaOH	Waller et al. (2018)
Oxazole	LZU-192	1,770	Rhombus	HCl, H_2SO_4, H_3PO_4, and NaOH	Waller et al. (2018)
Imidazole	LZU-501	815	Hexagonal	Common solvents, HCl, and NaOH	Wang et al. (2019)

We will investigate the characteristics of COFs in three key areas:

1. Pore design
2. Chemical stability
3. Scalability

4.2.1 Pore Design

Like other porous materials, COFs' pore structures have been extensively studied. As demonstrated by COF-6, CTF-1, TpPa-1, and Py-An COFs, the pore sizes of COFs can be precisely tuned by systematically arranging building units in a periodic and orderly manner.

Different polygonal geometries, including triangular, tetragonal, hexagonal, rhombic, and other shapes, can be observed in the pores of COFs. The pore size and shape can be precisely customized by altering the building units' length and structural arrangement and adding particular functional groups to the framework. For example, Banerjee et al. successfully synthesized a series of COFs with pore sizes ranging from 1.4 nm to 2.6 nm by systematically changing the length and geometry of linear building blocks (Dey et al., 2017). Similarly, Jiang et al. produced bifunctional building units containing boronic acid and aldehyde groups, which were then combined with two different reactive groups to form frameworks with hexagonal, tetragonal, and rhombic geometries (Chen et al., 2015).

Lavigne et al. added alkyl groups to the COF structure at the beginning of 2008. This caused pore diameter changes by directing the groups toward the center of the pores in a geometrically confined system (Tilford et al., 2008). The pore widths steadily decreased as the alkyl groups' length increased because the greater volumes filled the pores. The template-assisted method can be used to produce porous structures. For example, using polystyrene spheres as hard templates resulted in the creation of macro-TpBpy, which had a more defined macroporous structure and a larger surface area and total pore volume than the original TpBpy (Zhao et al., 2019). Furthermore, the macroporous diameters and template regions of the COF varied depending on the size and composition of polystyrene spheres used.

The copolymerization of all three components resulted in triple-pore COFs, with each unit containing two small triangular micropores and one large hexagonal mesopore. To illustrate the advantages of hierarchical porous COF structures, Ma et al. studied the immobilization of enzymes using a COF made of 4,4′-(ethyne-1,2-diyl) dibenzaldehyde (EDDA) as the connecting building unit and 4,4',4'',4'''-(ethene-1,1,2,2-tetrayl) tetraaniline (ETTA) as a flexible vertex piece (Pang et al., 2016, Sun et al., 2019). The dual-pore COF is made up of triangular micropores and hexagonal mesopores, with enzymes confined to the mesopores and reagents and products transported efficiently through the micropores. Hierarchical architectures provide better enzyme accessibility and increased denaturant resistance than nonporous COFs, opening the door for a sophisticated chemical approach to produce COFs with complex topologies and multifunctional characteristics.

Although COFs have a lot of potential for membrane applications, their effectiveness in membrane-based separation and nanofiltration processes is limited because their pore sizes frequently surpass those of the gas molecules that are being separated.

Consequently, many research teams have focused on developing COF membranes with smaller pore sizes. For instance, Ma et al. produced a COF membrane with a 0.6 nm pore size by changing the layer stacking arrangement (Li et al., 2020). The tunable and well-defined porous structures of COFs offer extensive potential for diverse practical applications. The dual-pore COF is made up of triangular micropores and hexagonal mesopores, with enzymes contained within the mesopores and reagents and products transported efficiently through the micropores. The dual-pore COF is made up of triangular micropores and hexagonal mesopores, with enzymes confined to the mesopores and reagents and products transported efficiently through the micropores.

4.2.2 Chemical Stability

The reversibility of formation reactions has a significant impact on the performance of various crystalline COFs. While many COFs exhibit remarkable thermal stability, the water generated as a by-product during the reversible process promotes reverse reactions, thereby accelerating the degradation of COFs (Hayat et al., 2024). Exposure to water and adverse conditions has a substantial impact on the structural integrity, crystallinity, and porosity of COFs, especially those that contain boron. Enhancing their hydrolytic stability through structural changes, strong linkage chemistry, or protective post-synthetic treatments is crucial for their successful use in practical applications.

Linkage conversion represents another intriguing approach. Amide-linked COFs can be synthesized by direct oxidation of imine linkages (Waller et al., 2016, Huang et al., 2021). Similarly, imine linkages are converted into oxazole- and thiazole-linked COFs by a series of linker exchanges and oxidative cyclization. The combined effects of irreversible enol-to-keto tautomerization and the reversible Schiff base reaction were responsible for the crystalline TpPa-1's exceptional chemical stability in both boiling water and 9 M HCl. However, it became unstable under alkaline conditions (Kandambeth et al., 2012, Waller et al., 2018). This instability results from an irreversible reaction under basic conditions that compromises the acid-sensitive imine linkages, whereas the keto form is still susceptible to conversion into the enol form.

The second strategy for increasing COF stability is linker modification, which entails functionalizing the framework and incorporating intramolecular or interlayer hydrogen bonds (Li et al., 2017). Early research found that functionalization with alkyl chains or pyridine provides effective steric hindrance, protecting boronate ester sites from hydrolysis while decreasing crystallinity and porosity. Subsequent research revealed that adding methyl sulfide or methoxy groups to the edges of phenyl rings in imine-linked COFs reduces the polarization effect of C-N bonds, improves interlayer interactions, and maintains high crystallinity and porosity (Huang et al., 2017). Kandambeth et al. (2013) proposed innovative strategies for achieving remarkable stability (Kandambeth et al., 2018).

The DhaTph COF remained structurally stable under acidic conditions, whereas the methoxy-substituted DmaTph COF, which lacks hydrogen bonding, degraded in 3 M HCl. COFs are resistant to strong acids and bases due to enhanced interlayer C-H. N hydrogen bonds with $-OCH_3$ functionality play a crucial role in enhancing the internal stability of COFs. These non-covalent interactions help anchor the framework and restrict the flexibility of imine linkages. This suggests that hydrogen-bond interactions within COFs can inhibit imine-N hydrolysis by shielding the nitrogen center from hydrolytic attack, thereby improving the overall material stability (Halder et al., 2018). Along with linker modification and linkage conversion, the blending strategy is frequently used to improve stability. When COFs are combined with functional materials like cross-linked polymers, CNTs, and graphene, they form COF-based composites with improved structural stability. These hybrids demonstrate excellent stability across various catalytic processes, particularly those involving metals.

4.2.3 Scalability

The majority of COFs are typically developed in sealed Pyrex tubes using a solvothermal method. Water can slowly diffuse out of the reaction medium due to the comparatively low internal pressure during this process, which facilitates the nucleation and crystallization of the COFs. Maintaining water equilibrium in reversible COF formation reactions presents significant challenges under standard conditions. When the precursor solution is heated in an open environment, water rapidly evaporates from the reaction mixture, disrupting the process and resulting in a loss of structural order in the synthesized COFs. Developing a straightforward and scalable process for producing COF materials is essential to overcoming the constraints of COF synthesis in industrial conditions. A number of alternative approaches, such as mechanochemical method, ionothermal method, room-temperature method, and microwave-assisted synthesis, have been studied to improve reaction conditions in addition to the conventional solvothermal method.

The preparation of COFs often requires meticulous screening of various solvents and multiple experimental iterations. The complexity of synthetic procedures and the restricted availability of resources present significant challenges to large-scale production. Consequently, most COFs are typically synthesized in milligram-scale quantities. To address the demands of industrial-scale production, it is essential to explore innovative and scalable synthetic strategies that align with environmentally sustainable practices. The widespread applicability of substrates and the common use of imine linkages have drawn significant attention from researchers toward the preparation of imine-based COFs. Jiang et al. (2014) introduced a vapor-assisted solid-state preparation method for imine-based COFs (Jiang et al., 2014). In this approach, oligomers formed by the direct grinding of monomers were exposed to small quantities of solvent vapor, leading to their conversion into bulk network polymers. Additionally, the composition of solvent vapor mixtures can be changed to control the morphology of COFs. Dichtel et al. replaced the traditional arylamine monomers with N-aryl benzophenone imine precursors to create the imine BND-TFB COF in 2017. These nonoxidizable precursors eliminated the requirement for additional protective steps, enabling the production of high-quality COFs through microwave-assisted heating. More recently, Zhang et al. developed a ventilation-vial method to prepare the 3D imine COF-300-V in a scalable manner that enhanced crystal nucleation and growth (Chen et al., 2019). This method achieved gram-scale yields, surpassing the production levels typically achieved through conventional solvothermal methods (Huang et al., 2021).

The complex reversible reactions, constrained by the design principles of COF, necessitate extended preparation methods. Catalysts in coupling reactions are widely regarded as an effective method for improving scalability and accelerating reaction kinetics. Banerjee et al. found that adding p-toluene sulfonic acid (PTSA) to the reaction mixture accelerated the synthesis of β-ketoenamine-linked TpPa-1 COF. This resulted in architectural stability and a large surface area in just one minute (Karak et al., 2017). Additionally, TpPa-1 can be synthesized on a large scale using a solvent-free approach with a twin-screw extruder. Because PTSA-amine salts are formed, mechanistic studies suggest that the reaction may be reversible. PTSA is essential for water retention, molecular organization, and reactivity control (Huang et al., 2021).

While these studies highlight the potential for scaling up COF material production, achieving true mass production remains a significant challenge. COFs are sensitive to chemicals because of their layer-by-layer stacked structures. Although adding functional groups can increase the framework's functionality and stability, doing so also uses resources and raises the cost of preparation.

4.2.4 Structural Regularity

Porosity alone does not always ensure the formation of crystalline structures in COF synthesis. Strong covalent bonds frequently lead to the development of disordered materials, which is evident in porous organic polymers, in contrast to amorphous organic polymers, achieving structural regularity in the targeted synthesis of COFs. Regular pore architectures can only be constructed if rigid building blocks are rationally designed and synthetic reactions are carefully chosen.

Even when these conditions are satisfied, the preparation of structurally ordered COF is not assured. This challenge has been demonstrated by Thomas et al. Specifically, the reversible cyclotrimerization reaction resulted in the formation of amorphous porous materials when rigid yet distorted nitrile-based building blocks were utilized. This observation emphasizes how important it is to make sure that building units' shapes and angles are compatible in order to promote the effective reversible formation of covalent bonds.

For example, if one of the building units is linear, it should react with another linear unit to form a linear material (Rambo et al., 2007). Conversely, when this linear unit interacts with components oriented at angles of 120° or 90°, the formation of hexagonal or tetragonal 2D COF structures is anticipated. Consequently, it forms the anticipated three-dimensional COF structure when coupled with a tetrahedral monomer (Uribe-Romo et al., 2009).

4.2.5 Functionality

The core principles outlined emphasize the importance of periodicity and porosity in the structural design and development of crystalline COF materials.

However, when developing functional COFs tailored to specific applications, it is critical to consider the integration of functional moieties within COFs.

The Pd(II)-coordinated COF-LZU1 material was synthesized via a simple post-synthetic modification of COF-LZU1 with Pd(OAc)- under ambient conditions. The resulting Pd/COF-LZU1 exhibits an excellent catalytic activity in the Suzuki–Miyaura coupling reaction. This method's key advantage is the homogeneous dispersion of functional sites, which contributes to the enhanced hydrothermal and chemical stability of the synthesized COFs.

To accomplish this, design principles governing periodicity and porosity must be integrated simultaneously while remaining aligned with the desired functional properties.

This presents difficulties for maintaining structural consistency in functional COFs as well as for synthesizing functional building blocks. Nonetheless, the bottom-up approach has proven to be extremely effective in producing functional COFs for various applications, including photoelectric devices and gas storage (Haase et al., 2020, Yeo et al., 2024, Yusran et al., 2024, 2024).

4.3 SYNTHESIS OF COFs

The basic idea behind COF synthesis, DCC, makes it easier to develop highly ordered crystalline structures with superior thermodynamic stability.

The precise atomic-level assembly of organic building blocks is made possible by COFs, which are organic, microporous crystalline materials that produce distinct nanopore architectures. The first COF was successfully created in 2005 by Cote et al. (2005) [59] by molecularly dehydrating 1,4-benzene diboronic acid.

One-dimensional (1D), 2D, and 3D COFs were then produced by carefully putting together various building blocks, each of which had unique functional characteristics. Table 4.2 shows the characteristics and synthetic methods for typical COFs; it is still critical to find viable methods for synthesizing COFs on a large scale.

TABLE 4.2
Porosity Features and Synthesis Process of COFs

Linkages	COF	BET Surface Area ($m^2\ g^{-1}$)	Pore Size (nm)	Process	References
Imine	LZU1	410	1.8	Solvothermal process	Wang et al. (2017)
	TpPa-1	535	1.25		Ma et al. (2017)
Boronate ester	COF-5	1,590	2.7	Microwave process	Bao et al. (2016)
Imine	TpBD	885	2.3	Room temperature process	Yang et al. (2015)
	COF (TAPB、TPA)	178.87	3.6		Lin et al. (2017)
	TbBd	196	2.8		Gao et al. (2017)
Triazine	CTF-1	791	1.2	Ionothermal process	Liu et al. (2012)

Nowadays, a popular and widely used method for growing on surfaces is solvothermal synthesis. Additionally, solvothermal conditions have been utilized to synthesize COFs using microwave processes (Xu et al., 2015, Xiao et al., 2023). Our main focus was on COFs that were successful in sample pretreatment and could be readily synthesized.

4.3.1 Solvothermal Process

The synthesis of COFs uses DCC, which produces well-ordered crystalline structures with remarkable thermodynamic stability.

Generally, a Pyrex tube containing the monomer and mixed solvent was degassed three times using the freeze-pump-thaw method. The Pyrex tube was sealed with a flame spray gun and placed in an oven set at a constant temperature for a specified period of time, which resulted in a considerable amount of precipitate in the system. After a number of purifications, COFs were finally produced. The properties of COFs in solvothermal reactions are strongly influenced by factors such as solvent, reaction temperature, and duration. The typical temperature range for the synthesis of COFs is 85°C–120°C, depending on the building units' chemical reactivity. Schiff base chemistry-produced COFs occasionally required days of high-temperature heating (around 120°C) (Bull et al., 2022).

Han et al. (2018) used a solvothermal method to successfully prepare a novel 3D chiral COF, which was then used as a chiral stationary phase in HPLC for enantioseparation of racemic alcohols. Also, Wang et al. (2017) used solvothermal methods to synthesize COF-LZU1, which demonstrated remarkable PAH extraction efficiency. Figure 4.3a depicts the imine groups that link COF-LZU1 together. The solvothermal process was also used to create COF(TpPa-1), a hydrophilic porous material for enriching N-linked glycopeptides (Ma et al., 2017). Paclitaxel was extracted from rat plasma samples using a solvothermal technique to produce the COF-1 modification (Chen et al., 2017).

Using a solvothermal technique, a recently developed hydrazine-linked chiral COF was made and utilized as a stationary phase in HPLC. Additionally, a solvothermal reaction in a capillary was used to synthesize COF-5 as a stationary phase for electrochromatography separation (Zhang et al., 2017, Vardhan et al., 2023). Notably, all the aforementioned COF synthesis methods were based on the solvothermal approach, differing primarily in applied temperatures.

4.3.2 Microwave Synthesis

A number of benefits over the solvothermal method, such as higher product purity, shorter reaction times, and the capacity to observe phase behavior while precisely controlling temperature and pressure, have been attributed to this approach, which has been extensively studied as an alternative strategy for COF synthesis.

In a study, Cooper et al. recently synthesized COF-5 using microwave irradiation and stirring for 20 minutes with a 200W power input. Compared to the traditional solvothermal method, which utilized 72 hours, this synthesis method showed a reaction rate 200 times faster.

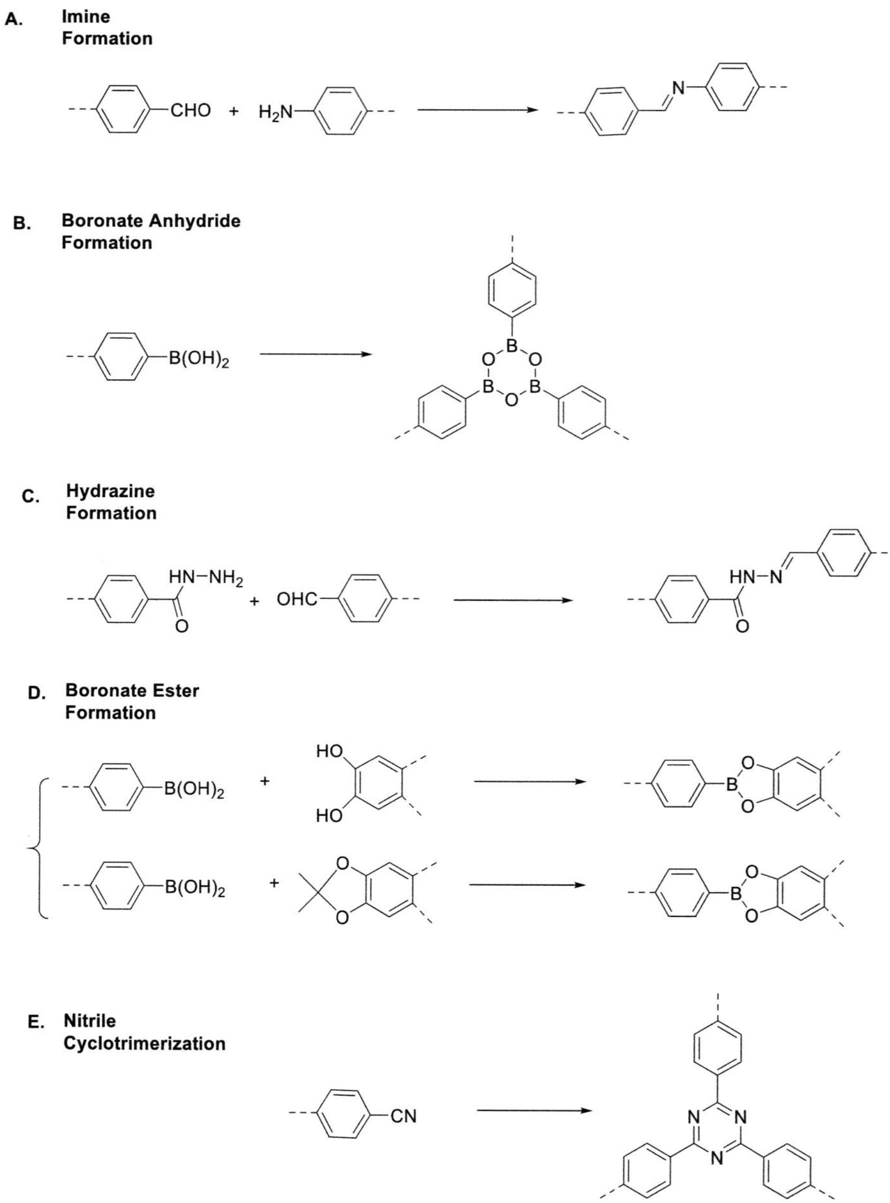

FIGURE 4.3 The connection of COF materials.

Moreover, microwave-assisted synthesis enhanced the surface area of COF-5 by effectively removing impurities, resulting in a BET surface area larger than that of its solvothermal synthesized counterpart. To eliminate impurities, the isolated COF-5 gray precipitate was heated in the reaction vessel with dry acetone using a 200 W power input. To eliminate the purple coloring in the supernatant brought on by the development of oxidized 2,3,6,7,10,11-hexahydroxytriphenylene (HHTP), the process was carried out twice (Abuzeid et al., 2021).

4.3.3 Mechanochemical Synthesis

This is a popular approach in chemical synthesis because it saves time, produces less waste, and employs less hazardous solvents. This approach has been shown to allow for multiple bonding by combining coordination and covalent bonds, resulting in the development of an open framework structure (Shinde et al., 2016). As part of the standard mechanochemical synthesis process, the reactants and solvent are combined in a mortar at room temperature with a pestle or zirconia balls. During grinding, morphological changes and the breaking and reforming of chemical bonds take place as the particle size decreases (Elseman et al., 2016, Lyu et al., 2017). This method can produce imine-linked, β-ketoenamine-linked, and hydrazine-linked COFs (Shinde et al., 2016). Three imine-linked COFs (TpPa-1, TpPa-2, and TpBD) were effectively prepared by Banerjee by condensing Tp with aromatic diamines like benzidine (BD) and 2,5-dimethyl-p-phenylenediamine at room temperature.

As the grinding time increased from 0 to 45 minutes, the product's color changed from orange to dark crimson. The COF's newly formed aromatic C=C and C-N bonds were confirmed by As the grinding time increased from (FT-IR) spectroscopy, which also revealed details about the mechanochemically synthesized material's covalent bond formation (Biswal et al., 2013). In a brief amount of time (45–180 minutes), other imine-linked COFs have been produced mechanochemically through the Schiff base reaction, including TpBpy-COF (Lv et al., 2019). Furthermore, COF layers were exfoliated as a result of the mechanochemical synthesis process (Biswal et al., 2013). In contrast, COFs made by the solvothermal process have better surface area, porosity, and crystallinity than those made by the mechanochemical method.

To produce COFs, Biswal et al. (2013) used solvent-free mechanochemical grinding at room temperature. These COFs showed exceptional stability in boiling water as well as in basic and acidic environments. The COFs made using the solvothermal method had a different layered morphology than the COFs made using the mechanochemical method. As the grinding time increased from 0 to 45 minutes, the product's color changed from orange to dark crimson. The COF's newly formed aromatic C=C and C-N bonds were confirmed by FT-IR spectroscopy, which also revealed details about the mechanochemically synthesized material's covalent bond formation (Uribe-Romo et al., 2009, Spitler et al., 2012, Biswal et al., 2013).

Crystalline COFs are produced at higher rates when the precursors are ground with the help of a liquid. By encouraging the proximity of precursors, the use of catalytic amounts of a liquid accelerates the rate of reaction. The liquid-assisted grinding technique was used to develop a crystalline hydrazone-linked COF and an unsubstituted hydrazone linkage with high crystallinity, which is a challenging task even with the solvothermal method (Das et al., 2014). Additionally, a simple liquid-assisted grinding method was used to synthesize COF-based 2D/2D heterojunction photocatalysts. Shinde et al. prepared COFs functionalized with bipyridine using both solvothermal (TpBpy-ST) and mechanochemical (TpBpy-MC) methods. These COFs were produced through a Schiff base reaction between 1,3,5-triformylphloroglucinol and

2,2′-bipyridine-5,5′-diamine. The final materials' chemical and thermal stability were comparable to those of cross-linked polymers (Shinde et al., 2016, Hou et al., 2019).

Mechanochemistry has become a very important technique because it produces no waste. Furthermore, mechanochemical approaches are faster, do not require hazardous solvents, and are less expensive than the solvothermal process. However, low porosity and crystallinity are the primary problems with materials made by mechanochemical processes, which restrict their usefulness in practical uses. The interfacial polymerization technique is also used in the production of COFs (Ma et al., 2019). Aside from a few exceptions, the majority of COFs must be synthesized in extremely harsh environments. Traditional methods of COF crystallization also require a lengthy reaction time. Mechanochemical, sonochemical, and microwave-assisted techniques are being used in an attempt to overcome these limitations. To produce COFs with high crystallinity, more research into simpler and more effective synthetic methods is required as there are still difficulties in COF preparation.

4.3.4 Sonochemical Synthesis

This method is less expensive, more energy efficient, and takes less time. During the sonochemical process, bubbles form and collapse in the solvent, resulting in localized high temperatures and pressures that accelerate COF crystallization (Głowniak et al., 2023). This method has several advantages, including a shorter reaction time, a large surface area, and products with small crystal sizes. COF-1 and COF-5 were efficiently produced in less than an hour using the sonochemical method. Additionally, the COFs have a maximum surface area of 2,122 $m^2\ g^{-1}$ and a mean crystal size that is 100 times smaller than the solvothermal method. Furthermore, COFs can be deposited onto materials such as graphene and carbon nanotubes using sonochemical synthesis (Yoo et al., 2016). In general, the combination of solvothermal synthesis and sonochemical synthesis facilitates the easy and quick formation of desired COFs.

4.3.5 Ionothermal Synthesis

In contrast to solvothermal synthesis, this process necessitates the use of molten salts and higher temperatures. It has mostly been used to synthesize Covalent Organic Frameworks (COFs) rather than Covalent Triazine Frameworks (CTFs), which are based on the trimerization of nitriles, because of the difficult reaction environment.

In their recent ionothermal synthesis of 3D COFs (3D-IL-COFs) at room temperature and pressure, Qiu et al. employed a room-temperature ionic liquid, 1-butyl-3-methylimidazolium bis((trifluoromethyl)sulfonyl) imide, as a catalyst and a solvent to facilitate the formation of Schiff bases. Notably, unlike the traditional solvothermal approach, which typically takes 3–7 days, this method allowed for the formation of 3D-IL-COFs in a significantly shorter reaction time.

Thomas created porous CTF-1 for the first time in 2008 using molten $ZnCl_2$ at temperatures between 400°C and 700°C (Kuhn et al., 2008). As an effective catalyst for their trimerization, molten $ZnCl_2$ can dissolve aromatic nitriles, which break down through C-H bond cleavage at temperatures higher than 400°C. Although it results in a loss of crystallinity and an amorphous structure, an increase in ZnCl2 concentration improves the porosity of CTF-1. Similarly, in 2010, CTF-2 was synthesized from 2,6-naphthalenedicarbonitrile using ionothermal synthesis at 400°C (Bojdys et al., 2010). Despite this, partial carbonization is unavoidable under the harsh conditions of ionothermal synthesis, producing CTF materials that appear as dark brown or even black powders. Combining ionothermal synthesis with additional techniques, like the microwave-assisted approach (Ren et al., 2012), not only significantly reduces reaction time, but the product also exhibits a luminous and non-carbonizing condition. Low carbonization and crystalline amorphous CTF made at a low temperature (40°C) were also achieved via ionothermal treatment (Kuecken et al., 2015).

4.3.6 Light-Induced Synthesis

Using a co-catalyst and simulating sunlight irradiation, a highly crystalline and conjugated COF with wavelengths ranging from 200 to 2,500 nm was successfully developed. The precursors, 1,2,4,5-benzenetetramine tetrahydrochloride and octahydrate (HCH), were dissolved in a range of solvent systems in an inert atmosphere to facilitate the reaction. Condensation was facilitated using acetic acid and water as co-catalysts. In the absence of light, hcc-COF transitioned into an amorphous state. The development of an extended conjugated structure, which facilitates improved charge transfer, is most likely the cause of the remarkable electrical conductivity observed in crystalline hcc-COF (Kim et al., 2019).

4.3.7 Room-Temperature Synthesis

Although COFs synthesized at room temperature exhibit an excellent acid–base resistance and high thermal stability, their practical applications remain limited. To date, only a few successful examples of high-performance COF synthesis using this approach have been reported. In 2015, at room temperature, Yang et al. (2015) created a simple and efficient solution-phase synthesis technique to create COF(TpBD), which demonstrated enhanced thermal stability. While this technique is faster than conventional solvothermal synthesis, it resulted in COF(TpBD) with moderate crystallinity and relatively low yields. A straightforward room-temperature method for developing COFs was prepared by Lin et al. (2017), who also showed how effective they are as a novel absorbent for protein exclusion and peptide enrichment from biological materials.

Similarly, their group used a simple room-temperature method to prepare the COF using BD and 1,3,5-triformylbenzene (Tb) as ligands. Even though these examples showed that COFs could be synthesized at room temperature, encouraging these synthetic pathways is still a major obstacle to increasing the potential applications of COFs (Li et al., 2018).

4.3.8 Knoevenagel Polycondensation Reaction

The first conjugated 2D COFs based on C=C linkages were reported by Zhuang et al. in 2016. This was accomplished by the Knoevenagel condensation process, which was previously believed to be impractical in the field of COFs. This common organic reaction serves as a catalyst base and transforms aldehydes or ketones into the desired olefins, usually in cyano-substituted and cis-configurations (Zhu et al., 2024). The primary monomers utilized in this investigation to generate a two-dimensional poly(phenylenevinylene) (2DPPV) framework via Knoevenagel polycondensation were 1,4-phenylenediacetonitrile and a tridentate aromatic aldehyde. The straightforward conversion of the heat-treated and activated sample into a porous carbon nanosheet structure offers an inexpensive, non-template approach to creating two-dimensional porous carbon materials suitable for energy-related applications. The potential applications of carbon materials and 2D polymers in energy conversion and storage, optoelectronics, sensing, and catalysis make this discovery noteworthy (Cao et al., 2019).

4.3.9 Photochemical Synthesis

Kim et al. were the first to report the photochemical synthesis of COF-5, which resulted in the generation of uniformly sized, sea-urchin-shaped COF-5. The discovery of novel synthetic pathways is highly valued in synthetic chemistry because it significantly improves the reaction efficiency. The photochemical approach yielded growth that was about 48 times greater. This method yielded approximately 75% and produced COF with a surface area of 2026.9 $m^2 g^{-1}$. The sea-urchin-shaped COF-5 has a large surface area due to its hierarchical structure (Kim et al., 2017). Using a similar method, a multilayer cladding, sea cucumber-shaped COF containing magnetic CNTs was synthesized using photochemistry. Magnetic properties were imparted by embedding magnetic nanoparticles into CNTs and depositing a COF shell around the nanotubes (Li et al., 2020). In a short amount of time, this method produces distinct morphology COFs more efficiently than the traditional synthesis method.

4.3.10 Other Methods

Furthermore, exploiting nanoparticles' magnetic characteristics for COF synthesis is a low-cost, low-toxicity approach. He et al. (2017) grafted COF onto amino-functionalized Fe_3O_4 nanoparticles to create a bouquet-shaped magnetic porous nanocomposite (He et al., 2017). In a similar manner, a two-step process consisting of mechanochemical grinding and crystallization was used to create a magnetic COF composite (Liao et al., 2017). Yan et al. recently reported a simple solution-phase method for generating spherical imine-based COFs at room temperature and atmospheric pressure. Because of its simplicity, this methodology may be a good choice for large-scale production if it can be generalized (Yang et al., 2015).

4.4 CONCLUSION

The systematically arranged pore structures of COFs, a novel class of crystalline porous materials, contribute to their high porosity, large surface area, and low density. These substances are a promising class of crystalline porous polymers that have attracted a lot of interest.

COFs perform better than other porous materials, such as MOFs and inorganic zeolites, due to their consistent design and structural adaptability. Because of their low density, high specific surface area, and adaptable features and functions, they make excellent platforms for a range of applications. They are useful in many different areas, such as drug delivery, electronic devices, luminescence, chemical sensing, gas storage, heterogeneous catalysis, and energy storage and conversion. This chapter summarizes the structural properties and synthesis methods for carbon-organic frameworks.

REFERENCES

Abuzeid, Hesham R., Ahmed FM EL-Mahdy, and Shiao-Wei Kuo. "Covalent organic frameworks: design principles, synthetic strategies, and diverse applications." *Giant* 6 (2021): 100054.

Altaf, Amna, Nadeem Baig, Manzar Sohail, Muhammad Sher, Anwar Ul-Hamid, and Muhammad Altaf. "Covalent organic frameworks: advances in synthesis and applications." *Materials Today Communications* 28 (2021): 102612.

Bao, Tao, Pingxiu Tang, Deying Kong, Zhenkun Mao, and Zilin Chen. "Polydopamine-supported immobilization of covalent-organic framework-5 in capillary as stationary phase for electrochromatographic separation." *Journal of Chromatography A* 1445 (2016): 140–148.

Biswal, Bishnu P., Suman Chandra, Sharath Kandambeth, Binit Lukose, Thomas Heine, and Rahul Banerjee. "Mechanochemical synthesis of chemically stable isoreticular covalent organic frameworks." *Journal of the American Chemical Society* 135, no. 14 (2013): 5328–5331.

Bojdys, Michael J., Jekaterina Jeromenok, Arne Thomas, and Markus Antonietti. "Rational extension of the family of layered, covalent, triazine-based frameworks with regular porosity." *Advanced Materials* 22, no. 19 (2010): 2202–2205.

Bull, O. S., I. Bull, G. K. Amadi, and C. O. Odu. "Covalent organic frameworks (COFs): a review." *Journal of Applied Sciences and Environmental Management* 26, no. 1 (2022): 145–179.

Cao, Shuai, Bing Li, Rongmei Zhu, and Huan Pang. "Design and synthesis of covalent organic frameworks towards energy and environment fields." *Chemical Engineering Journal* 355 (2019): 602–623.

Chen, Xiong, Matthew Addicoat, Enquan Jin, Hong Xu, Taku Hayashi, Fei Xu, Ning Huang, Stephan Irle, and Donglin Jiang. "Designed synthesis of double-stage two-dimensional covalent organic frameworks." *Scientific Reports* 5, no. 1 (2015): 14650.

Chen, Yichong, Zhao-Lin Shi, Lei Wei, Beibei Zhou, Jing Tan, Hao-Long Zhou, and Yue-Biao Zhang. "Guest-dependent dynamics in a 3D covalent organic framework." *Journal of the American Chemical Society* 141, no. 7 (2019): 3298–3303.

Chen, Yuling, and Zilin Chen. "COF-1-modified magnetic nanoparticles for highly selective and efficient solid-phase microextraction of paclitaxel." *Talanta* 165 (2017): 188–193.

Cote, Adrien P., Annabelle I. Benin, Nathan W. Ockwig, Michael O'Keeffe, Adam J. Matzger, and Omar M. Yaghi. "Porous, crystalline, covalent organic frameworks." *Science* 310, no. 5751 (2005): 1166–1170.

Dalapati, Sasanka, Matthew Addicoat, Shangbin Jin, Tsuneaki Sakurai, Jia Gao, Hong Xu, Stephan Irle, Shu Seki, and Donglin Jiang. "Rational design of crystalline supermicroporous covalent organic frameworks with triangular topologies." *Nature Communications* 6, no. 1 (2015): 7786.

Das, Gobinda, Digambar Balaji Shinde, Sharath Kandambeth, Bishnu P. Biswal, and Rahul Banerjee. "Mechanosynthesis of imine, β-ketoenamine, and hydrogen-bonded imine-linked covalent organic frameworks using liquid-assisted grinding." *Chemical Communications* 50, no. 84 (2014): 12615–12618.

Dey, Kaushik, Manas Pal, Kanhu Charan Rout, ShebeebKunjattu H, Anuja Das, Rabibrata Mukherjee, Ulhas K. Kharul, and Rahul Banerjee. "Selective molecular separation by interfacially crystallized covalent organic framework thin films." *Journal of the American Chemical Society* 139, no. 37 (2017): 13083–13091.

Elseman, Ahmed Mourtada, Mohamed M. Rashad, and Ali M. Hassan. "Easily attainable, efficient solar cell with mass yield of nanorod single-crystalline organo-metal halide perovskite based on a ball milling technique." *ACS Sustainable Chemistry & Engineering* 4, no. 9 (2016): 4875–4886.

Feng, Juanjuan, Jiaqing Feng, Xiangping Ji, Chunying Li, Sen Han, Haili Sun, and Min Sun. "Recent advances of covalent organic frameworks for solid-phase microextraction." *TrAC Trends in Analytical Chemistry* 137 (2021): 116208.

Gao, Chaohong, Guo Lin, Zhixian Lei, Qiong Zheng, Jiashi Lin, and Zian Lin. "Facile synthesis of core–shell structured magnetic covalent organic framework composite nanospheres for selective enrichment of peptides with simultaneous exclusion of proteins." *Journal of Materials Chemistry B* 5, no. 36 (2017): 7496–7503.

Głowniak, Sylwia, Barbara Szczęśniak, Jerzy Choma, and Mietek Jaroniec. "Recent developments in sonochemical synthesis of nanoporous materials." *Molecules* 28, no. 6 (2023): 2639.

Guan, Xinyu, Fengqian Chen, Qianrong Fang, and Shilun Qiu. "Design and applications of three dimensional covalent organic frameworks." *Chemical Society Reviews* 49, no. 5 (2020): 1357–1384.

Haase, Frederik, and Bettina V. Lotsch. "Solving the COF trilemma: towards crystalline, stable and functional covalent organic frameworks." *Chemical Society Reviews* 49, no. 23 (2020): 8469–8500.

Halder, Arjun, Meena Ghosh, Abdul Khayum M, Saibal Bera, Matthew Addicoat, Himadri Sekhar Sasmal, Suvendu Karak, Sreekumar Kurungot, and Rahul Banerjee. "Interlayer hydrogen-bonded covalent organic frameworks as high-performance supercapacitors." *Journal of the American Chemical Society* 140, no. 35 (2018): 10941–10945.

Han, Xing, Jinjing Huang, Chen Yuan, Yan Liu, and Yong Cui. "Chiral 3D covalent organic frameworks for high performance liquid chromatographic enantioseparation." *Journal of the American Chemical Society* 140, no. 3 (2018): 892–895.

Hao, Shuang, and Zhiqian Jia. "Covalent organic frameworks membranes: fabrication and applications." *Microporous and Mesoporous Materials* (2023): 112742.

Hayat, Asif, Saleem Raza, Mohammed A. Amin, Zeeshan Ajmal, Majed M. Alghamdi, Adel A. El-Zahhar, Hamid Ali et al. "Developing new-generation covalent organic frameworks as sustainable catalysts: synthesis, properties, types and solar energy production." *Materials Science and Engineering: R: Reports* 157 (2024): 100771.

He, Sijing, Tao Zeng, Saihua Wang, Hongyun Niu, and Yaqi Cai. "Facile synthesis of magnetic covalent organic framework with three-dimensional bouquet-like structure for enhanced extraction of organic targets." *ACS Applied Materials & Interfaces* 9, no. 3 (2017): 2959–2965.

Hegazy, H. H., Siva Sankar Sana, Tholkappiyan Ramachandran, Yedluri Anil Kumar, Dasha Kumar Kulurumotlakatla, Hisham SM Abd-Rabboh, and Seong Cheol Kim. "Covalent organic frameworks in supercapacitors: unraveling the pros and cons for energy storage." *Journal of Energy Storage* 74 (2023): 109405.

Hou, Yuxia, Cheng-Xing Cui, Enhui Zhang, Ji-Chao Wang, Ying Li, Yuping Zhang, Yuquan Zhang, Qing Wang, and Jianzhuang Jiang. "A hybrid of gC 3 N 4 and porphyrin-based covalent organic frameworks via liquid-assisted grinding for enhanced visible-light-driven photoactivity." *Dalton Transactions* 48, no. 40 (2019): 14989–14995.

Huang, Jing, Xiaojuan Liu, Wei Zhang, Zhifeng Liu, Hua Zhong, Binbin Shao, Qinghua Liang, Yang Liu, and Qingyun He. "Functionalization of covalent organic frameworks by metal modification: construction, properties and applications." *Chemical Engineering Journal* 404 (2021): 127136.

Huang, Ning, Lipeng Zhai, Hong Xu, and Donglin Jiang. "Stable covalent organic frameworks for exceptional mercury removal from aqueous solutions." *Journal of the American Chemical Society* 139, no. 6 (2017): 2428–2434.

Jaid, Ghaidaa M., Adnan A. AbdulRazak, Hicham Meskher, Saad Al-Saadi, and Qusay F. Alsalhy. "Metal-organic frameworks (MOFs), covalent organic frameworks (COFs), and hydrogen-bonded organic frameworks (HOFs) in mixed matrix membranes." *Materials Today Sustainability* 25 (2024): 100672.

Jiang, Cheng, Mi Tang, Shaolong Zhu, Jidong Zhang, Yanchao Wu, Yuan Chen, Cong Xia, Chengliang Wang, and Wenping Hu. "Constructing universal ionic sieves via alignment of two-dimensional covalent organic frameworks (COFs)." *Angewandte Chemie International Edition* 57, no. 49 (2018): 16072–16076.

Jiang, Shu-Yan, Shi-Xian Gan, Xi Zhang, Hui Li, Qiao-Yan Qi, Fu-Zhi Cui, Jian Lu, and Xin Zhao. "Aminal-linked covalent organic frameworks through condensation of secondary amine with aldehyde." *Journal of the American Chemical Society* 141, no. 38 (2019): 14981–14986.

Jiang, Yi, Wei Huang, Jianying Wang, Qi Wu, Haijiang Wang, Lili Pan, and Xikui Liu. "Green, scalable and morphology controlled synthesis of nanofibrous covalent organic frameworks and their nanohybrids through a vapor-assisted solid-state approach." *Journal of Materials Chemistry A* 2, no. 22 (2014): 8201–8204.

Jin, Enquan, Juan Li, Keyu Geng, Qiuhong Jiang, Hong Xu, Qing Xu, and Donglin Jiang. "Designed synthesis of stable light-emitting two-dimensional sp2 carbon-conjugated covalent organic frameworks." *Nature communications* 9, no. 1 (2018): 4143.

Kandambeth, Sha rath, Arijit Mallick, Binit Lukose, Manoj V. Mane, Thomas Heine, and Rahul Banerjee. "Construction of crystalline 2D covalent organic frameworks with remarkable chemical (acid/base) stability via a combined reversible and irreversible route." *Journal of the American Chemical Society* 134, no. 48 (2012): 19524–19527.

Kandambeth, Sharath, Digambar Balaji Shinde, Manas K. Panda, Binit Lukose, Thomas Heine, and Rahul Banerjee. "Enhancement of chemical stability and crystallinity in porphyrin-containing covalent organic frameworks by intramolecular hydrogen bonds." *Angewandte Chemie* 49 (2013): 13290–13294.

Karak, Suvendu, Sharath Kandambeth, Bishnu P. Biswal, Himadri Sekhar Sasmal, Sushil Kumar, Pradip Pachfule, and Rahul Banerjee. "Constructing ultraporous covalent organic frameworks in seconds via an organic terracotta process." *Journal of the American Chemical Society* 139, no. 5 (2017): 1856–1862.

Khayum M, Abdul, Vidyanand Vijayakumar, Suvendu Karak, Sharath Kandambeth, Mohitosh Bhadra, Karthika Suresh, Nikhil Acharambath, Sreekumar Kurungot, and Rahul Banerjee. "Convergent covalent organic framework thin sheets as flexible supercapacitor electrodes." *ACS Applied Materials & Interfaces* 10, no. 33 (2018): 28139–28146.

Kim, Soyoung, and HeeCheul Choi. "Light-promoted synthesis of highly-conjugated crystalline covalent organic framework." *Communications Chemistry* 2, no. 1 (2019): 60.

Kim, Soyoung, Chibeom Park, Minkyung Lee, Intek Song, Jungah Kim, Minhui Lee, Jaehoon Jung, Yousoo Kim, Hyunseob Lim, and HeeCheul Choi. "Rapid photochemical synthesis of sea-urchin-shaped hierarchical porous COF-5 and its lithography-free patterned growth." *Advanced Functional Materials* 27, no. 32 (2017): 1700925.

Kuecken, Sophie, Johannes Schmidt, Linjie Zhi, and Arne Thomas. "Conversion of amorphous polymer networks to covalent organic frameworks under ionothermal conditions: a facile synthesis route for covalent triazine frameworks." *Journal of Materials Chemistry A* 3, no. 48 (2015): 24422–24427.

Kuhn, Pierre, Markus Antonietti, and Arne Thomas. "Porous, covalent triazine-based frameworks prepared by ionothermal synthesis." *Angewandte Chemie International Edition* 47, no. 18 (2008): 3450–3453.

Li, Jian, Xiang Zhou, Jing Wang, and Xiufen Li. "Two-dimensional covalent organic frameworks (COFs) for membrane separation: a mini review." *Industrial & Engineering Chemistry Research* 58, no. 34 (2019): 15394–15406.

Li, Li-Hua, Xiao-Lin Feng, Xiao-Hui Cui, Yun-Xiang Ma, San-Yuan Ding, and Wei Wang. "Salen-based covalent organic framework." *Journal of the American Chemical Society* 139, no. 17 (2017): 6042–6045.

Li, Na, Junjie Du, Di Wu, Jichao Liu, Ning Li, Zhiwei Sun, Guoliang Li, and Yongning Wu. "Recent advances in facile synthesis and applications of covalent organic framework materials as superior adsorbents in sample pretreatment." *TrAC Trends in Analytical Chemistry* 108 (2018): 154–166.

Li, Xinle, Chongqing Yang, Bing Sun, Songliang Cai, Ziman Chen, YongqinLv, Jian Zhang, and Yi Liu. "Expeditious synthesis of covalent organic frameworks: a review." *Journal of Materials Chemistry A* 8, no. 32 (2020): 16045–16060.

Li, Yang, Qianxun Wu, Xinghua Guo, Meicheng Zhang, Bin Chen, Guanyi Wei, Xing Li, Xiaofeng Li, Shoujian Li, and Lijian Ma. "Laminated self-standing covalent organic framework membrane with uniformly distributed subnanopores for ionic and molecular sieving." *Nature Communications* 11, no. 1 (2020): 599.

Liao, Yaozu, Jiahuan Li, and Arne Thomas. "General route to high surface area covalent organic frameworks and their metal oxide composites as magnetically recoverable adsorbents and for energy storage." *ACS Macro Letters* 6, no. 12 (2017): 1444–1450.

Lin, Guo, Chaohong Gao, Qiong Zheng, Zhixian Lei, Huijuan Geng, Zian Lin, Huanghao Yang, and Zongwei Cai. "Room-temperature synthesis of core–shell structured magnetic covalent organic frameworks for efficient enrichment of peptides and simultaneous exclusion of proteins." *Chemical Communications* 53, no. 26 (2017): 3649–3652.

Liu, Jingliang, Enmin Zong, Heyun Fu, Shourong Zheng, Zhaoyi Xu, and Dongqiang Zhu. "Adsorption of aromatic compounds on porous covalent triazine-based framework." *Journal of Colloid and Interface Science* 372, no. 1 (2012): 99–107.

Liu, Xiaolu, Hongwei Pang, Xuewei Liu, Qian Li, Ning Zhang, Liang Mao, Muqing Qiu, Baowei Hu, Hui Yang, and Xiangke Wang. "Orderly porous covalent organic frameworks-based materials: superior adsorbents for pollutants removal from aqueous solutions." *The Innovation* 2, no. 1 (2021).

López-Magano, Alberto, Saba Daliran, Ali Reza Oveisi, Rubén Mas-Ballesté, Amarajothi Dhakshinamoorthy, José Alemán, Hermenegildo Garcia, and Rafael Luque. "Recent advances in the use of covalent organic frameworks as heterogenous photocatalysts in organic synthesis." *Advanced Materials* 35, no. 24 (2023): 2209475.

Lv, Hongzhou, Xiaoli Zhao, Hongyun Niu, Sijing He, Zhi Tang, Fengchang Wu, and John P. Giesy. "Ball milling synthesis of covalent organic framework as a highly active photocatalyst for degradation of organic contaminants." *Journal of Hazardous Materials* 369 (2019): 494–502.

Lyu, Honghong, Bin Gao, Feng He, Cheng Ding, Jingchun Tang, and John C. Crittenden. "Ball-milled carbon nanomaterials for energy and environmental applications." *ACS Sustainable Chemistry & Engineering* 5, no. 11 (2017): 9568–9585.

Ma, Mingxuan, Yonghao Yang, Zhonghua Huang, Fuhong Huang, Quanliang Li, and Hongyu Liu. "Recent progress in the synthesis and applications of covalent organic framework-based composites." *Nanoscale* 16, no. 4 (2024): 1600–1632.

Ma, Xionghui, Chaohai Pang, Shuhuai Li, Yuhao Xiong, Jianping Li, Jinhui Luo, and Yan Yang. "Synthesis of Zr-coordinated amide porphyrin-based two-dimensional covalent organic framework at liquid-liquid interface for electrochemical sensing of tetracycline." *Biosensors and Bioelectronics* 146 (2019): 111734.

Ma, Yu-Fang, Fang Yuan, Xiao-Hui Zhang, Ying-Lin Zhou, and Xin-Xiang Zhang. "Highly efficient enrichment of N-linked glycopeptides using a hydrophilic covalent-organic framework." *Analyst* 142, no. 17 (2017): 3212–3218.

Pang, Zhong-Fu, Shun-Qi Xu, Tian-You Zhou, Rong-Ran Liang, Tian-Guang Zhan, and Xin Zhao. "Construction of covalent organic frameworks bearing three different kinds of pores through the heterostructural mixed linker strategy." *Journal of the American Chemical Society* 138, no. 14 (2016): 4710–4713.

Qian, Cheng, Anyang Qiu, Qihang Huang, Jiatong Liu, Qiang Gao, Hongwei Wu, and Yaozu Liao. "Dynamic Imine Exchange Enables Access to Nanoporous Covalent Organic Frameworks with High Crystallinity for the Heck Reaction." *ACS Applied Nano Materials* 7, no. 8 (2024): 9668–9677.

Raghav, Sapna, Jyoti Raghav, Ma'aruf Abdulmumin Muhammad, Madhur Babu Singh, Pallavi Jain, Dinesh Kumar. "Recent advancements in synthesis and applications of covalent organic frameworks." *Applied Materials Today*, 44(2025): 102694.

Rambo, Brett M., and John J. Lavigne. "Defining self-assembling linear oligo (dioxaborole) s." *Chemistry of Materials* 19, no. 15 (2007): 3732–3739.

Ren, Shijie, Michael J. Bojdys, Robert Dawson, Andrea Laybourn, Yaroslav Z. Khimyak, Dave J. Adams, and Andrew I. Cooper. "Porous, fluorescent, covalent triazine-based frameworks via room-temperature and microwave-assisted synthesis." *Advanced Materials* 24, no. 17 (2012): 2357–2361.

Shinde, Digambar Balaji, Harshitha Barike Aiyappa, Mohitosh Bhadra, Bishnu P. Biswal, Pritish Wadge, Sharath Kandambeth, Bikash Garai, Tanay Kundu, Sreekumar Kurungot, and Rahul Banerjee. "A mechanochemically synthesized covalent organic framework as a proton-conducting solid electrolyte." *Journal of Materials Chemistry A* 4, no. 7 (2016): 2682–2690.

Singh, Vikram, Seungwook Jang, Niraj K. Vishwakarma, and Dong-Pyo Kim. "Intensified synthesis and post-synthetic modification of covalent organic frameworks using a continuous flow of microdroplets technique." *NPG Asia Materials* 10, no. 1 (2018): e456–e456.

Spitler, Eric L., John W. Colson, Fernando J. Uribe-Romo, Arthur R. Woll, Marissa R. Giovino, Abraham Saldivar, and William R. Dichtel. "Lattice expansion of highly oriented 2D phthalocyanine covalent organic framework films." *Angewandte Chemie* 124, no. 11 (2012): 2677–2681.

Sun, Qi, Briana Aguila, Pui Ching Lan, and Shengqian Ma. "Tuning pore heterogeneity in covalent organic frameworks for enhanced enzyme accessibility and resistance against denaturants." *Advanced Materials* 31, no. 19 (2019): 1900008.

Tilford, R. William, Sam J. Mugavero, Perry J. Pellechia, and John J. Lavigne. "Tailoring microporosity in covalent organic frameworks." *Advanced Materials* 20, no. 14 (2008): 2741–2746.

Uribe-Romo, Fernando J., Joseph R. Hunt, Hiroyasu Furukawa, Cornelius Klock, Michael O'Keeffe, and Omar M. Yaghi. "A crystalline imine-linked 3-D porous covalent organic framework." *Journal of the American Chemical Society* 131, no. 13 (2009): 4570–4571.

Vardhan, Harsh, Grace Rummer, Angela Deng, and Shengqian Ma. "Large-scale synthesis of covalent organic frameworks: challenges and opportunities." *Membranes* 13, no. 8 (2023): 696.

Waller, Peter J., Steven J. Lyle, Thomas M. Osborn Popp, Christian S. Diercks, Jeffrey A. Reimer, and Omar M. Yaghi. "Chemical conversion of linkages in covalent organic frameworks." *Journal of the American Chemical Society* 138, no. 48 (2016): 15519–15522.

Waller, Peter J., Yasmeen S. AlFaraj, Christian S. Diercks, Nanette N. Jarenwattananon, and Omar M. Yaghi. "Conversion of imine to oxazole and thiazole linkages in covalent organic frameworks." *Journal of the American Chemical Society* 140, no. 29 (2018): 9099–9103.

Wang, Keke, Yuanzhe Tang, Qin Jiang, Youshi Lan, Hongliang Huang, Dahuan Liu, and Chongli Zhong. "A thiophene-containing covalent triazine-based framework with ultramicropore for CO_2 capture." *Journal of Energy Chemistry* 26, no. 5 (2017): 902–908.

Wang, Peng-Lai, San-Yuan Ding, Zhi-Cong Zhang, Zhi-Peng Wang, and Wei Wang. "Constructing robust covalent organic frameworks via multicomponent reactions." *Journal of the American Chemical Society* 141, no. 45 (2019): 18004–18008.

Wang, Rong, and Zilin Chen. "A covalent organic framework-based magnetic sorbent for solid phase extraction of polycyclic aromatic hydrocarbons, and its hyphenation to HPLC for quantitation." *Microchimica Acta* 184, no. 10 (2017): 3867–3874.

Wei, Pi-Feng, Ming-Zhu Qi, Zhi-Peng Wang, San-Yuan Ding, Wei Yu, Qiang Liu, Li-Ke Wang, Huai-Zhen Wang, Wan-Kai An, and Wei Wang. "Benzoxazole-linked ultrastable covalent organic frameworks for photocatalysis." *Journal of the American Chemical Society* 140, no. 13 (2018): 4623–4631.

Xiao, Jing, Jia Chen, Juewen Liu, Hirotaka Ihara, and Hongdeng Qiu. "Synthesis strategies of covalent organic frameworks: an overview from nonconventional heating methods and reaction media." *Green Energy & Environment* 8, no. 6 (2023): 1596–1618.

Xu, Hong, Jia Gao, and Donglin Jiang. "Stable, crystalline, porous, covalent organic frameworks as a platform for chiral organocatalysts." *Nature chemistry* 7, no. 11 (2015): 905–912.

Yang, Cheng-Xiong, Chang Liu, Yi-Meng Cao, and Xiu-Ping Yan. "Facile room-temperature solution-phase synthesis of a spherical covalent organic framework for high-resolution chromatographic separation." *Chemical Communications* 51, no. 61 (2015): 12254–12257.

Yang, Yizhou, and Karl Börjesson. "Electroactive covalent organic frameworks: a new choice for organic electronics." *Trends in Chemistry* 4, no. 1 (2022): 60–75.

Yeo, Chien Ing, Yee Seng Tan, Hafiz Taimoor Ahmed Awan, Abdul Hanan, Weng Pin Wong, Rashmi Walvekar, Bey Hing Goh, and Mohammad Khalid. "A review on the advancements in covalent organic frameworks for photocatalytic reduction of carbon dioxide." *Coordination Chemistry Reviews* 521 (2024): 216167.

Yoo, JongTae, Sung-Ju Cho, Gwan Yeong Jung, Su Hwan Kim, Keun-Ho Choi, Jeong-Hoon Kim, Chang Kee Lee, Sang Kyu Kwak, and Sang-Young Lee. "COF-net on CNT-net as a molecularly designed, hierarchical porous chemical trap for polysulfides in lithium–sulfur batteries." *Nano Letters* 16, no. 5 (2016): 3292–3300.

Yuan, Shushan, Xin Li, Junyong Zhu, Gang Zhang, Peter Van Puyvelde, and Bart Van der Bruggen. "Covalent organic frameworks for membrane separation." *Chemical Society Reviews* 48, no. 10 (2019): 2665–2681.

Yusran, Yusran, Bo Miao, Shilun Qiu, and Qianrong Fang. "Functional covalent organic frameworks: design principles to potential applications." *Accounts of Materials Research* 5, no. 10 (2024): 1263–1278.

Zhang, Bing, Mufeng Wei, Haiyan Mao, Xiaokun Pei, Sultan A. Alshmimri, Jeffrey A. Reimer, and Omar M. Yaghi. "Crystalline dioxin-linked covalent organic frameworks from irreversible reactions." *Journal of the American Chemical Society* 140, no. 40 (2018): 12715–12719.

Zhang, Kai, Song-Liang Cai, Yi-Lun Yan, Zi-Hao He, Hui-Mei Lin, Xiao-Ling Huang, Sheng-Run Zheng, Jun Fan, and Wei-Guang Zhang. "Construction of a hydrazone-linked chiral covalent organic framework–silica composite as the stationary phase for high performance liquid chromatography." *Journal of Chromatography A* 1519 (2017): 100–109.

Zhao, Xiaojia, Pradip Pachfule, Shuang Li, Thomas Langenhahn, Mengyang Ye, Christopher Schlesiger, Sebastian Praetz, Johannes Schmidt, and Arne Thomas. "Macro/microporous covalent organic frameworks for efficient electrocatalysis." *Journal of the American Chemical Society* 141, no. 16 (2019): 6623–6630.

Zhu, Yaqin, Lizhen Chen, Junjie Pan, Shaohua Jiang, Jiaxiu Wang, Guoying Zhang, and Kai Zhang. "Recent advances in COF-derived carbon materials: synthesis, properties, and applications." *Progress in Materials Science* (2024): 101373.

5 History and Classification of Covalent Organic Frameworks

Jasvinder Kaur, Neha Choudhary, and Shivanshi Tandon

5.1 INTRODUCTION

Covalent organic frameworks (COFs) are a new generation of crystalline, porous organic materials with great structural and functional diversity. COFs are different from traditional porous materials such as zeolites and metal-organic frameworks (MOFs) because they are made up solely of light organic elements such as carbon, hydrogen, oxygen, nitrogen, and boron held together by strong covalent bonds. This cross-linking architecture gives rise to greater stability and allows for tunability on the molecular scale. Organic building blocks are designed to form COFs with specific pore sizes, shapes, and functionalities, with much flexibility for various applications (X. Li et al. 2020; Bukhari et al. 2023). COFs are most often synthesized through reticular chemistry, in which predesigned monomers create covalent bonding to establish extended, periodic lattices. COFs are two types of networks: two-dimensional (2D) and three-dimensional (3D) networks, synthesized on the basis of alternate modes of molecular connection. 2D COFs have a π–π stacked layered architecture, while 3D COFs form interthreaded or knotted networks with increased surface area and mechanical stability. These frameworks are highly thermally and chemically stable and can also be modified postsynthesis, thus significantly expanding their application fields (Raghav et al. 2025; Song et al. 2019; Meng and Mirica 2021).

COFs have attracted great interest in various scientific fields due to their highly ordered tunable porosity such as for gas storage and separation, catalysis, optoelectronics, drug delivery, and environmental remediation. Precise atomic-level design allows the possibility to adapt the chemical environment for specific applications; thus, COFs are a powerful tool in material science. Further research with COFs holds the promise to revolutionize several fields such as energy storage, biomedicine, and green chemistry. It will, once again, ensure their relevance to modern chemistry and materials science (Song et al. 2019; M. Ma et al. 2024).

The importance of COFs in organic material chemistry lies in their extraordinary combination of crystallinity, porosity, and structural diversity. Due to the modular nature of the building blocks, researchers can easily fine-tune pore sizes, shapes, and functionalities to fit applications in the fields of gas storage, catalysis, drug delivery, sensing, and energy storage. There is further versatility on such materials since COFs

DOI: 10.1201/9781003605386-5

allow for postsynthetic modification (PSM), which further enables functionalization that can accommodate specific needs ranging from biomedical research to environmental sustainability. Moreover, COFs opened frontiers in reticular chemistry as they concentrate more on material design using atomic-level precision to revolutionize how chemists will approach the synthesis of materials (L. Zhu and Zhang 2017; Lohse et al. 2018).

This chapter will thus cover the fundamental aspects of COFs by analyzing their historical background and classification. Research on COFs has really gone a long way since it was first developed. New methods of synthesis with different functional group variations have amplified their utility and applicability in many fields of study. Their history will reveal the development path from the mere concept to more recent breakthroughs and applications, thus summarizing some of the important discoveries (Ge et al. 2024; Essalhi et al. 2025).

In addition, this chapter will classify COFs based on basic classification parameters, including dimensionality, linkage chemistry, and functionalization. Dimensional classification differentiates between 2D and 3D frameworks, which have different structural characteristics and applications. Another critical classification criterion is linkage chemistry, where different covalent linkages, such as boronate esters, imines, and triazine bonds, define the stability and reactivity of the framework. In addition, functional groups within COFs determine the physicochemical properties of these materials, making them useful for certain applications.

Important literature examples throughout the chapter will represent the design principles, structural diversity, and some practical applications of COFs. By the end of this chapter, the readers will get more insight into basic aspects of COFs, how they were derived historically, and different approaches that classify their growth in scientific investigations and industrial practice.

5.2 HISTORICAL BACKGROUND OF COFs

The COFs are among the breakthroughs of porous materials with this new class of crystalline organic polymers being covalently bonded. COFs were developed based on increasing demand for the preparation of ultra-high molecular, chemically stable porous materials, whose structures are precisely modifiable to suit specific applications. This section covers the early progress leading to the identification of COFs, with some of the key milestones and pivotal research that have established COFs as an important class of material in reticular chemistry (Waller et al. 2015).

5.2.1 Early Developments

5.2.1.1 Initial Concepts in Microporous Polymers

The origins of COFs stem from the extensive work on microporous materials, which have traditionally been dominated by inorganic frameworks like zeolites and MOFs (Kirlikovali et al. 2023). Even though these materials were structurally rigid and possessed high surface areas, requirements for low-density, chemically varied, and tunable organic systems resulted in research in the area of microporous polymers. The initial organic porous materials, such as hypercrossliked polymers(HCPs), conjugated microporous polymers (CMPs), and porous aromatic frameworks (PAFs), were porous but not crystalline and structurally well-defined. The requirement for a well-defined, covalently linked crystalline organic structure eventually led to the creation of COFs with the objective of merging high porosity with ordered structures (Yusran et al. 2020).

5.2.1.2 Discovery of COFs: Milestones and Pivotal Research

The discovery of COFs by Yaghi and his team in UCLA in 2005 was one of the finest breakthroughs in materials science. Initial COFs, i.e., COF-1 and COF-5, utilized boronate ester linkages, which have proved that it is feasible to form covalently bonded crystalline frameworks. COFs were designed with extremely accurate pore structures and high predictability based on reticular chemistry (Côté et al. 2005).

A series of landmark breakthroughs include the introduction of imine-linked COFs in 2007, with enhanced chemical stability, followed by the synthesis of triazine-linked COFs in 2012, achieving thermal and chemical robustness. Between 2014 and 2016, the possibility of functionalization of pore walls through PSM paved a new window of application for COF in catalysis, sensing, and energy storage. Since 2018, access to very stable COFs based on novel linkages, initiated from β-ketoenamine and hydrazone-based frameworks, has been facilitated by dynamic covalent chemistry (X. Ma and Scott 2018).

These breakthroughs have accelerated COF research in the production of highly tunable materials with excellent porosity, stability, and chemical specificity. Advances in synthetic strategies have further extended their industrial and scientific applications.

5.2.2 Evolution of COF Research

5.2.2.1 Advancements in Synthetic Techniques

Advances in synthetic methodologies impelled the development of COFs as they allowed more control over structure, porosity, and functional properties. Improvements of such alternatives like microwave-assisted, ionothermal, and mechanochemical synthesis helped reduce efficiency, scale the synthetic process, and reduce environmental burden relative to conventional solvothermal synthesis (Mosupi et al. 2025). One of the keys to this breakthrough was the adoption of DCC; this prevents error on covalent bond formation, self-healing properties, and allows for reversible covalent bond formation. Catalyst-assisted polymerization has opened up a vast horizon for COF applications by allowing diverse functional groups, and their PSMs have enhanced selectivity for the applications of gas separation and molecular sensing. These developments have significantly broadened the utility of COFs in catalysis, energy storage, and drug delivery (Yusen Li et al. 2020; Machado et al. 2021).

5.2.2.2 Role of Reticular Chemistry in COF Development

Reticular chemistry, pioneered by Omar M. Yaghi, has revolutionized the arena of COFs through the controlled synthesis of long, periodic frameworks with strong covalent bonds. This has enabled a geometry of the framework based on a predictable topology, making COFs very tunable materials with large surface areas, thermal stability, and unique electronic characteristics. The strictly organic structures of COFs, unlike conventional porous materials, greatly increase their utility for a wide range of applications. Another related area in chemistry is reticular chemistry, which has witnessed the design of COFs to possess hierarchical porosity for uses in catalysis and adsorption. Faster development in the discovery and optimization of new structures of COFs can be achieved through computational design approaches (B. Zhang et al. 2019; Gropp et al. 2020).

5.2.2.3 Increasing Focus on Structural and Functional Versatility

In the last two decades, there have been tremendous advancements in COFs that have shifted from maximum surface area and porosity to the integration of functional groups that increase chemical reactivity and selectivity. Recent uses of COFs have involved the introduction of heteroatoms like nitrogen, oxygen, sulfur, and metals, thereby enhancing their catalytic, photocatalytic, and energy storage capabilities (Hao and Jia 2023). One of the key structural advances has been the shift from 2D COFs, with layered π-π stacking, to 3D COFs of more rigid, cross-linked architectures, which have led to better stability and greater volumetric surface areas. Such changes have broadened the utility of COFs in gas storage, supercapacitors, and battery electrodes. More recently, COFs have emerged as responsive materials that can change their properties upon external stimuli such as light, pH, and temperature. Such adaptability opens the fields of controlled drug delivery and environmental remediation. In a nutshell, the synthesis and reticular chemistry of COFs have thus placed them at the center of the future versatile class of nanotechnology materials, materials science, and sustainable chemistry (Q. Zhu et al. 2020).

5.2.3 Impact of COFs in Modern Chemistry

COFs have revolutionized modern chemistry because they have allowed for the easy design of highly porous, crystalline materials with precise control of structure. These have a modular architecture that is superior in chemical stability and therefore are very light, unlike the traditional porous materials zeolites and MOFs (Martín-Illán et al. 2023).

COFs have been recognized for a number of strengths: they find a foundation in reticular chemistry that allows predictable self-assembly of molecular building blocks into topologies of desired, thus broad applications in the domains of catalysis, energy storage, gas separation, drug delivery, environmental remediation, and so forth. There are also further PSMs providing added scope for adaptability and functional diversity (Haase and Lotsch 2020).

Integration with AI-driven design and nanotechnology sets the future of material science. Computational modeling speeds up the discovery of new structures of COFs, increasing the potential application areas. This kind of interdisciplinarity in research will bring many researchers together, bridging the gaps between chemistry, materials science, and engineering in cutting-edge scientific developments (Pokrajac et al. 2021; Zatsu et al. 2024).

5.2.3.1 Applications Influencing COF Research Trajectories

COFs evolve incessantly owing to their widespread application, influencing synthesis and further modifications in structures. Their exceptionally large surface areas with tunable pore sizes enable them to store gas for effective separation and even for hydrogen fuel storage and carbon capture. Environmentally, their effectiveness in the removal of polluting agents through purification of water was observed without a loss in stability under challenging conditions, mainly removal of pollutants, heavy metals, and radioactive contaminations (Zhao et al. 2021). COFs are promising in biomedical research for drug delivery and biosensing due to their biocompatibility and controlled release properties, thus opening up possibilities in antimicrobial coatings and biocatalysis. COFs are also emerging in optoelectronics as their π-conjugated structures allow for good charge transport and are thus ideal for organic LEDs, solar cells, and transistors. Continued interest in COFs pushes the framework of scientific research for new synthetic strategies and mass production, thus ensuring its long-term applicability in various fields (H. Kaur et al. 2024).

5.2.3.2 Influence of Interdisciplinary Fields like Catalysis, Sensing, and Energy Storage

COFs have been found to be of great interest in several scientific disciplines, such as catalysis, sensing, and energy storage, due to their high surface area, tunable pore environments, and chemical stability. In catalysis, COFs act as heterogeneous catalysts, especially in photocatalysis and electrocatalysis, catalyzing reactions such as water splitting and CO_2 reduction with improved selectivity and efficiency. It uses COFs in the identification of gases, pollutants, and biomarkers by functionalizing them with recognition sites such that detection can be carried out selectively and sensitively. The COFs have had applications in supercapacitors, lithium-ion batteries, and fuel cells based on improved charge storage and cycling stability owing to their high conductivity and porous structures. Interdisciplinary research in COF continues to drive innovations and is therefore key to advancing modern materials science and industrial applications (Yeo et al. 2024) (Figure 5.1).

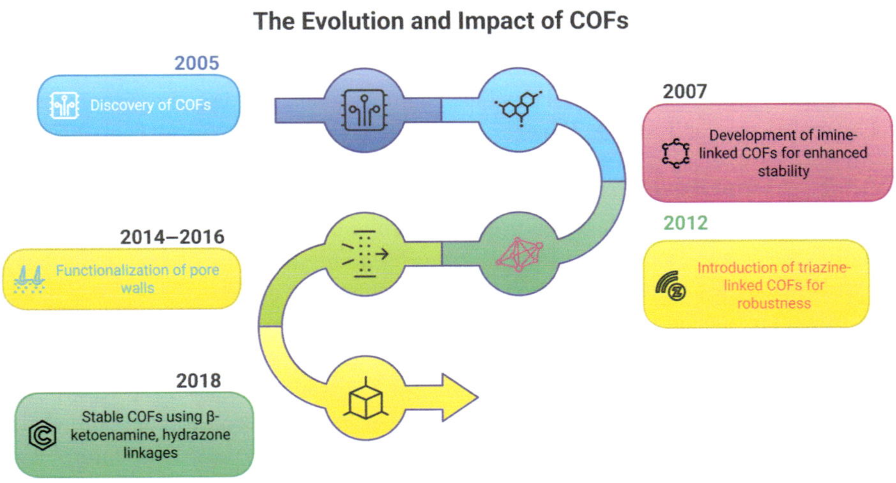

FIGURE 5.1 The evolution and impact of COFs.

5.3 GENERAL FEATURES OF COVALENT ORGANIC FRAMEWORKS

COFs have characteristics of structural and chemical properties unique to them that differentiate them from conventional porous materials. Their well-ordered structure, tunable porosity, and extensive stability make them potentially suitable for several applications, which include catalysis, gas storage, drug delivery, and energy storage. At the most basic level, such properties are defined by crystallinity, dimensionality, and atomic-level design and synthesis of COFs; yet afterward synthesis, alterations in pore environments are available. This chapter details the structural properties and well-defined frameworks and, on their synthesis, does disclose quite precise control (Song et al. 2019; AlHazaimeh et al. 2024).

5.3.1 Structural Properties

5.3.1.1 Crystallinity and Dimensionality

COFs possess a highly ordered arrangement of covalently linked organic linkers, leading to crystallinity, and are distinguished from amorphous porous polymers. The improved mechanical stability and integrity are hence imparted for uses involving homogeneous pore channels, like molecular sieving and selective adsorption (Dubey et al. 2024).

COFs have two main types based on their dimensionality.

2D COFs have layered structures comprising planar sheets that are stacked with π-π interactions or van der Waals forces holding them together. This layered structure presents well-defined channels, thereby finding applications in semiconductor charge transport and optoelectronics and separation technologies involving membranes. The surface areas are extremely high, and their pore networks are accessible as well (Fu et al. 2022; Kuai and Wang 2024).

3D COFs improve pore accessibility, mechanical strength, and surface areas by increasing covalent bonding in three directions. The frameworks have flexibility in gas storage as well as catalysis for energy storage devices. Additionally, 3D COFs show superior chemical and thermal stability, making them extremely appropriate for future industrial applications. Tuning the dimensionality of COFs enables materials to be engineered with specific technological demands in mind, such as for use in molecular separation and electrocatalysis (Q. Zhu et al. 2020; Sheng et al. 2024; Yusran et al. 2024).

5.3.1.2 Atomic-Level Precision in Design and Synthesis

Reticular chemistry allows atomic-scale control in COFs by allowing systematic building block selection and assembly, giving rise to highly defined pore architecture and functionalities. COFs have much lower defect content compared to porous materials, which results in very controllable properties in terms of pore size distributions, surface chemistry, and electronic properties. Functional group introduction at the time of synthesis offers more flexibility toward applications in selective adsorption, sensing, and photocatalysis. PSM can also sharpen the structure without compromising its integrity. The size of crystallinity, dimensionality, and molecular control render COFs ultra-versatile for innovative scientific and industrial applications (Yaghi 2016; Rath et al. 2024).

5.3.2 Chemical Features of Covalent Organic Frameworks

5.3.2.1 High Chemical Stability and Functionalization Potential

The high chemical stability of COFs arises from the presence of strong covalent bonds and, consequently, an insensitivity toward chemical and thermal degradation. Robust linkages like boronate esters, imines, triazines, and β-ketoenamines allow the structure to perform under harsh conditions, such as catalysis, gas storage, and drug delivery. One of the significant advantages of COFs is that they are highly functionalizable and thus their properties, such as hydrophilicity, electronic behavior, and catalytic activity, can be controlled very precisely. The functional groups such as hydroxyl, amino, and sulfonic acid can be added for selective interaction. Further, COFs show stability in different solvents, acids, and bases. This makes them permanently usable and serves as a better substitute than traditional porous materials. This includes zeolites and MOFs, among others (Evans et al. 2021; Yaqin Li et al. 2022b).

5.3.2.2 Postsynthetic Modifications and Applications

Enhancing the functionality of the initial synthesis, PSMs afford structural changes while retaining integrity by modification introduced for functionalities, such as sulfonic acid groups for proton conductivity and metal ions for catalytic efficiency in organic reactions. PSMs have also improved stability and processability through polymer grafting, ligand exchange, and conductive material doping and opened wider applications for these materials in drug delivery, gas separation, and energy storage. Furthermore, selective pore wall modifications allow for targeted molecular recognition for both sensors and adsorbents. In summary, PSMs improve COFs by greatly enhancing their flexibility for catalytic, environmental, and biomedical applications (Viciano-Chumillas et al. 2022; Khojastehnezhad et al. 2024).

5.3.3 Porosity and Accessibility

COFs are very porous materials whose crystalline structure is precisely engineered, allowing control over pore size, high surface areas, and tunable functionalities. Their porosity is a function of the choice of monomer, the covalent linkages used, and framework dimensionality as 2D or 3D. Even after solvent removal, COFs retain permanent porosity, and hence, they have found applications in gas storage, molecular separation, and catalysis. Highly accessible pores enable efficient molecular interactions, and synthetic modifications may allow for further enhancing their usability for a wide range of scientific and industrial purposes (S. Li et al. 2022; Patial et al. 2023).

5.3.3.1 Intrapore Chemical Environments

The chemical environment in COF pores is highly important for the interaction with guest molecules because it would allow, through rational design and PSM, precise control over them. Unlike other porous materials like zeolites and MOFs, COFs can incorporate functional groups, heteroatoms, or charged moieties to enhance selective adsorption and reactivity. The intrapore environment can actually be tuned hydrophilic, hydrophobic, acidic, or basic depending on the approach of synthesis. For example, imine-linked COFs with high nitrogen levels capture effectively, while boronate ester-linked COFs carry electron-rich molecules. Further functionalization by incorporation of metal nanoparticles or catalytic centers enhances COFs in terms of catalytic applications, sensing, and drug delivery applications. This versatility helps COFs to act as selective adsorbents and separation materials for water purification, gas separation, and controlled release drug delivery (Abuzeid et al. 2021; Meng and Mirica 2021; Essalhi et al. 2025).

5.3.3.2 Implications of High Pore Accessibility for Practical Applications

High pore accessibility in COFs confers significant benefit toward numerous practical application domains. Well-ordered pore networks of these materials contribute to easy molecule diffusion throughout the adsorption-based application such as gas storage, in which the highly expansive surface area and adjustable pore sizes contribute to gas capture, such as hydrogen, methane, and carbon dioxide (Y. Zhu et al. 2025). Advantages in catalysis arise from improving the efficiency of a reaction with accessible active sites for heterogeneous photocatalysis. Their open channels further allow for the encapsulation of drugs and controlled release for targeted drug delivery systems. In environmental remediation, COFs act as efficient adsorbents for the removal of pollutants, heavy metals, and organic contaminants. It means COFs are one of the most versatile materials for applications in fields related to energy, environment, healthcare, and catalysis (L. Zhu and Zhang 2017; Naghdi et al. 2023).

5.4 CLASSIFICATION OF COVALENT ORGANIC FRAMEWORKS

Classification of COFs is necessary in understanding the structural diversity, chemical properties, and potential applications. Classifications of COFs can be made according to several criteria, which include their dimensionality: whether they are 2D or 3D, linkage chemistry, and incorporation of functional groups. The important classification among these involves dimensionality, as it plays a crucial role in determining the overall properties, stability, and potential applications of COFs (Akinnawo 2024) (Figure 5.2).

5.4.1 Dimensionality-Based Classification

Broadly, COFs can be distinguished as 2D and 3D depending on the frameworks of their building blocks. Its dimensionality affects the crystallinity and porosity, but at the same time, it is a factor in the mechanical stability, strength, and functionality of a material, presented in Table 5.1.

5.4.1.1 2D Covalent Organic Frameworks

2D COFs are composed of planar molecular layers bonded covalently in two dimensions, in the x and y axes, but weak non-covalent interactions such as π–π stacking, van der Waals forces, and hydrogen bonding are present along the z-axis. The resulting structure boasts high surface area, crystallinity, and defined pores, where efficient charge transport and adsorption can be obtained. Chemical environments within pores can be tailored by monomer selection and PSMs (X. Xu et al. 2021; Ji et al. 2024).

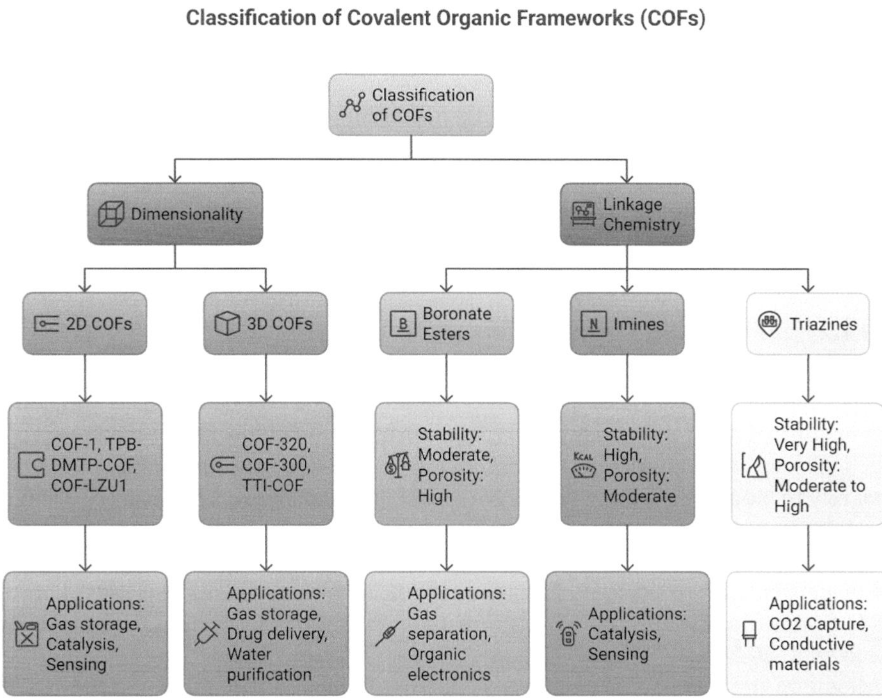

FIGURE 5.2 Classification of covalent organic frameworks (COFs).

TABLE 5.1
Comparison Between 2D and 3D COFs

Feature	2D COFs	3D COFs	References
Structural growth	Extends in the x-y plane, stacked via weak interactions	Extends in x-y-z, covalently bonded throughout	Bhattacharjee et al. (2024)
Stability	Moderate (can be affected by external forces)	Higher mechanical and chemical stability	Moosavi et al. (2018)
Porosity	High, but lower than 3D COFs	Very high due to interpenetrated structures	Kang et al. (2020)
Gas storage	Good	Excellent due to higher pore volume	B. Li et al. (2016)
Catalysis	Effective, but surface-limited	More active sites, better accessibility	J. Ding et al. (2022)
Drug delivery	Used for small-molecule delivery	More suitable for sustained-release formulations	Adepu et al. (2021)
Water purification	Moderate efficiency	Highly efficient due to enhanced stability and porosity	Raffi et al. (2024)

Some of the most notable 2D COFs include highly porous boronate ester-based COF-1; imine-linked TPB-DMTP-COF, with applications in both photocatalysis and energy storage; and the imine-linked, aqueously stable COF-LZU1, whose application is toward gas adsorption (Song et al. 2019; Yang Li et al. 2021).

2DCOFsexhibitwideapplicationsincludinggasstorageandseparation(H_2,CO_2,CH_2), catalysis, oxidation, hydrogenation, photocatalysis, energy storage supercapacitors, lithium-ion batteries, fuel cells, and sensing detecting pollutants and biomolecules. They are among the most versatile materials due to their tunable porosity and stability (Khalil et al. 2024).

5.4.1.2 3D Covalent Organic Frameworks (COFs)

3D COFs are materials where there are covalent bonds extended in three dimensions, including the x, y, and z axes. It is found that 3D COFs show higher mechanical stability, higher robustness, and larger surface area compared to the 2D COFs due to strong covalent bonding throughout the entire structure and unlike 2D COFs that rely on noncovalent interactions for stacking of the layers. The resulting frameworks have better porosity, more significant pore volumes, and improved chemical and mechanical strength. Therefore, it can be applied in harsher environments, for instance, acidic or basic, to use its great properties. Furthermore, its rigid structure adds to the functionality in applications like adsorption, separation, and catalysis (Guo et al. 2024).

Some of the remarkable examples of 3D COFs include the boronate ester-based COF-320, a diamond-like structure that makes it highly porous and promising for gas storage, the imine-linked COF-300, and the nitrogen-rich triazine-based TTI-COF with photocatalytic hydrogen evolution properties (Yang et al. 2024). These COFs have yielded different applications because of their properties, enhanced stability, and tunable functionality. They are exhaustively used as materials in gas storage and carbon capture, drug delivery, catalysis, and water purification. The 3D COFs adsorb gases, such as CO_2, H_2, and CH_2, with very high stability and microporosity, making them great candidates for utilization in environmental and clean energy applications, targeted drug delivery systems, chemical transformations, and removal of contaminants from water sources (Xiaolu Liu et al. 2021).

The dimensionality plays a key role in determining the overall chemical stability, structural integrity, and practical applications of these materials. Though 2D COFs have shown a wider surface area and tunable porosity together with efficient charge transport, 3D COFs offer higher structural rigidity, excellent gas storage, and greater mechanical stability. The choice of use between 2D and 3D COFs will rest on the intended application: 2D COFs for electronic and catalytic applications and 3D COFs are instead more generally suited for adsorption, filtration, and environmental applications. Advances in COF synthesis and functionalization continue to be developed to expand utility across a myriad of scientific fields (Essalhi et al. 2025).

5.4.2 Linkage Chemistry-Based Classification

Classification of COFs according to linkage chemistry is critical in the understanding of structural diversity and functional versatility of COFs. Overall stability, porosity, chemical functionality, and applications of the COF are dependent on choosing linking chemistry. Boronate esters, imines, and triazines are commonly used connections, and each has advantages in the assembly of COFs. This chapter is devoted to the complete synthesis, characteristics, and exemplary examples of the three major linking chemistries.

5.4.2.1 Boronate Ester Linkages

Usually based on the boronate ester linkages, the COFs are designed by means of the reaction between boronic acids and diols. Such a reaction generates very structurally rigid, highly porous in nature stable, covalent bonds between boron and oxygen atoms. Usually utilizing solvents and bases such as sodium or potassium carbonate to neutralize acid byproducts, synthesis consists of condensation reactions between boronic acids and diols. The solvothermal procedure is a quite often used synthesis technique producing well-ordered crystalline materials (Kirchner et al. 2024). Among the most well-known examples are COF-5, a 2D hexagonal structure with significant surface area, and COF-10, a stable 2D porous structure perfect for use in catalysis and energy storage. A few key benefits of boronate ester-linked COFs are chemical stability, crystallinity, and functional group insertion potential. Although theoretically open to postsynthetic changes, the nature of the electron shortfall imposed by the boron atom also makes these materials possibly beneficial for selective adsorption and catalysis (Hegazy et al. 2023).

5.4.2.2 Imine Linkages

Given their stable but reversible bonding chemistry, the synthesis of COFs depends on the Imines or Schiff bases somewhat critically. Since they are generated by the condensation of an amine group and a carbonyl group, imine linkages are preferred as simple synthesis, stability, and tunability. Usually under acid or base catalysis at solvothermal conditions, the synthesis of these linkages is achieved by reactions between main amines and carbonyl molecules such as aldehydes or ketones. Strong but reversible such imine bonds allow postsynthetic changes to modify the characteristics of the COFs (Huang et al. 2024). Among COFs with imine connections are TPB-COF and COF-1. Both of these expose the varied character of these materials and demonstrate great porosity and stability as well as outstanding gas adsorption capabilities. Tunable imine linkages enable integration of many functional groups, hence improving the reactivity of the COF toward uses in catalysis, gas storage, and sensing. Furthermore, the reversibility of these bonds enables postsynthesis material change, hence enabling the construction of COFs with particular features fit for various uses (Akinnawo 2024; Huang et al. 2024).

5.4.2.3 Triazine Linkages

Triazine linkages are a kind of exceptionally strong and flexible connection displayed by COFs. Among its advantages are increased chemical activity, exceptional electrical properties, and remarkable stability. Usually, the condensation of carbon-rich monomers with the triazine derivatives nitrogen-rich heterocycles with three nitrogen atoms in a six-membered ring obtains this form of linkage (Akinnawo 2024). The triazine-based COFs are strong with additional features.

For all advanced uses, these are thus excellently suited. The example of COF-300, synthesized from a triazine derivative and a carbon-based monomer, to create a stable 2D structure fairly captures this class of COFs. This material is suitable for catalysis since it exhibits remarkable stability under very hostile circumstances, including basic or acidic ones. Another example is COF-500, whose high porosity stable 3D framework created by triazine linkages is suitable for gas and liquid adsorption (Yeo et al. 2024).

Because of their great stability and conductivity together with their nitrogen richness, these COFs have great potential in catalysis, energy storage (such as batteries and supercapacitors), and capture of carbon dioxide. Improved catalysis and sensing efficiency come from the important interactions between COFs and the substrates. Because of their good thermal stability and strong electron-withdrawing characteristics, these materials are quite interesting for use in electronics and optoelectronics. Triazine-linked COFs' porous structure and further functionalizing possibilities improve their adaptability and enable a broad spectrum of uses in gas storage, catalysis, and other domains. This might so combine triazine linkages with other chemistries, including boronate esters and imines, that can give exact control over the characteristics of COFs and so offer fascinating opportunities for future material designs (Alenazi et al. 2025).

5.4.3 Functional Group Incorporation

Indeed, the incorporation of functional groups into COFs underlines the improvement of the latter characteristics and, hence, additional stretching of their applications. Known alternately as presynthetic and postsynthetic modification, alteration takes place either before or during the synthesis of COFs. Using predetermined functional groups, COFs could be designed for catalysis, gas storage, separation, drug delivery, and optoelectronics (Machado et al. 2021).

5.4.3.1 Functionalization Strategies During and After Synthesis

Presynthetic functionalization is the process by which functional groups are added to monomers, or building blocks, before COFs are assembled. This guarantees homogeneous distribution of the functional groups in the COF structure, therefore providing control over such factors as porosity, hydrophilicity, electrical properties, and chemical reactivity. Amine, hydroxyl, and thiol groups can increase chemical affinity toward a particular molecule; carboxyl and sulfonic acid groups are added to increase proton conductivity or catalytic efficiency for use in catalytic applications. Fluorinated groups such as $-CF_3$ (Machado et al. 2021; Yusran et al. 2024) can also alter the pore environment in applications of gas separation.

Postsynthetic functionalization (PSM) offers an advantage over other methods by adding capabilities to the preformed COFs without altering their crystalline structure. The method is quite helpful in including features sensitive to the environment of the synthesis of COFs. Generally, PSM is accomplished via covalent attachment of small molecules or metal complexes to reactive spots on the COFs. Such a method helps to incorporate catalytic centers or tweak hydrophilic and hydrophobic characteristics.

By sequential addition of various functional groups, PSM helps synthesize multifunctional COFs, which can be especially intended for gas adsorption and catalysis (Rejali et al. 2023; Khojastehnezhad et al. 2024).

5.4.3.2 Tailoring COFs for Specific Applications

Tailoring the COFs through their functionalization is a way toward significantly enhancing application areas, giving substantial advances on catalysis and drug delivery besides gas storage or sensing technologies. The introduction of functional groups allows for flexibility with either synthetic modifications or postsynthetic methodologies to fulfill such specific needs in these diverse application areas (Martín-Illán et al. 2023; Yeo et al. 2024).

Catalysis is a highly surface area material with tunable porosity that makes it very suitable for heterogeneous catalysis. The functional groups like -SO_3H, -NH_2, and metal-coordinating ligands added to COFs enhance their catalytic potential for organic transformations, such as oxidation, hydrogenation, and esterification. The COFs functionalized with metal-binding groups, such as carboxylates or phosphonates, exhibit exceptional performance in several reactions (Y. Zhang and Riduan 2012).

It is also interesting that biocompatible functional groups, in COFs, have improved drug dispersion, stability, and pH-responsiveness, allowing for controlled release in selective environments such as tumor tissues. More recently, functionalized COFs were used for theranostic applications by introducing a drug-delivery function combined with imaging functions (Jia et al. 2020).

Functionalization of the functional groups by amines (-NH_2) or hydroxyl groups (-OH) in COFs makes an excellent gas storage and separation material, which is very effective for selectively adsorbing gases like CO_2 and H_2. The amine-functionalized COF has remarkable efficiency in capturing carbon, and metal-functionalized COFs are most promising for hydrogen storage because of strong binding interaction (Dautzenberg et al. 2023).

In addition, COFs have been applied to sensing and optoelectronics due to their tunable electronic properties. Nitro (-NO_2) and amines (-NH_2) functional groups enable COFs to sense metal ions, explosives, and biological molecules. The redox-active groups are currently being studied for energy storage, including supercapacitors and batteries (Blätte et al. 2024).

In short, functionalization is a vital process that is opening up areas of application in COFs; it is the key to finding exciting possibilities within material science. It opens future avenues for developments in health, environmental sustainability, and energy storage. This would be highly compatible with current research in nanomedicines and green chemistry in which functionalized materials have the promise to advance solutions that are scientific and practical at the same time.

5.5 NOTABLE EXAMPLES IN LITERATURE

COFs are highly versatile for their structural potential, tunable porosity, and high stability. Many experiments have been successful in demonstrating applications of COFs in gas storage, catalysis, energy storage, and drug delivery. Some of the essential studies on COFs are related to the synthesis and application with respect to designs of frameworks varying with different types of linkages and dimensionalities (Z. Wang et al. 2024).

5.5.1 Key Studies Showcasing the Design and Applications of COFs

One of the early seminal publications on COFs was made by Côté et al. in 2005, where they achieved the synthesis of the first crystalline COFs, COF-1 and COF-5, by self-condensation of boronic acids. Frameworks showed high thermal stability with surface area and well-defined porosity that favored applications in gas storage. This pioneering work introduced the concept of reticular chemistry in organic framework materials and paved the way for further investigations of COFs (Côté et al. 2005).

Ding and Wang (2012) further designed imine-linked COFs, like TpPa-1, by applying Schiff base chemistry between aldehydes and amines. Such COFs displayed good chemical stability and resistance to hydrolysis, an important limitation that the earlier boronate-based COFs could not overcome. Due to this strong linkage, imine-based COFs have been extensively applied for catalytic purposes, adsorption, and environmental remediation (S. Y. Ding and Wang 2012).

Recently, Dalapati et al. (2013) synthesized β-ketoenamine-linked COF, TpBD by the reaction between 1,3,5-triformylphloroglucinol (Tp) and benzidine (BD). The synthesized COFs showed strong acid and base resistance, rendering them excellent prospects for harsh industrial applications. Moreover, they displayed a high CO_2 uptake capacity, thus useful in carbon capture and separation technologies (Dalapati et al. 2013).

Liu et al. (2018) have proposed another fantastic development in the form of triazine-linked COFs, like CTF-1, which have been synthesized by ionothermal condensation of aromatic nitriles. Because of their high nitrogen content, electron-rich nature, and high conductivity, they have been tested for photocatalytic hydrogen evolution and energy storage (Z. Liu et al. 2018; Xigui Liu et al. 2019). The recent development in COFs expands their scope of application in the biomedical field. Xu et al. (2020) synthesized a biocompatible porphyrin-based COF for photodynamic cancer therapy. Its peculiar π-conjugated structure permitted the COF to efficiently absorb light and produce ROS, which further resulted in increased anticancer efficacy (N. Xu et al. 2020).

Later, Guo et al. (2024) designed metal-doped COFs for enzyme-mimetic catalysis, where Mn(II)-modified COFs showed peroxidase-like activity for biosensing. This opened a bright future opportunity for the application of COFs in biomimetic and medical fields toward being incorporated into diagnostic and therapeutic systems (Z. Li et al. 2021).

5.5.2 Comparative Analysis of COFs with Different Linkages and Dimensions

COFs are categorized based on their dimensionality, either as 2D or 3D structures, and linkage chemistry, including boronate, imine, triazine, and β-ketoenamine.

TABLE 5.2
Linkage Chemistry-Based Comparison

COF Type	Linkage Chemistry	Key Properties	Applications	Examples	References
Boronate ester COFs	Boronic acid condensation	High porosity, but moisture-sensitive	Gas storage, separation	COF-1, COF-5	Akinnawo (2024)
Imine COFs	Schiff base reaction	Good stability, tunable pore structure	Catalysis, environmental remediation	TpPa-1, COF-300	Prakash et al. (2024)
Triazine COFs	Ionothermal polymerization of nitriles	High thermal stability, electron-rich	Photocatalysis, energy storage	CTF-1	G. Li et al. (2021)
β-Ketoenamine COFs	Aldehyde-ammonia condensation	Exceptional acid and base resistance	CO_2 capture, industrial catalysis	TpBD	Yaqin Li et al. (2022a)
Metal-doped COFs	Postsynthetic metalation	Enhanced catalytic and sensing properties	Biosensors, enzyme-mimetic catalysis	Mn(II)-COF	K. Y. Wang et al. (2023)

A comparative analysis of the different COFs is presented with respect to their structural and functional properties, dependent on the nature of the linkage and dimensionality, as represented in Table 5.2.

2D COFs, such as COF-1, COF-5, and TpPa-1, typically have layered structures of π–π-stacking interactions. These frameworks, through reversible condensation reactions, provide a significant surface area, tunable pore sizes, and excellent crystallinity. There are strengths; one is their high pore volumes that ensure the best opportunities for gas adsorption, gas separation, as well as being excellent catalytically active, though in others they are hindered by aggregation that leads to poor electrical stability (Kandambeth et al. 2022; Koner et al. 2024).

On the contrary, 3D COFs, including COF-300 (imine-linked) and CTF-1 (triazine-linked), show interpenetrating networks that confer improved mechanical strength, chemical durability, and 3D connectivity of pores, which are good for energy storage and catalysis. The better charge transport with nonlayered structure and good accessibility to the active sites gives them an excellent edge. It is the key challenge in terms of achieving good crystallinity of 3D COFs as their interpenetrated network is very complicated (Bhambri et al. 2022).

It has been determined in Table 5.2 that linking chemistry is extremely influential in ascertaining stability, functionality, and potential use in COFs. Generally, boronate-based COFs have excellent surface areas; however, there are limitations set forth by this fact, i.e., tendency toward hydrolysis, rendering the material unsustainable over a protracted period. Conversely, the imine as well as the β-ketoenamine framework possess greater chemoresistance in addition to increased flexibility, for the catalysis as well as the adsorption-based applications. The electronic and photonic fields excel in triazine-based COFs, while the biomedical sensing and catalysis hold much promise with metal-doped COFs. This section delivers critical advances on the research with COFs, and how this selection of dimensionality and chemistry of linkage dictates the physicochemical properties as well as the functionalities of the COFs. New structures with enhanced stability and reactivity tailored for new functions will be advanced for applications that are further spread in the fields of energy, environment, and healthcare.

5.6 CHALLENGES AND FUTURE DIRECTIONS

Despite the great achievements in COF research, there are many challenges that need to be addressed before these structures can be broadly applied and practically used. Such challenges are generally related to synthesis, structural stability, and possible future developments. Overcoming such concerns is key to unlocking the full potential of COFs in various scientific and industrial domains.

5.6.1 Synthetic Challenges – Limitations in Reproducibility and Scalability

This probably remains one of the biggest obstacles in COF research: namely, the difficulty of reproducibility and scalability in synthesis. Whereas MOFs have well-defined synthetic protocols, problems with batch-to-batch variability are rather common for COFs due to their covalent bonding nature and extreme sensitivity to reaction conditions. Polymerization reactions for building COFs, for example, through boronate ester formation or imine condensation, are very sensitive to temperature control, solvent conditions, and the choice of catalyst in order to obtain high crystallinity and yields. Furthermore, the necessity for long reaction times and special conditions for reaction, such as the solvothermal or ionothermal methods, poses a significant limitation for scaling-up production. Such shortcomings limit the commercial production of COFs because the repetition of synthesis commonly results in faults, decreased crystallinity, and porosity values. Thus, it is much needed to fabricate reliable yet efficient synthetic methodologies with high reproduction values that help maintain the desirability aspects of COFs (Martinez-Bulit et al. 2022; Vardhan et al. 2023).

5.6.2 Structural Challenges – Overcoming Issues with Crystallinity and Stability

Another significant challenge for the long-term usability of COFs is their structural stability. Unlike the traditional porous materials, COFs are built by dynamic covalent bonds, which, despite being highly tunable, may also cause issues associated with the degradation of structure over time. The crystallinity of COFs is crucial to their porosity and functionality; however, most of the reported COFs possess amorphous or semicrystalline structures because of imperfect polymerization. Moreover, the stability of COFs under various environmental conditions, including acidic or basic media, restricts their functionalities (Jaid et al. 2024). For example, boronate ester-linked COFs are very sensitive to hydrolysis and, therefore, not suitable for aqueous applications. To improve structural stability, researchers have been very active in developing new linkage chemistries, including C–C coupling or β-ketoenamine linkages, that give better resistance to environmental degradation. Consequently, truly overcoming those challenges will require innovative design strategies that make COFs crystalline and robust but modularity and functional properties are not degraded (Abuzeid et al. 2021).

5.6.3 Future Prospects

5.6.3.1 New Linkage Chemistries and Functional Materials

Researchers are also trying to look for new kinds of covalent bonds, which would result in improved stability and functionality with overcoming the deficiencies that accompany COF synthesis and stability. Exploration in carbon–carbon-linked COFs, π-conjugated COFs, and dynamic covalent frameworks will pave the way to new types of materials, highly stable and functionally excellent. That would not be enough to say that; heteroatom incorporation, such as nitrogen, sulfur, or phosphorus, in the COF framework will further improve the electronic properties to be more tuned for applications in energy storage and optoelectronics. Improvements in computational chemistry and machine learning are also helping rationalize the design of COFs with tailored properties, accelerating the discovery of new frameworks with improved performance (Prakash et al. 2024; Yeo et al. 2024).

5.6.3.2 Prospects for Advanced Applications in Emerging Fields

With continuous advancements in COF chemistry, their applications are now going into cutting-edge fields such as drug delivery, environmental remediation, energy storage, and catalysis. Functionalized COFs with specific active sites have thus shown promise in heterogeneous catalysis, enabling selective chemical transformations. Furthermore, the porosity and surface area of COFs are being tuned to be used in gas storage and separation, including hydrogen and carbon dioxide capture. Attention for biomedical sciences toward COFs as biocompatible and drug-loaded material is offered to prospects on targeted drug delivery systems. Another integration of electronic and photonic devices with COFs is being opened up into novel sensor technologies and next-generation energy materials. Further, while the research work in COFs is being scaled up, cross-disciplinary cooperation will be pivotal in translating bench-scale discoveries to commercially viable applications that have potential impact (J. Kaur et al. 2023; Shao et al. 2024).

In summary, COFs are well-established, promising materials with very unique structural and functional properties; however, many synthetic and structural challenges must still be overcome. Further research that refines the synthesis techniques to improve stability will unlock the true potential of COFs for their real-world applications.

5.7 CONCLUSION

Covalent organic frameworks, abbreviated as COFs, are a transformative class of crystalline porous materials offering revolutionary precision in the design of structure and functionalization. These materials show great promise in a whole host of fields, including catalysis, energy storage, drug delivery, and environmental remediation. The development of COFs has been motivated by new synthetic strategies, especially reticular chemistry that provides tunable porosity, stability, and diversity in functionality. Evolved from them is the increasing application of covalent bonding principles toward producing robust, tunable materials with atomic-level control.

COFs can also be categorized based on the dimensionality in which they are prepared, linkage chemistry, and functional group incorporation. 2D COFs consist of layered structures with in-plane connectivity; in 3D COFs, the networks extend in all directions. Due to its structure, these have higher stability and mechanical properties. The linkage chemistry – the use of boronate esters, imines, or triazine-based frameworks – provides the variable contributions to stability and reactivity of the networks. Functional group incorporation remains a means of precisely tuning properties, thereby qualifying COFs for particular applications. Therefore, COFs continue to face serious issues, mainly including scalability control of crystallinity and long-time stability. New linkage chemistries, hybrid frameworks, and PSMs can unlock applications for COFs. Ongoing research and innovation will eventually bring COFs into the material family that makes an impact on energy storage, biomedical fields, and environmental sustainability and advances the development of materials science and nanotechnology.

REFERENCES

Abuzeid, Hesham R., Ahmed F.M. EL-Mahdy, and Shiao Wei Kuo. 2021. "Covalent Organic Frameworks: Design Principles, Synthetic Strategies, and Diverse Applications." *Giant* 6 (June): 100054. doi:10.1016/J.GIANT.2021.100054.

Adepu, Shivakalyani, Seeram Ramakrishna, Rita Costa-Pinto, and Ana Leite Oliveira. 2021. "Controlled Drug Delivery Systems: Current Status and Future Directions." *Molecules* 26 (19): 5905. doi:10.3390/MOLECULES26195905.

Akinnawo, Solomon Oluwaseun. 2024. "Covalent Organic Frameworks: Design, Synthesis, Characterization, and Applications." *ChemPhysMater* 3 (1): 36–63. doi:10.1016/J.CHPHMA.2023.08.003.

Alenazi, Maha H., Aasif Helal, Mohd Yusuf Khan, Amjad Khalil, Abuzar Khan, Muhammad Usman, and Md Hasan Zahir. 2025. "Covalent Organic Frameworks (COFs) for CO_2 Utilizations." *Carbon Capture Science & Technology* 14 (March): 100365. doi:10.1016/J.CCST.2025.100365.

AlHazaimeh, Taima, Manal Raheem, and Amani Al-Othman. 2024. "Covalent Organic Frameworks (COFs): Characteristics and Applications for Lithium-Ion Batteries (LIBs) and Sodium Ion Batteries (SIBs)." *International Journal of Thermofluids* 21 (February): 100531. doi:10.1016/J.IJFT.2023.100531.

Bhambri, Himanshi, Sadhika Khullar, Sakshi, and Sanjay K. Mandal. 2022. "Nitrogen-Rich Covalent Organic Frameworks: A Promising Class of Sensory Materials." *Materials Advances* 3 (1): 19–124. doi:10.1039/D1MA00506E.

Bhattacharjee, Rameswar, Henry Jervis, Megan E. McCormack, Marina A. Petrukhina, and Miklos Kertesz. 2024. "Structure and Bonding in π-Stacked Perylenes: The Impact of Charge on Pancake Bonding." *Journal of the American Chemical Society* 146 (15): 10465–10477. doi: https://doi.org/10.1021/jacs.3c14065.

Blätte, Dominic, Frank Ortmann, and Thomas Bein. 2024. "Photons, Excitons, and Electrons in Covalent Organic Frameworks." *Journal of the American Chemical Society* 146 (47): 32161–32205. doi:10.1021/JACS.3C14833.

Bukhari, Syed Nasir Abbas, Naveed Ahmed, Muhammad Wahab Amjad, Muhammad Ajaz Hussain, Mervat A. Elsherif, Hasan Ejaz, and Nasser H. Alotaibi. 2023. "Covalent Organic Frameworks (COFs) as Multi-Target Multifunctional Frameworks." *Polymers* 15 (2): 267. doi:10.3390/POLYM15020267.

Côté, Adrien P., Annabelle I. Benin, Nathan W. Ockwig, Michael O'Keeffe, Adam J. Matzger, and Omar M. Yaghi. 2005. "Porous, Crystalline, Covalent Organic Frameworks." *Science* 310 (5751): 1166–1170. doi:10.1126/SCIENCE.1120411.

Dalapati, Sasanka, Shangbin Jin, Jia Gao, Yanhong Xu, Atsushi Nagai, and Donglin Jiang. 2013. "An Azine-Linked Covalent Organic Framework." *Journal of the American Chemical Society* 135 (46): 17310–17313. doi: https://doi.org/10.1021/ja4103293.

Dautzenberg, Ellen, Guanna Li, and Louis C.P.M. de Smet. 2023. "Aromatic Amine-Functionalized Covalent Organic Frameworks (COFs) for CO_2/N_2 Separation." *ACS Applied Materials and Interfaces* 15 (4): 5118–5127. doi: https://doi.org/10.1021/acsami.2c17672.

Ding, Jia, Hashim Jalil Khan, Gholamreza Vahedi Sarrigani, Paul Fitzgerald, Amirali Ebrahimi Ghadi, Olivier Lefebvre, Chao Meng, et al. 2022. "Enhancing the Active Site Accessibility of Cobalt-Silica Catalysts for Improved Fenton-like Performance." *Chemical Engineering Journal* 432 (March): 134435. doi:10.1016/J.CEJ.2021.134435.

Ding, San Yuan, and Wei Wang. 2012. "Covalent Organic Frameworks (COFs): From Design to Applications." *Chemical Society Reviews* 42 (2): 548–568. doi:10.1039/C2CS35072F.

Dubey, Prashant, Vishal Shrivastav, Tribani Boruah, Giorgio Zoppellaro, Radek Zbořil, Aristides Bakandritsos, and Shashank Sundriyal. 2024. "Unveiling the Potential of Covalent Organic Frameworks for Energy Storage: Developments, Challenges, and Future Prospects." *Advanced Energy Materials* 14 (24): 2400521. doi:10.1002/AENM.202400521.

Essalhi, Mohamed, El Hassan Mahmoud, Ali Tayeb, Rawan A. Al-Qahtani, Ahmad Salam Farooqi, and Mahmoud Abdelnaby. 2025. "Structural Design of Covalent Organic Frameworks and Their Recent Advancements in Carbon Capture Applications: A Review." *Carbon Capture Science & Technology* 14 (March): 100370. doi:10.1016/J.CCST.2025.100370.

Evans, Austin M., Matthew R. Ryder, Woojung Ji, Michael J. Strauss, Amanda R. Corcos, Edon Vitaku, Nathan C. Flanders, Ryan P. Bisbey, and William R. Dichtel. 2021. "Trends in the Thermal Stability of Two-Dimensional Covalent Organic Frameworks." *Faraday Discussions* 225: 226–240. doi:10.1039/D0FD00054J.

Fu, Shuai, Enquan Jin, Hiroki Hanayama, Wenhao Zheng, Heng Zhang, Lucia Di Virgilio, Matthew A. Addicoat, et al. 2022. "Outstanding Charge Mobility by Band Transport in Two-Dimensional Semiconducting Covalent Organic Frameworks." *Journal of the American Chemical Society* 144 (16): 7489–7496. doi: https://doi.org/10.1021/jacs.2c02408.

Ge, Shengbo, Kexin Wei, Wanxi Peng, Runzhou Huang, Esther Akinlabi, Hongyan Xia, Muhammad Wakil Shahzad, Xuehua Zhang, Ben Bin Xu, and Jianchun Jiang. 2024. "A Comprehensive Review of Covalent Organic Frameworks (COFs) and Their Derivatives in Environmental Pollution Control." *Chemical Society Reviews* 53 (23): 11259–11302. doi:10.1039/D4CS00521J.

Gropp, Cornelius, Stefano Canossa, Stefan Wuttke, Felipe Gándara, Qiaowei Li, Laura Gagliardi, and Omar M. Yaghi. 2020. "Standard Practices of Reticular Chemistry." *ACS Central Science* 6 (8): 1255–1273. doi: 10.1021/acscentsci.0c00592.

Guo, Zi'ang, Zeyue Zhang, and Junliang Sun. 2024. "Topological Analysis and Structural Determination of 3D Covalent Organic Frameworks." *Advanced Materials* 36 (19): 2312889. doi:10.1002/ADMA.202312889.

Haase, Frederik, and Bettina V. Lotsch. 2020. "Solving the COF Trilemma: Towards Crystalline, Stable and Functional Covalent Organic Frameworks." *Chemical Society Reviews* 49 (23): 8469–8500. doi:10.1039/D0CS01027H.

Hao, Shuang, and Zhiqian Jia. 2023. "Covalent Organic Frameworks Membranes: Fabrication and Applications." *Microporous and Mesoporous Materials* 361 (November): 112742. doi:10.1016/J.MICROMESO.2023.112742.

Hegazy, H. H., Siva Sankar Sana, Tholkappiyan Ramachandran, Yedluri Anil Kumar, Dasha Kumar Kulurumotlakatla, Hisham S.M. Abd-Rabboh, and Seong Cheol Kim. 2023. "Covalent Organic Frameworks in Supercapacitors: Unraveling the Pros and Cons for Energy Storage." *Journal of Energy Storage* 74 (December): 109405. doi:10.1016/J.EST.2023.109405.

Huang, Tianhong, Weifeng Zhang, Shuai Yang, Liping Wang, and Gui Yu. 2024. "Imine-Linked Covalent Organic Frameworks: Recent Advances in Design, Synthesis, and Application." *SmartMat* 5 (6): e1309. doi:10.1002/SMM2.1309.

Jaid, Ghaidaa M., Adnan A. AbdulRazak, Hicham Meskher, Saad Al-Saadi, and Qusay F. Alsalhy. 2024. "Metal-Organic Frameworks (MOFs), Covalent Organic Frameworks (COFs), and Hydrogen-Bonded Organic Frameworks (HOFs) in Mixed Matrix Membranes." *Materials Today Sustainability* 25 (March): 100672. doi:10.1016/J.MTSUST.2024.100672.

Ji, Chunqing, Chengjun Kang, Bidhan Chandra Patra, and Dan Zhao. 2024. "Flexible Covalent Organic Frameworks: Design, Synthesis, and Applications." *CCS Chemistry* 6 (4): 856–881. doi:10.31635/CCSCHEM.023.202303415.

Jia, Yutao, Lina Zhang, Binnan He, Yulong Lin, Meng Li, and Jing Wang. 2020. "8-Hydroxyquinoline Functionalized Covalent Organic Framework as a PH Sensitive Carrier for Drug Delivery." *Materials Science and Engineering: C* 117 (December): 111243. doi:10.1016/J.MSEC.2020.111243.

Kandambeth, Sharath, Vinayak S. Kale, Osama Shekhah, Husam N. Alshareef, and Mohamed Eddaoudi. 2022. "2D Covalent-Organic Framework Electrodes for Supercapacitors and Rechargeable Metal-Ion Batteries." *Advanced Energy Materials* 12 (4): 2100177. doi:10.1002/AENM.202100177.

Kang, Xing, Xiaowei Wu, Xing Han, Chen Yuan, Yan Liu, and Yong Cui. 2020. "Rational Synthesis of Interpenetrated 3D Covalent Organic Frameworks for Asymmetric Photocatalysis." *Chemical Science* 11 (6): 1494–1502. doi:10.1039/C9SC04882K.

Kaur, Harjot, Samarjeet Singh Siwal, Reena V. Saini, and Vijay Kumar Thakur. 2024. "Covalent-Organic Framework-Based Materials in Theranostic Applications: Insights into Their Advantages and Challenges." *ACS Omega* 9 (6): 6235–6252. doi: https://doi.org/10.1021/acsomega.3c08456.

Kaur, Jasvinder, Vivek Sharma, and Ram K. Gupta. 2023. "Taming of Organic Pollutants Using Covalent-Organic-Framework-Based Materials: Challenges and Future Prospects." *Nano LIFE* 13 (04). doi:10.1142/S1793984423300054.

Khalil, Islam E., Prasenjit Das, and Arne Thomas. 2024. "Two-Dimensional Covalent Organic Frameworks: Structural Insights across Different Length Scales and Their Impact on Photocatalytic Efficiency." *Accounts of Chemical Research* 57 (21): 3138. doi:10.1021/ACS.ACCOUNTS.4C00491.

Khojastehnezhad, Amir, Ali Samie, Anna Bisio, Hani M. El-Kaderi, and Mohamed Siaj. 2024. "Impact of Postsynthetic Modification on the Covalent Organic Framework (COF) Structures." *ACS Applied Materials and Interfaces*. https://doi.org/10.1021/acsami.4c14030.

Kirchner, Philipp H., Louis Schramm, Svetlana Ivanova, Kazutaka Shoyama, Frank Würthner, and Florian Beuerle. 2024. "A Water-Stable Boronate Ester Cage." *Journal of the American Chemical Society* 146 (8): 5305–5315. doi: 10.1021/jacs.3c12002.

Kirlikovali, Kent O., Sylvia L. Hanna, Florencia A. Son, and Omar K. Farha. 2023. "Back to the Basics: Developing Advanced Metal-Organic Frameworks Using Fundamental Chemistry Concepts." *ACS Nanoscience Au* 3 (1): 37–45. doi: 10.1021/acsnanoscienceau.2c00046.

Koner, Kalipada, Himadri Sekhar Sasmal, Dinesh Shetty, and Rahul Banerjee. 2024. "Thickness-Driven Synthesis and Applications of Covalent Organic Framework Nanosheets." *Angewandte Chemie International Edition* 63 (31): e202406418. doi:10.1002/ANIE.202406418.

Kuai, Yanbing, and Yuhua Wang. 2024. "Innovative COF@MXene Composites for High Performance Energy Applications." *Carbon Neutrality* 3 (1): 1–35. doi:10.1007/S43979-024-00110-X.

Li, Bin, Hui Min Wen, Wei Zhou, Jeff Q. Xu, and Banglin Chen. 2016. "Porous Metal-Organic Frameworks: Promising Materials for Methane Storage." *Chem* 1 (4): 557–580. doi:10.1016/J.CHEMPR.2016.09.009.

Li, Guiliang, Jianrong Ye, Yi Shen, Qile Fang, and Fu Liu. 2021. "Covalent Triazine Frameworks Composite Membrane (CdS/CTF-1) with Enhanced Photocatalytic in-Situ Cleaning and Disinfection Properties for Sustainable Separation." *Chemical Engineering Journal* 421 (October): 127784. doi:10.1016/J.CEJ.2020.127784.

Li, Shimei, Jian Zou, Longfei Tan, Zhongbing Huang, Ping Liang, and Xianwei Meng. 2022. "Covalent Organic Frameworks: From Linkages to Biomedical Applications." *Chemical Engineering Journal* 446 (October): 137148. doi:10.1016/J.CEJ.2022.137148.

Li, Xinle, Songliang Cai, Bing Sun, Chongqing Yang, Jian Zhang, and Yi Liu. 2020. "Chemically Robust Covalent Organic Frameworks: Progress and Perspective." *Matter* 3 (5): 1507–1540. doi:10.1016/J.MATT.2020.09.007.

Li, Yang, Meghdad Karimi, Yun Nan Gong, Nan Dai, Vahid Safarifard, and Hai Long Jiang. 2021. "Integration of Metal-Organic Frameworks and Covalent Organic Frameworks: Design, Synthesis, and Applications." *Matter* 4 (7): 2230–2265. doi:10.1016/J.MATT.2021.03.022.

Li, Yaqin, Maosong Liu, Jinjun Wu, Junbo Li, Xianglin Yu, and Qichun Zhang. 2022a. "Highly Stable β-Ketoenamine-Based Covalent Organic Frameworks (COFs): Synthesis and Optoelectrical Applications." *Frontiers of Optoelectronics* 15 (1): 1–42. doi:10.1007/S12200-022-00032-5.

Li, Yaqin, Maosong Liu, Jinjun Wu, Junbo Li, Xianglin Yu, and Qichun Zhang. 2022b. "Highly Stable β-Ketoenamine-Based Covalent Organic Frameworks (COFs): Synthesis and Optoelectrical Applications." *Frontiers of Optoelectronics* 15 (1): 38. doi:10.1007/S12200-022-00032-5.

Li, Yusen, Weiben Chen, Guolong Xing, Donglin Jiang, and Long Chen. 2020. "New Synthetic Strategies toward Covalent Organic Frameworks." *Chemical Society Reviews* 49 (10): 2852–2868. doi:10.1039/D0CS00199F.

Li, Zhixi, Jun Guo, Yue Wan, Yutian Qin, and Meiting Zhao. 2021. "Combining Metal-Organic Frameworks (MOFs) and Covalent-Organic Frameworks (COFs): Emerging Opportunities for New Materials and Applications." *Nano Research* 15 (4): 3514–3532. doi:10.1007/S12274-021-3980-0.

Liu, Xiaolu, Hongwei Pang, Xuewei Liu, Qian Li, Ning Zhang, Liang Mao, Muqing Qiu, Baowei Hu, Hui Yang, and Xiangke Wang. 2021. "Orderly Porous Covalent Organic Frameworks-Based Materials: Superior Adsorbents for Pollutants Removal from Aqueous Solutions." *The Innovation* 2 (1): 100076. doi:10.1016/J.XINN.2021.100076.

Liu, Xigui, Danlian Huang, Cui Lai, Guangming Zeng, Lei Qin, Han Wang, Huan Yi, et al. 2019. "Recent Advances in Covalent Organic Frameworks (COFs) as a Smart Sensing Material." *Chemical Society Reviews* 48 (20): 5266–5302. doi:10.1039/C9CS00299E.

Liu, Zhongshan, Hongwei Wang, Junjie Ou, Lianfang Chen, and Mingliang Ye. 2018. "Construction of Hierarchically Porous Monoliths from Covalent Organic Frameworks (COFs) and Their Application for Bisphenol A Removal." *Journal of Hazardous Materials* 355 (August): 145–153. doi:10.1016/J.JHAZMAT.2018.05.022.

Lohse, Maria S, Thomas Bein, M S Lohse, and T Bein. 2018. "Covalent Organic Frameworks: Structures, Synthesis, and Applications." *Advanced Functional Materials* 28 (33): 1705553. doi:10.1002/ADFM.201705553.

Ma, Mingxuan, Yonghao Yang, Zhonghua Huang, Fuhong Huang, Quanliang Li, and Hongyu Liu. 2024. "Recent Progress in the Synthesis and Applications of Covalent Organic Framework-Based Composites." *Nanoscale* 16 (4): 1600–1632. doi:10.1039/D3NR05797F.

Ma, Xingjian, and Timothy F. Scott. 2018. "Approaches and Challenges in the Synthesis of Three-Dimensional Covalent-Organic Frameworks." *Communications Chemistry* 1 (1): 1–15. doi:10.1038/s42004-018-0098-8.

Machado, Tiago F., M. Elisa Silva Serra, Dina Murtinho, Artur J.M. Valente, and Mu Naushad. 2021. "Covalent Organic Frameworks: Synthesis, Properties and Applications—An Overview." *Polymers* 13 (6) 970. doi:10.3390/POLYM13060970.

Martinez-Bulit, Pablo, Alessandro Sorrenti, David Rodriguez San Miguel, Michele Mattera, Yonca Belce, Yanming Xia, Shenglin Ma, Mu Hua Huang, Salvador Pané, and Josep Puigmartí-Luis. 2022. "In Flow-Based Technologies: A New Paradigm for the Synthesis and Processing of Covalent-Organic Frameworks." *Chemical Engineering Journal* 435 (May): 135117. doi:10.1016/J.CEJ.2022.135117.

Martín-Illán, Jesús, David Rodríguez-San-Miguel, and Félix Zamora. 2023. "Evolution of Covalent Organic Frameworks: From Design to Real-World Applications." *Coordination Chemistry Reviews* 495 (November): 215342. doi:10.1016/J.CCR.2023.215342.

Meng, Zheng, and Katherine A. Mirica. 2021. "Covalent Organic Frameworks as Multifunctional Materials for Chemical Detection." *Chemical Society Reviews* 50 (24): 13498–13558. doi:10.1039/D1CS00600B.

Moosavi, Seyed Mohamad, Peter G. Boyd, Lev Sarkisov, and Berend Smit. 2018. "Improving the Mechanical Stability of Metal-Organic Frameworks Using Chemical Caryatids." *ACS Central Science* 4 (7): 832–839. doi: 10.1021/acscentsci.8b00157.

Mosupi, Keaoleboga, Mike Masukume, Guoming Weng, Nicholas M. Musyoka, and Henrietta W. Langmi. 2025. "Recent Advances in Fe-Based Metal-Organic Frameworks: Structural Features, Synthetic Strategies and Applications." *Coordination Chemistry Reviews* 529 (April): 216467. doi:10.1016/J.CCR.2025.216467.

Naghdi, Shaghayegh, Masoumeh Moheb Shahrestani, Mohammad Zendehbad, Hoorieh Djahaniani, Hossein Kazemian, and Dominik Eder. 2023. "Recent Advances in Application of Metal-Organic Frameworks (MOFs) as Adsorbent and Catalyst in Removal of Persistent Organic Pollutants (POPs)." *Journal of Hazardous Materials* 442 (January): 130127. doi:10.1016/J.JHAZMAT.2022.130127.

Patial, Shilpa, Vatika Soni, Abhinandan Kumar, Pankaj Raizada, Tansir Ahamad, Xuan Minh Pham, Quyet Van Le, Van Huy Nguyen, Sourbh Thakur, and Pardeep Singh. 2023. "Rational Design, Structure Properties, and Synthesis Strategies of Dual-Pore Covalent Organic Frameworks (COFs) for Potent Applications: A Review." *Environmental Research* 218 (February): 114982. doi:10.1016/J.ENVRES.2022.114982.

Pokrajac, Lisa, Ali Abbas, Wojciech Chrzanowski, Goretty M. Dias, Benjamin J. Eggleton, Steven Maguire, Elicia Maine, et al. 2021. "Nanotechnology for a Sustainable Future: Addressing Global Challenges with the International Network4Sustainable Nanotechnology." *ACS Nano* 15 (12): 18608–18623. doi: 10.1021/acsnano.1c10919.

Prakash, Kamal, Rakesh Deka, and Shaikh M. Mobin. 2024. "A Review on Covalent Organic Frameworks: Exploration of Their Growing Potential as Porous Materials in Photocatalytic Applications." *Inorganic Chemistry Frontiers* 11 (20): 6711–6752. doi:10.1039/D4QI01480D.

Raffi, Amirul Afiat, Noor Fadilah Yusof, Khairul Hamimah Abas, Mohd Hafiz Dzarfan Othman, Juhana Jaafar, Mohd Akhmal Muhamad Sidek, and Mukhlis A Rahman. 2024. "Enhanced Water Purification Performance of Composite PVA–ZnO–Al_2O_3

Hollow Fiber Membrane: A Breakthrough Approach for High-Temperature Stability, Low-Pressure Water Separation." *Chemical Engineering Journal* 499 (November): 155770. doi:10.1016/J.CEJ.2024.155770.

Raghav, Sapna, Jyoti Raghav, Ma'aruf Abdulmumin Muhammad, Madhur Babu Singh, Pallavi Jain, Dinesh Kumar. 2025. "Recent advancements in synthesis and applications of covalent organic frameworks." *Applied Materials Today* 44(June): 102694. doi:10.1016/j.apmt.2025.102694

Rath, Bibhuti Bhusan, Simon Krause, and Bettina Valeska Lotsch. 2024. "Active Site Engineering in Reticular Covalent Organic Frameworks for Photocatalytic CO_2 Reduction." *Advanced Functional Materials* 34 (43): 2309060. doi:10.1002/ADFM.202309060.

Rejali, Narges Abdolhossein, Mohammad Dinari, and Yong Wang. 2023. "Post-Synthetic Modifications of Covalent Organic Frameworks (COFs) for Diverse Applications." *Chemical Communications* 59 (78): 11631–11647. doi:10.1039/D3CC03091A.

Shao, Gaofeng, Xiaogu Huang, Xiaodong Shen, Changxia Li, and Arne Thomas. 2024. "Metal–Organic Framework and Covalent–Organic Framework-Based Aerogels: Synthesis, Functionality, and Applications." *Advanced Science* 11 (48): 2409290. doi:10.1002/ADVS.202409290.

Sheng, Kai, Junyong Zhu, Lei Ge, Jiale Du, Tingting Luo, Lei Jiang, Wenheng Jing, Shi Peng Sun, Yatao Zhang, and Bart Van der Bruggen. 2024. "Three-Dimensional Covalent Organic Frameworks for Advanced Membrane Separations." *Advanced Membranes* 4 (January): 100109. doi:10.1016/J.ADVMEM.2024.100109.

Song, Yanpei, Qi Sun, Briana Aguila, Shengqian Ma, Y Song, Q Sun, B Aguila, and S Ma. 2019. "Opportunities of Covalent Organic Frameworks for Advanced Applications." *Advanced Science* 6 (2): 1801410. doi:10.1002/ADVS.201801410.

Vardhan, Harsh, Grace Rummer, Angela Deng, and Shengqian Ma. 2023. "Large-Scale Synthesis of Covalent Organic Frameworks: Challenges and Opportunities." *Membranes* 13 (8): 696. doi:10.3390/MEMBRANES13080696.

Viciano-Chumillas, Marta, Xiangyu Liu, Antonio Leyva-Pérez, Donatella Armentano, Jesús Ferrando-Soria, and Emilio Pardo. 2022. "Mixed Component Metal-Organic Frameworks: Heterogeneity and Complexity at the Service of Application Performances." *Coordination Chemistry Reviews* 451 (January): 214273. doi:10.1016/J.CCR.2021.214273.

Waller, Peter J., Felipe Gándara, and Omar M. Yaghi. 2015. "Chemistry of Covalent Organic Frameworks." *Accounts of Chemical Research* 48 (12). American Chemical Society: 3053–3063. doi:10.1021/ACS.ACCOUNTS.5B00369/ASSET/IMAGES/MEDIUM/AR-2015–00369G_0009.GIF.

Wang, Kun Yu, Jiaqi Zhang, Yu Chuan Hsu, Hengyu Lin, Zongsu Han, Jiandong Pang, Zhentao Yang, Rong Ran Liang, Wei Shi, and Hong Cai Zhou. 2023. "Bioinspired Framework Catalysts: From Enzyme Immobilization to Biomimetic Catalysis." *Chemical Reviews* 123 (9): 5347–5420. doi: 10.1021/acs.chemrev.2c00879.

Wang, Zhaobo, Chen Wang, Qiming Hu, Xuefei Lei, Rui Guo, Junhua You, Ye Tian, Fan Yang, and Meili Lv. 2024. "Covalent-Metal Organic Frameworks: Preparation and Applications." *Chemical Engineering Journal* 483 (March): 149217. doi:10.1016/J.CEJ.2024.149217.

Xu, Nanfeng, Yingxue Diao, Xihao Qin, Zhengtao Xu, Hanzhong Ke, and Xunjin Zhu. 2020. "Donor–Acceptor Covalent Organic Frameworks of Nickel(II) Porphyrin for Selective and Efficient CO_2 Reduction into CO." *Dalton Transactions* 49 (44): 15587–15591. doi:10.1039/D0DT03205K.

Xu, Xiaomin, Zirui Lou, Simin Cheng, Philip C.Y. Chow, Norbert Koch, and Hui Ming Cheng. 2021. "Van Der Waals Organic/Inorganic Heterostructures in the Two-Dimensional Limit." *Chem* 7 (11): 2989–3026. doi:10.1016/J.CHEMPR.2021.08.013.

Yaghi, Omar M. 2016. "Reticular Chemistry—Construction, Properties, and Precision Reactions of Frameworks." *Journal of the American Chemical Society* 138 (48): 15507–15509. doi:10.1021/JACS.6B11821.

Yang, Shu, Keke Zhi, Zhimin Zhang, Rukiya Kerem, Qiong Hong, Lei Zhao, Wenbo Wu, Lulu Wang, and Duozhi Wang. 2024. "Nitrogen-Rich Triazine-Based Covalent Organic Frameworks as Efficient Visible Light Photocatalysts for Hydrogen Peroxide Production." *Nanomaterials* 14 (7): 643. https://doi.org/10.3390/nano14070643.

Yeo, Chien Ing, Yee Seng Tan, Hafiz Taimoor Ahmed Awan, Abdul Hanan, Weng Pin Wong, Rashmi Walvekar, Bey Hing Goh, and Mohammad Khalid. 2024. "A Review on the Advancements in Covalent Organic Frameworks for Photocatalytic Reduction of Carbon Dioxide." *Coordination Chemistry Reviews* 521 (December): 216167. doi:10.1016/J.CCR.2024.216167.

Yusran, Yusran, Xinyu Guan, Hui Li, Qianrong Fang, and Shilun Qiu. 2020. "Postsynthetic Functionalization of Covalent Organic Frameworks." *National Science Review* 7 (1): 170–190. doi:10.1093/NSR/NWZ122.

Yusran, Yusran, Bo Miao, Shilun Qiu, and Qianrong Fang. 2024. "Functional Covalent Organic Frameworks: Design Principles to Potential Applications." *Accounts of Materials Research* 5 (10) 1263–1278. doi: 10.1021/accountsmr.4c00195.

Zatsu, Vilhouphrenuo, Angel Elizabeth Shine, Joel M. Tharakan, Dayanand Peter, Thottiam Vasudevan Ranganathan, Saqer S. Alotaibi, Robert Mugabi, Abdullatif Bin Muhsinah, Muhammad Waseem, and Gulzar Ahmad Nayik. 2024. "Revolutionizing the Food Industry: The Transformative Power of Artificial Intelligence-a Review." *Food Chemistry: X* 24 (December): 101867. doi:10.1016/J.FOCHX.2024.101867.

Zhang, Bing, Haiyan Mao, Roc Matheu, Jeffrey A. Reimer, Sultan A. Alshmimri, Saeed Alshihri, and Omar M. Yaghi. 2019. "Reticular Synthesis of Multinary Covalent Organic Frameworks." *Journal of the American Chemical Society* 141 (29): 11420–11424. doi: 10.1021/jacs.9b05626.

Zhang, Yugen, and Siti NurhannaRiduan. 2012. "Functional Porous Organic Polymers for Heterogeneous Catalysis." *Chemical Society Reviews* 41 (6): 2083–2094. doi:10.1039/C1CS15227K.

Zhao, Xiaojia, Pradip Pachfule, and Arne Thomas. 2021. "Covalent Organic Frameworks (COFs) for Electrochemical Applications." *Chemical Society Reviews* 50 (12): 6871–6913. doi:10.1039/D0CS01569E.

Zhu, Lijuan, and Yue Biao Zhang. 2017. "Crystallization of Covalent Organic Frameworks for Gas Storage Applications." *Molecules : A Journal of Synthetic Chemistry and Natural Product Chemistry* 22 (7): 1149. doi:10.3390/MOLECULES22071149.

Zhu, Qiang, Xue Wang, Rob Clowes, Peng Cui, Linjiang Chen, Marc A. Little, and Andrew I. Cooper. 2020. "3D Cage Cofs: A Dynamic Three-Dimensional Covalent Organic Framework with High-Connectivity Organic Cage Nodes." *Journal of the American Chemical Society* 142 (39): 16842–16848. doi: 10.1021/jacs.0c07732.

Zhu, Yaqin, Lizhen Chen, Junjie Pan, Shaohua Jiang, Jiaxiu Wang, Guoying Zhang, and Kai Zhang. 2025. "Recent Advances in COF-Derived Carbon Materials: Synthesis, Properties, and Applications." *Progress in Materials Science* 148 (February): 101373. doi:10.1016/J.PMATSCI.2024.101373.

6 Remediation of Radioactive Contaminants in Aquatic Environment Using Covalent Organic Frameworks

Amjad Ali, Rajesh Kumar, and Serajul Haque Faizi

6.1 INTRODUCTION

Environment is the home for the perpetuation of all kinds of living organisms, and water plays a vital role in sustainable living. In the last few decades, modernization and excessive need and curiosity of humans have exploited the natural resources. That caused the uncontrolled emission of wastes from manufacturing industries, which contaminates the water and air, leading to major cause of detrimental effects on living things at all levels. The major reason for water contamination and environmental pollution is the fast and continuous expansion of industry and agriculture (Figure 6.1). The industries of battery production, metal plating, paper manufacturing, tanning and mining make indiscriminate discharge of spent nuclear fuel, nucleology, radioactive medical facilities, nuclear weapons, etc. along with agricultural inputs like fertilizers, pesticides and radioisotopes in agriculture have significantly contributed to environmental contamination (Gendy et al., 2021). Among the most concerning pollutants, heavy metal ions and radionuclides play a critical role in water pollution, posing serious risks to ecosystems and human health (Wang and Zhuang, 2019). However, nuclear energy has been widely developed and utilized as a clean energy. However, significant environmental and health concerns are posed due to their radioactive wastes (Te, Th, U, Nd, Tc, Ra, etc.) generated from nuclear power plants. The harmful effects of radionuclides such as long half-life of radionuclides and high toxicity pose serious risks to health (radioactive iodine can also be hazardous) (Michaelides and Michaelides, 2020). Gradually, the absorption capability of soil has saturated because the radio pollutants have a very long half-life and take 100–1,000 years to disintegrate, causing accumulation; hence, excess radio-pollutants may percolate to water bodies that ultimately enter the food chain and affect the flora and fauna. Their effect starts when the concentration surpasses threshold limits. A few

DOI: 10.1201/9781003605386-6

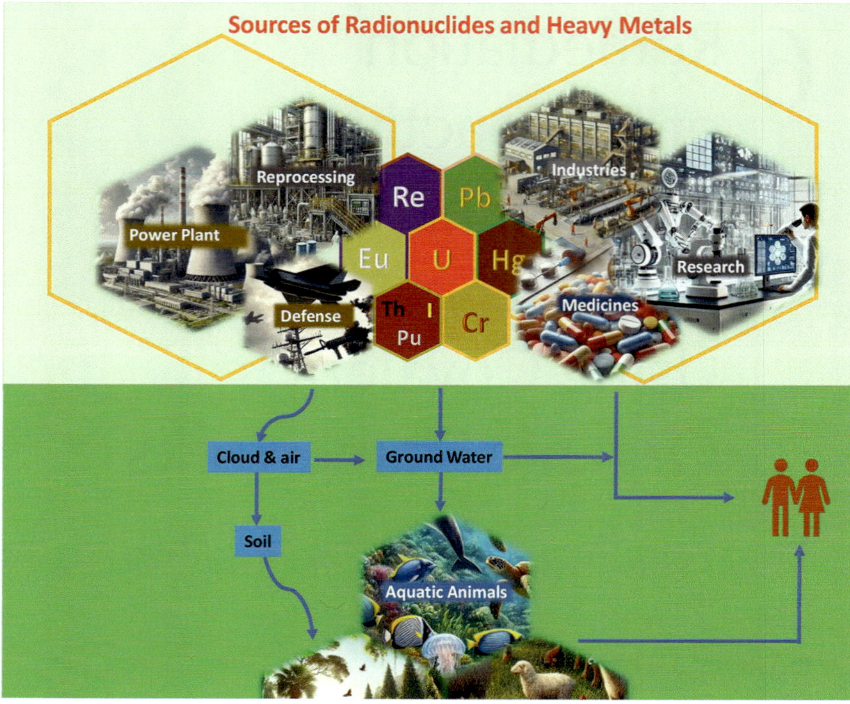

FIGURE 6.1 Some sources of heavy metals and radioactive pollutants.

well-known examples of hazardous effects of heavy metal ions and RADPOLs in excess amounts are as follows:

1. Lead (Pb) ions may affect the brain and kidney functions and the nervous system (Wu et al., 2023).
2. Mercury (Hg) ions may cause amnesia, elevate heart rates, damage kidney functions, etc. (Liu et al., 2023).
3. Chromium (Cr) ions may cause skin irritations and seriously affect the liver and kidney functions (Irshad et al., 2023; Rasheed, 2022).
4. Uranium (U^{6+}) may cause irreversible damage to nervous, reproductive, circulatory and digestive systems (Wani et al., 2022).
5. Europium (Eu^{3+}) may cause neoplastic transformations, leading to tumorigenesis and the development of cancer. Death in extreme cases (Wani et al., 2022).

To mitigate the severe health consequences (Schilz et al., 2022), it is essential to remove radio-wastes from the environment and water bodies. There are several technologies/methods employed for the remediation of toxic heavy metal ions and radioactive pollutants (RADPOLs):

1. Ion exchange (Abo-Farha et al., 2009),
2. Membrane filtration (Dialynas and Diamadopoulos, 2009),

3. Chemical precipitation (Bhatnagar and Sillanpaa, 2010),
4. Phytoremediation (Singh et al., 2023; Gul et al., 2022; Yan et al., 2021),
5. Reverse osmosis (Chen et al., 2021; Cai et al., 2020),
6. Chemical/solvent extraction and leaching (Bayode et al., 2024),
7. Adsorption (Fu and Wang, 2011; Peligro et al., 2016).

Most of these systems are not efficient enough due to the lack of regeneration and self-cleaning properties. Over time, these techniques have been developed to address these issues and minimize the risk. Among them, adsorption is a more efficient, effective and economical approach for the redressal of environment due to its low cost, easy availability of materials, simplicity and high efficiency. It can be used for remedial of RADPOLs even at extremely low concentration. Several materials have also been developed to use as adsorbents to capture pollutants and metal ions, which are as follows:

1. Zeolites (Abbas, 2022),
2. Activated charcoals (Srivastava et al., 2008) and
3. Metal-organic frameworks (MOFs) (Gao et al., 2021).

However, there are limitations for the use of these materials on a larger scale due to their small pore sizes, disordered structures and poor selectivity. Hence, their efficiency gets reduced in pollutant removal from environment. Since selective adsorption is a crucial factor (Chakraborty et al., 2022; Zeng et al., 2022), materials with high adsorption capacity and selectivity can be considered a better choice for heavy metal ions and radionuclides removal from polluted water. To provide these requirements, covalent organic frameworks (COFs) fulfil the purpose, as they are

1. Highly efficient organic materials for wastewater treatment (Wang and Zhuang, 2019);
2. Composed of lightweight elements such as hydrogen (H), boron (B), carbon (C), nitrogen (N) and oxygen (O) (Gendy et al., 2021);
3. Have crystalline porous structures linked by covalent bonds; and
4. Their pore sizes can be adjusted or manipulated according to the use by varying the size of linkers.

COFs consist of predesigned organic building blocks linked by strong covalent bonds, imparting 1D, 2D, or 3D structural arrangements with desired functionalities and porosity (Figure 6.2). Reticular organic chemistry enables the creation of permanently ordered porous frameworks with tailored shapes and sizes, which is the most determining and efficient feature for the purification of water. In the direction of the synthesis of various COFs, the first COF was successfully synthesized by Yaghi et al. in 2005, using a solvothermal method (Côté et al., 2005). They reported a simple method for the synthesis and crystallization of micro- and mesoporous crystalline COFs. The efficient synthesis of the first two members of COFs is shown in Scheme 6.1. COF-1 $[(C_3H_2BO)_6{\cdot}(C_9H_{12})]$ is a simple one pot condensation reaction under mild condition formed by self-molecular assembly (Scheme 6.1a) and COF-5 $(C_9H_4BO_2)$

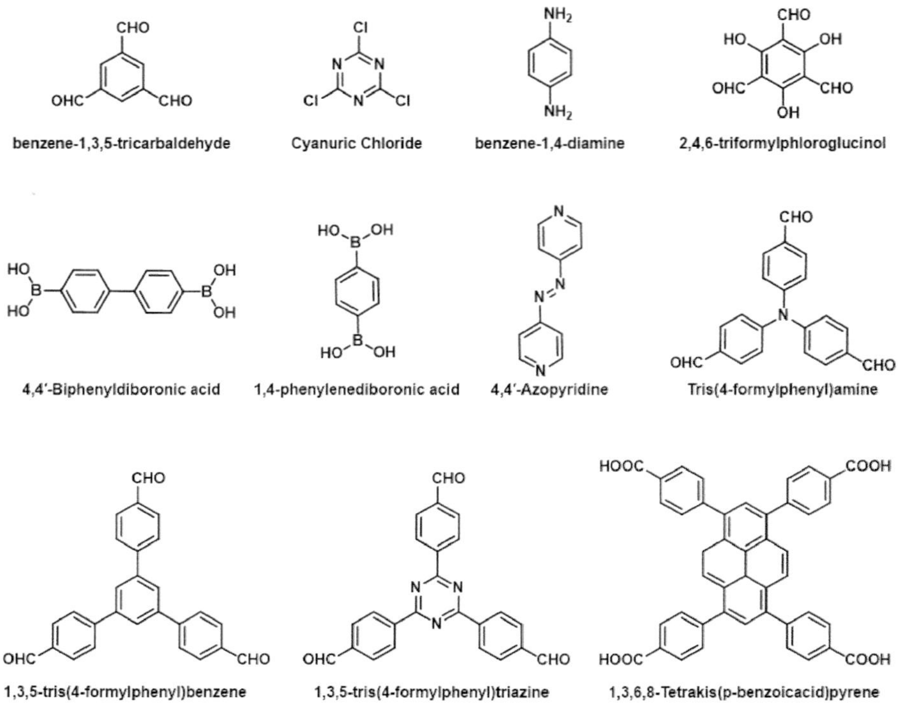

FIGURE 6.2 Some frequently used COF building block linkers.

(Scheme 6.1b) is synthesized at 120°C in 72 h; both have pore sizes 15 Å and 27 Å, respectively, and have low density and high rigidity with exceptional thermal stabilities of up to 600°C. They possess a definite surface area with permanent porosity more than that of well-known zeolites and porous silicates. Borate links are moisture sensitive, making COFs with borate linkers unstable (Kumar et al., 2021; Emmanuel et al., 2024). Lavinge group reported a handy method to synthesize a stable COF-18. It was refluxed for three days under continuous nitrogen flow to yield the COF-18 (Tilford et al., 2006). In another report, Zhang reported a hydrazone-connected chiral COF@SiO_2 (Zhang et al., 2017). In the process of synthesis of COFs, reversible reactions play a crucial role in constructing well-organized lattice architectures of COFs. This enables COF systems to reorganize, self-repair and modify structural construction, which is essential for maintaining highly ordered frameworks (Spitler et al., 2011; Cao et al., 2019). Two key factors in COF design and fabrication are the selection of building units (topological design) and the synthesis technique (Smith et al., 2016; Cao et al., 2019). Although current synthesis methods have limitations, the ongoing development of novel synthetic techniques will significantly expand fabrication possibilities (Ding and Wang, 2013; Dey et al., 2017; Cao et al., 2019; Raghav et al. 2025). COF composites can be synthesized through three main approaches:

1. Integrating nanoparticles (NPs) onto prefabricated COFs, or
2. Growing COFs onto presynthesized NPs, or

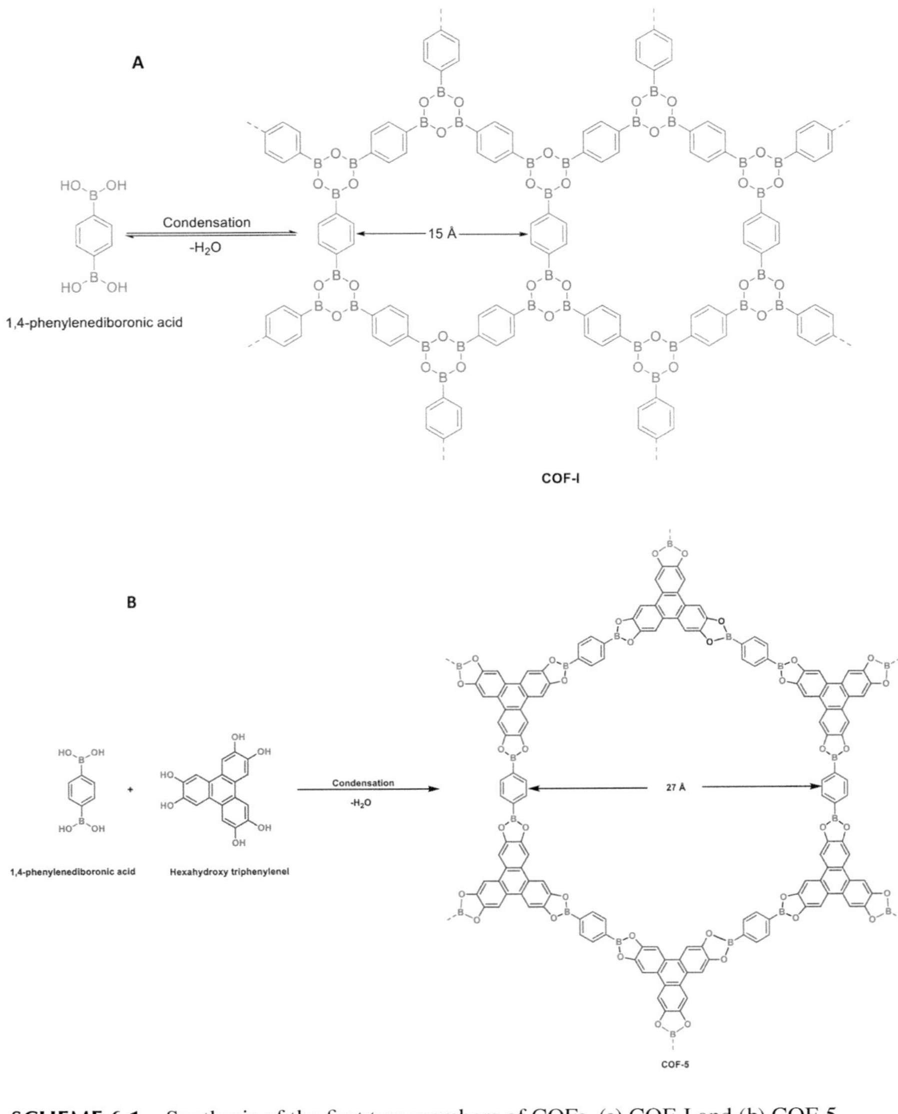

SCHEME 6.1 Synthesis of the first two members of COFs, (a) COF-I and (b) COF-5.

3. Embedding NPs into COF building blocks before assembly, allowing simultaneous COF and NP formation (Kumar et al., 2021).

These methods employ diverse techniques such as (a) solvothermal, which is performed in a closed vessel/system (autoclave) with non-aqueous solvent at a certain temperature and pressure. It has broad applications and can be used in bulk synthesis, for example, COF-1, COF-5 (Côté et al., 2005), and 2D films on SLG (Single Layer Graphene) (Corcos et al., 2019). (b) Ionothermal, a technique used to synthesize COFs at higher temperatures. Here, metal salt in the molten state or ionic liquids (ILs) are used as solvents, e.g., CFT-1 (Bojdys et al., 2010). (c) Mechanochemical, which is a solvent-free direct mixing and grinding of monomers to synthesize COFs. It is a simple, fast and economical method, e.g. TpPa-1 and TpPa-2 TpBD (Campbell et al., 2009). (d) Interface synthesis, which creates uniform COF films with high mechanical properties on the interface zone. This is advantageous over the traditional method, particularly in terms of flexibility, control and structure modification. However, limited building blocks are challenging, e.g. TFPA-TAPA Film (Wu et al., 2023). (e) Top-down approach, preexisting large COFs are converted to smaller 2D nanosheets. Here, different techniques such as ultrasonic, chemical and mechanical are employed to weaken COFs' strong π-π interactions, e.g. COF nanosheets (Berlanga et al., 2011). (f) Bottom-up approach, which involves generation of COFs on a specific surface with desired properties. Here, interface polymerization, surface reaction and surface-assisted synthesis methods are used to grow COF layers, e.g. COFs film synthesizes on SiO_2/SLG (Colson et al., 2015). (g) Flow synthesis method is employed to synthesize high-quality COF films. A consistent reaction environment involves the flow rate, monomer blending, temperature and duration, e.g. 2D COF thin films grown (Colson et al., 2015). (h) microwave-assisted, which uses microwave radiation to synthesize unstable materials. However, it cuts reaction time, increases yield and accelerates production of COFs, i.e. COF-5 (Cooper et al., 2009). (i) Sonochemical method uses ultrasonic radiation (20 kHz–10 MHz) to prepare COFs. It is fast and environmentally friendly and substitutes the solvothermal method, e.g. COF-1 and COF-5 (Yang et al., 2012). (j) hydrothermal, (k) solvent-free direct heating, vapour-assisted and micelle-assisted synthesis (Emmanuel et al., 2024). The solvothermal method is a widely used method and was first used for COF fabrication in 2005 (Emmanuel et al., 2024). It has been used to facilitate the synthesis of COFs based on boronate, imine, azines, borosilicate, esters, imides aquaraines, boroxine, hydrazones, phenazines and borazine (Kumar et al., 2021). This process involves activation of reactants in a Pyrex tube under vacuum using a freeze-pump-thaw technique, followed by sealing and heating of the tube at 80–120°C for 2–9 days. The resulting precipitate is washed with organic solvents like acetone and vacuum-dried (Kumar et al., 2021). The orderly porous COFs have antibacterial activity, definite surface area and biocompatibility. It is physically, chemically and thermally very stable; it is lightweight due to low density and high porosity, photochemically active due to a highly conjugated system; it has a tuneable structure due to adjustable pore size and it can easily be functionalized. Moreover, COFs have promising applications in gas purifications (Han et al., 2008; Zhang et al., 2019), catalysis (Hu et al., 2019), treatment of water (Dey et al., 2017; Wang et al., 2017) and adsorption of pollutant and separation (Wang et al., 2018; Gendy et al., 2021).

6.2 REMOVAL OF RADIOACTIVE POLLUTANTS USING ADSORPTION TECHNIQUE BY COF

The advancement of technology in almost all the dimensions for the effortless and efficient work have increased the dependence on clean energy source, i.e. nuclear energy.

Sometimes, clean sources indirectly affect the environment. For example, nuclear power plants, industries processing radioactive ores, and medical wastes

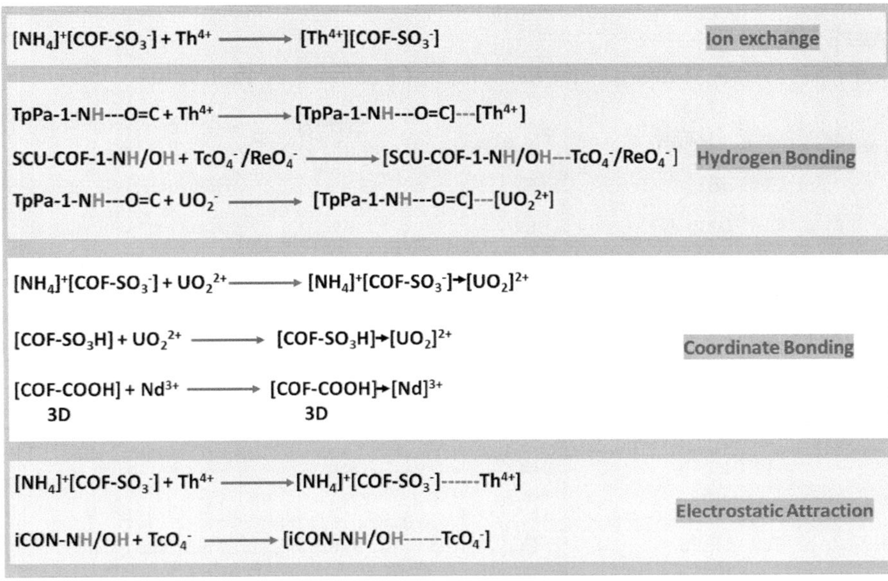

SCHEME 6.2 Various interactions between COFs and RADPOLs (Gendy et al., 2021).

and agriculture are the primary source of radio-wastes. The nuclear waste water primarily contains U (Uranium), Eu (Europium), Th (Thorium) and I (Iodine), and these are highly toxic and detrimental to the environment and health. Adsorption of RADPOLs by COFs is greatly affected by temperature, pH and time. The interaction of RADPOLs with COFs may involve an ion exchange, hydrogen bonding, coordination or electrostatic interactions. Some examples are shown in Scheme 6.2. These are key factors that affect the binding of RADPOLs to the COFs. Chitosan, a cheap and easily available ingredient, was used by Zhu's group and modified C2 and C6 disubstituted amidoxime functionalized chitosan foams (CMNSC-AO foams). Freshly prepared CMNSC-AO form showed an excellent sorption efficacy for U^{6+} of over 450 mg g^{-1}. Furthermore, a fresh sharp peak for CMNSC-AO-U form in FT-IR for the O=U=O stretching was observed at 900 cm^{-1}, and SEM-EDS analysis reveals that COF retained the good pore structure. In addition, COFs with multi-site functional groups $-NH_2$ and -NOH show an enhanced sorption rate. Adsorption of Pu^{4+} and U^{6+} by COF-IHEP1 is around 160 mg g^{-1}. From the density functional theory, it is found that RADPOLs are adsorbed in the network of COF-IHEP1; here, two P=O oxygen atoms from neighbouring layer and two oxygen from H_2O are coordinated with U^{6+}. In another example, remediation of Th^{4+} using COF$\left[NH_4\right]^+\left[COF-SO_3^-\right]$ employs ion exchange technique; here, cation exchange between $[NH_4]^+$ and Th^{4+} takes place with an adsorption capability of 395 mg g^{-1}. Table 6.1 shows different COFs with efficiency and active sites/interaction types (Gendy et al., 2021).

COFs are best for the remediation of radioactive wastes from environment and wastewater. The high porosity, lightweight, tuneable pores and can be functionalized with different groups make it better for the remediation.

TABLE 6.1
COFs with Active Sites for the Remedial of Radioactive Pollutants

S. No.	COFs	Adsorbates	Capacity q (mg g^{-1})	Active Sites/Interaction Type
1	COF-HBI	UO_2^{2+}	211	Benzimidazole on surface
2	COF-TpAb-AO	UO_2^{2+}	408	Amidoxime-binding site
3	*o*-GS-COF	UO_2^{2+}	243	Nitrogen atoms
4	$[NH_4]^+[COF-SO_3^-]$	UO_2^{2+}	869	$-SO_3^-$ units
5	COF-TpPa-1	UO_2^{2+}	152	May be -NH---O=C
6	COF-PDAN-AO	UO_2^{2+}	256	Amidoxime group
7	TFPT-BTAN-AO	UO_2^{2+}	427	Amidoxime group
8	TFPT-BTAN	UO_2^{2+}	350	-CN and sp^2 N
9	COF-IHEP1	UO_2^{2+}, Pu^{4+}	160	Phosphate group
10	COF-IHEP2	UO_2^{2+}, Pu^{4+}	140	Phosphate group
11	COF-HAP	UO_2^{2+}	510	Carboxylic group
12	NDA-TN-AO	UO_2^{2+}	6.07	Electrostatic attraction
13	COF-R5	UO_2^{2+}	253.7	Hydrogen bonding
14	COF–SO_3H	UO_2^{2+}	360	H-bonding
15	$[NH_4]^+[COF-SO_3^-]$	Th^{4+}	395	Ion exchange
16	Py-TFImI-25 COF	Th^{4+}	–	Benzimidazole
17	BC@LDH@HAP	Eu^{3+}	714	Phosphate and carbonate
18	TAPB-BPDA COF	I	988	Benzene and N-atom

6.3 FUTURE PERSPECTIVE

COFs are found to be a very fascinating and potentially important material for the removal of RADPOLs from the environment and water bodies. Although COFs perform better than usual sorption properties than activated charcoals, zeolites, MOFs, etc., the use of most of the COFs is limited to lab performances; hence, their large-scale applicability is not much explored. The extensive research on COFs must be done thoroughly to enhance performance, durability for large scale, and industrial use for the benefit of humans. Additionally, studies of radioactive pollutant adsorbed COFs must be done for their extended use, leaching tendency, recycling and subsequent effects, because it is important before employing COFs on an industrial scale use or it may be detrimental. Moreover, for real-life applications, efficient management of multiple reuse of COFs is must because their expulsion in the environment will not fulfil the purpose of their use. More work is needed in the direction of introducing more binding scaffolds with COF monomers, in making films or forms (3D architecture) as separation and reusability of COFs are issues to tackle with. This can be pointed out that the majority of the research studies are done on uranium removal, and the focus on exploration of COFs for the removal of other RADPOLs such as Th, Rd, Sr, I and Am must be brought into consideration. Moreover, during the synthesis of COFs, the monomer proportions must be controlled because unreacted monomers are less stable and can affect the overall stability and efficiency. Although COFs are very efficient for the removal of RADPOLs in comparison to activated carbons, zeolites and MOFs, the high cost, scalability and common usability are challenging. So, researchers are continuously working to make it cost-effective and easily available.

6.4 CONCLUSION

COFs being lightweight and highly porous with specific pore sizes can be tuned according to the need, the surface can be functionalized with specific groups according to their binding capabilities with heavy metals or radionuclides, and crystallinity makes them very suitable and efficient adsorbents for the removal of RADPOLs from aquatic environment. COFs have a very high range of adsorption capabilities for RADPOLs that ranges from 41.8 to 2362.4 mg g^{-1} for polluted water. The functionalization of COFs with H-bonding, electrostatic, coordination (hydroxyl, carbonyl, oxime, imidazole moiety, etc.), and ion exchange groups on the surface make them more efficient due to an increase in binding sites. Additionally, it is found from multiple adsorption–desorption cycles (up to 4–9 cycles) with more than 77% efficiency, which shows its potential for scale-up and industrial use. Employing COFs for wastewater treatment can be beneficial for the environment and human well-being, as it has high potential to mitigate contamination issues efficiently. The issues to be addressed are management and disposal of multiple times used COFs and their cost of production.

REFERENCES

Abbas, M., 2022. Removal of methylene blue pollutant from the textile industry by adsorption onto Zeolithe: kinetic and thermodynamic study. *J. Eng. Fibers Fabr.* 17, 1–11.

Abo-Farha, S.A., Abdel-Aal, A.Y., Ashour, I.A., Garamon, S.E., 2009. Removal of some heavy metal cations by synthetic resin purolite C100. *J. Hazard. Mater.* 169, 190–194.

Berlanga, I., Ruiz-González, M.L., González-Calbet, J.M., Fierro, J.L.G., Mas-Ballesté, R., Zamora, F., 2011. Delamination of layered covalent organic frameworks, *Small* 7, 1207–1211.

Bhatnagar, A., Sillanpaa, M., 2010. Utilization of agro-industrial and municipal waste materials as potential adsorbents for water treatment-a review. *Chem. Eng. J.* 157, 277–296.

Bojdys, M.J., Jeromenok, J., Thomas, A., Antonietti, M., 2010. Rational extension of the family of layered, covalent, triazine-based frameworks with regular porosity. *Adv. Mater.* 22, 2202.

Cai, Y.-H., Yang, X.J., Schafer, A.I., 2020. Removal of naturally occurring strontium by nanofiltration/reverse osmosis from groundwater. *Membranes (Basel)* 10, 321.

Campbell, N.L., Clowes, R., Ritchie, L.K., Cooper, A.I., 2009. Rapid microwave synthesis and purification of porous covalent organic frameworks. *Chem. Mater.* 21, 204–206.

Cao, S., Li, B., Zhu, R., Pang, H., 2019. Design and synthesis of covalent organic frameworks towards energy and environment fields. *Chem. Eng. J.* 355, 602–623.

Chakraborty, R., Asthana, A., Singh, A.K., Jain, B., Susan, A.B.H., 2022. Adsorption of heavy metal ions by various low-cost adsorbents: a review. *Int. J. Environ. Anal. Chem.* 102, 342–379.

Chen, B., Yu, S., Zhao, X., 2021. The separation of radionuclides and silicon from boron-containing radioactive wastewater with modified reverse osmosis membranes. *Process Saf. Environ. Prot.* 146, 639–646.

Colson, J.W., Mann, J.A., DeBlase, C.R., Dichtel, W.R., 2015. Patterned growth of oriented 2D covalent organic framework thin films on single-layer graphene. *J. Polym. Sci. A-Polym. Chem.* 53, 378–384.

Corcos, A.R., Levato, G.A., Jiang, Z.W., Evans, A.M., Livingston, A.G., Mari˜nas, B.J., et al., 2019. Reducing the pore size of covalent organic frameworks in thin-film composite membranes enhances solute rejection. *ACS Mater. Lett.* 1, 440–446.

Côté, A.P., Benin, A.I., Ockwig, N.W., O'Keeffe, M., Matzger, A.J., Yaghi, O.M., 2005. Porous, crystalline, covalent organic frameworks. *Science* 310, 1166–1170.

Dey, K., Pal, M., Rout, K.C., Kunjattu, H.S., Das, A., Mukherjee, R., Kharul, U.K., Banerjee, R., 2017. Selective molecular separation by interfacially crystallized covalent organic framework thin films. *J. Am. Chem. Soc.* 139, 13083–13091.

Dey, S., Bhunia, A., Breitzke, H., Groszewicz, P.B., Buntkowsky, G., Janiak, C., 2017. Two linkers are better than one: enhancing CO_2 capture and separation with porous covalent triazine-based frameworks from mixed nitrile linkers. *J. Mater. Chem. A* 5, 3609–3620.

Dialynas, E., Diamadopoulos, E., 2009. Integration of a membrane bioreactor coupled with reverse osmosis for advanced treatment of municipal wastewater. *Desalination* 238, 302–311.

Ding, S.-Y., Wang, W., 2013. Covalent organic frameworks (COFs): from design to applications. *Chem. Soc. Rev.* 42, 548–568.

Emmanuel, S.S., Olawoyin, C.O., Adesibikan, A.A., Nafiu, S.A., Bayode, A.A., 2024. Solvothermally and non-solvothermally fabricated covalent organic frameworks (COFs) for eco-friendly remediation of radiocontaminants in aquatic environments: a review. *J. Organomet. Chem.* 1005, 122984.

Fu, F., Wang, Q., 2011. Removal of heavy metal ions from wastewaters: a review. *J. Environ. Manage.* 92, 407–418.

Gao, M., Liu, G., Gao, Y., Chen, G., Huang, X., Xu, X., Wang, J., Yang, X., Xu, D., 2021. Recent advances in metal-organic frameworks/membranes for adsorption and removal of metal ions. *TrAC Trends Anal. Chem.* 137, 116226.

Gendy, E.A., Ifthikar, J., Ali, J., Oyekunle, D.T., Elkhlifia, Z., Shahib, I.I., Khodair, A.I., Chen, Z., 2021. Removal of heavy metals by covalent organic frameworks (COFs): a review on its mechanism and adsorption properties. *J. Environ. Chem. Eng.* 9, 105687.

Gul, I., Ahmad, I., Shah, G.M., 2022. Phytoremediation of radioactive contaminated sites. *Adv. Bioremediation Phytoremediation Sustain. Soil Manag. Princ. Monit.* 171–182.

Han, S.S., Furukawa, H., Yaghi, O.M., Goddard III, W.A., 2008. Covalent organic frameworks as exceptional hydrogen storage materials. *J. Am. Chem. Soc.* 130, 11580–11581.

Hu, X., Long, Y., Fan, M., Yuan, M., Zhao, H., Ma, J., Dong, Z., 2019. Two-dimensional covalent organic frameworks as self-template derived nitrogen-doped carbon nanosheets for eco-friendly metal-free catalysis. *Appl. Catal. B Environ.* 244, 25–35.

Irshad, M.A., Sattar, S., Nawaz, R., Al-Hussain, S.A., Rizwan, M., Bukhari, A., Waseem, M., Irfan, A., Inam, A., Zaki, M.E.A., 2023. Enhancing chromium removal and recovery from industrial wastewater using sustainable and efficient nanomaterial: a review. *Ecotoxicol. Environ. Saf.* 263, 115231.

Kumar, S., Kulkarni, V.V., Jangir, R., 2021. Covalent-organic framework composites: a review report on synthesis methods. *ChemistrySelect* 6, 11201–11223.

Liu, C., Li, Y., Gai, X., Xiang, Z., Jiang, W., He, S., Liu, Y., Xiao, H., 2023. Advances in green materials derived from wood for detecting and removing mercury ions in water. *Environ. Pollut.* 335, 122351.

Michaelides, E.E., Michaelides, D.N., 2020. Impact of nuclear energy on fossil fuel substitution. *Nucl. Eng. Des.* 366, 110742.

Peligro, F.R., Pavlovic, I., Rojas, R., Barriga, C., 2016. Removal of heavy metals from simulated wastewater by in situ formation of layered double hydroxides. *Chem. Eng. J.* 306, 1035–1040.

Raghav, Sapna, Raghav, J., Muhammad, M. A., Singh, M. B., Jain, P., Dinesh Kumar, D., 2025. Recent advancements in synthesis and applications of covalent organic frameworks. *Appl. Mater. Today*, 44, 102694.

Rasheed, T., 2022. Covalent organic frameworks as promising adsorbent paradigm for environmental pollutants from aqueous matrices: perspective and challenges. *Sci. Total Environ.* 833, 155279.

Schilz, J.R., Dashner-Titus, E.J., Simmons, K.A., Erdei, E., Bolt, A.M., MacKenzie, D.A., Hudson, L.G., 2022. The immunotoxicity of natural and depleted uranium: from cells to people. *Toxicol. Appl. Pharmacol.* 454, 116252.

Singh, G.B., Bhawna, S.F., Shewale, P., Dahiya, R., Aggarwal, A., Manju, F., Arya, S.K. 2023. Phytoremediation of radioactive elements, possibilities and challenges: special focus on agricultural aspects. *Int. J. Phytoremed.* 25, 1–8.

Smith, B.J., Overholts, A.C., Hwang, N., Dichtel, W.R., 2016. Insight into the crystallization of amorphous imine-linked polymer networks to 2D covalent organic frameworks. *Chem. Commun.* 52, 3690–3693.

Spitler, E.L., Giovino, M.R., White, S.L., Dichtel, W.R., 2011. A mechanistic study of Lewis acid-catalyzed covalent organic framework formation. *Chem. Sci.* 2, 1588–1593.

Srivastava, V.C., Mall, I.D., Mishra, I.M., 2008. Adsorption of toxic metal ions onto activated carbon - study of sorption behaviour through characterization and kinetics. *Chem. Eng. Process.* 47, 1275–1286.

Tilford, R.W. Gemmill, W.R., zur Loye, H.-C., Lavigne, J.J., 2006. Facile synthesis of a highly crystalline, covalently linked porous boronate network. *Chem. Mater.* 18, 5296–5301.

Wang, C., Li, Z., Chen, J., Li, Z., Yin, Y., Cao, L., Zhong, Y., Wu, H., 2017. Covalent organic framework modified polyamide nanofiltration membrane with enhanced performance for desalination. *J. Membr. Sci.* 523, 273–281.

Wang, J., Zhuang, S., 2019. Covalent organic frameworks (COFs) for environmental applications. *Coord. Chem. Rev.* 400, 213046.

Wang, W., Deng, S., Ren, L., Li, D., Wang, W., Vakili, M., Wang, B., Huang, J., Wang, Y., Yu, G., 2018. Stable covalent organic frameworks as efficient adsorbents for high and selective removal of an aryl-organophosphorus flame retardant from water. *ACS Appl. Mater. Interfaces* 10, 30265–30272.

Wani, A.A., Shahadat, M., Ali, S.W., Ahammad, S.Z., Uddin, M.K., 2022. Recent advances and future perspectives of polymer-based magnetic nanomaterials for detection and removal of radionuclides: a review. *J. Mol. Liq.* 365, 119976.

Wu, D., Hu, Y., Cheng, H., Ye, X., 2023. Detection techniques for lead ions in water: a review. *Molecules* 28, 3601.

Wu, D.C., Zhang, Q.S., Wang, X.Y., Zhang, B., 2023. Interface-confined synthesis of a non-planar redox-active covalent organic framework film for synaptic memristors. *Nanoscale* 15, 2726–2733.

Yan, L., Van Le, Q. Sonne, C., Yang, Y., Yang, H., Gu, H., Ma, N.L., Lam, S.S., Peng, W., 2021. Phytoremediation of radionuclides in soil, sediments and water. *J. Hazard. Mater.* 407, 124771.

Yang, S.-T., Kim, J., Cho, H.-Y., Kim, S., Ahn, W.-S., 2012. Facile synthesis of covalent organic frameworks COF-1 and COF-5 by sonochemical method. *Rsc Adv.* 2, 10179–10181.
Zhang, K., Cai, S.-L., Yan, Y.-L., He, Z.-H., Lin, H.-M., Huang, X.-L., Zheng, S.-R., Fan, J., Zhang, W.-G., 2017. Construction of a hydrazone-linked chiral covalent organic framework–silica composite as the stationary phase for high performance liquid chromatography. *J. Chromatogr. A* 1519, 100–109.
Zhang, W.-C., Zhang, S.-T., Sun, Z., Sun, W.-F., Zhang, L.-Y., Shi, Y.-B., 2019. First-principles study of molecular carbon dioxide adsorption in covalent organic framework-112. *Sci. Adv. Mater.* 11, 317–324.

7 Absorption and Photocatalytic Degradation of Environmental Pollution Using Covalent Organic Framework

Deepti Chaudhary, Anjali Yadav, Vishwajit Chavda, and Meenu Devi

7.1 INTRODUCTION

The quick growth of the agricultural, pharmaceutical, and chemical manufacturing companies has resulted in a rising release of synthetic substances into aquatic habitats, considerably exceeding water bodies' natural self-purification capacity. Hazardous heavy metal ions (e.g., Hg^{2+}, Cd^{2+}, $Cu^{2+,}$ and Pb^{2+}), radionuclides (e.g.,^{57}Co,^{59}Fe,^{65}Zn,^{79}Se,^{89}Sr,^{99}Tc,^{128}Pd,$^{129}I^{137}Cs$,^{153}Eu,^{137}Cs, and ^{235}U) in nuclear waste, and organic pollutants (e.g., rhodamine B, Bisphenol A, CR, polycyclic aromatic hydrocarbons, methyl orange, methylene blue, sulfamerazine, and pharmaceutical ibuprofen) in industrial wastewater are the major pollutants in aqueous solutions that develop in living organisms and pose major threats to humans and other species through the food chain (Liu et al., 2021). As a result, traditional chemical and physical treatment procedures are frequently useless in addressing them. This pollution is a severe danger to sustainable development. For example, organic dyes alone account for over 100,000 commercially accessible kinds, with an annual production of almost 10 million tonnes. Along with the widely recognized risks of cancer and toxicity, dyes in aquatic environments also contribute to water discolouration and limit oxygenation on the water surface, significantly impacting aquatic plants and animals. Consequently, it is highly significant to control and remediate wastewater pollution. In particular, treating organic contaminants in wastewater has gained international attention since it improves aquatic ecosystem health and has significant financial advantages (Bulgariu et al., 2019, Hu et al., 2022). Prolonged interaction with organic contaminants in water, whether intentional or not, increases the hazards of various health problems, including endocrine disruption, cancer, diabetes, allergies, learning

DOI: 10.1201/9781003605386-7

disabilities, asthma, neurological disorders, cardiovascular diseases, and failure of the reproductive system. As a result, developing efficient treatment technologies to eliminate these pollutants from contaminated wastewater has attracted wide attention (Asayesh-Ardakani et al., 2024). Although metal-organic frameworks (MOFs) exhibit various shapes, large surface areas, increased porosity, and abundant metal sites, their thermal and chemical stability remains a challenge because of the weak nature of coordination bonds. Covalent organic frameworks (COFs) have grown in popularity as a fast-expanding topic of chemical research over the last three years. Additionally, to their highly stable structures, COFs have outstanding light absorption and an ideal bandgap, making them intriguing semiconductor photocatalytic materials. Figure 7.1 shows the COFs used to degrade and adsorb various contaminants from wastewater treatment.

COFs exhibit special features that make them potential candidates for next-generation heterogeneous photocatalysts; nonetheless, some obstacles still prevent their advanced application in wastewater treatment. COF can create electronic transitions when excited by external energy sources, resulting in photogenerated carriers such as photogenerated electrons and holes. Because of the differences in the outer environment, photogenerated carriers can react with persulfate, the air in the atmosphere, and water to develop photogenerated holes, superoxide radicals, hydroxyl radicals, sulphate radicals, and singlet oxygen. However, in traditional photocatalytic degradation methods, hydroxyl free radicals, superoxide free radicals, and

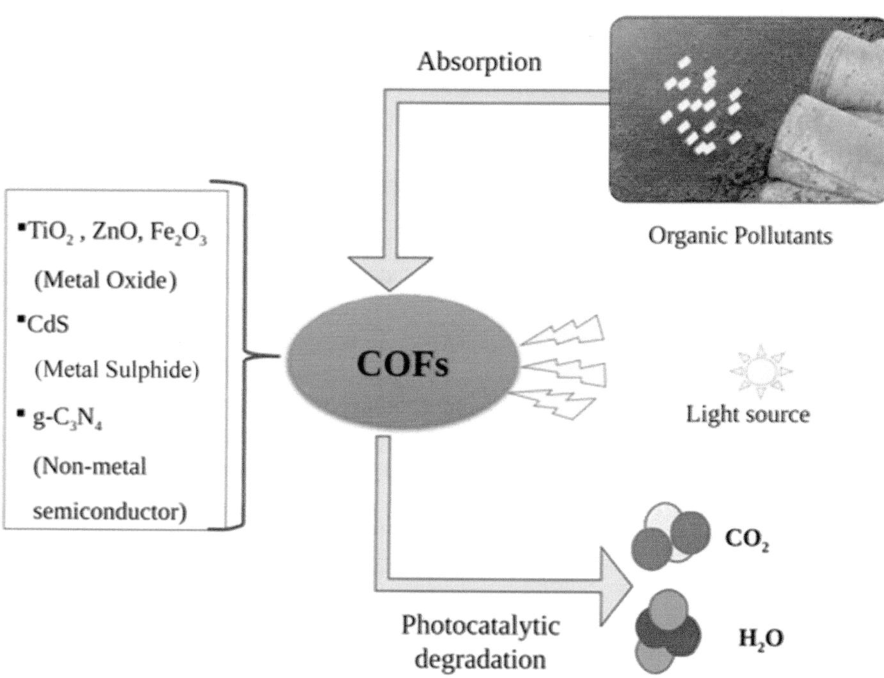

FIGURE 7.1 Various COFs are used to degrade and adsorb various pollutants from wastewater.

photogenerated holes significantly affect the degradation process. It is well known that light absorption, separation of electron-hole pairs, and electron transfer processes are crucial for the photocatalytic method (Hu et al.,2022). The interaction mechanisms between COFs and organic pollutants were π–π stacking, hydrogen bonding, hydrophobic, hydrophilic, and electrostatic interactions (Ma et al., 2023).

Recently, nanomaterials have received increasing attention for eliminating organic contaminants from water. Nanomaterials, like carbon-based materials and metal oxides, show exceptional absorption capacity and anti-microbial properties for wastewater treatment. The unique properties of nanoparticles (NPs), such as catalysis, high absorption capacity, and reactivity, have gained attention for their effectiveness in wastewater treatment (Liu et al., 2021). Nanomaterials, including nano-photocatalysts, nano-sorbents, and nano-membranes, provide diverse tools in wastewater treatment, which are recent advancements in nanomaterials. NPs are classified into two important categories: carbon-based NPs and metal-oxide NPs. Applying COF-based nanomaterials, such as photocatalysis, adsorption, fuel cells, and supercapacitors, has broad prospects.

COFs are crystalline, porous organic polymers (POPs). COFs are emerging as versatile materials in the fight against environmental pollution, particularly through their capabilities in absorption and photocatalytic degradation (Ritchie et al., 2010, Asayesh-Ardakani et al., 2024). COFs have a large specific surface area, tuneable pore structures, and great thermal and solvent stability, making them effective at capturing various pollutants, including heavy metal ions, endocrine-disrupting compounds, pharmaceutical toxic pollutants, and dye (Chen et al., 2021, Liu et al., 2021, Zhang et al.,2021).

Many water treatment technologies, like chemical, physical, and biological approaches, have been advanced for removing various emerging contaminants (He et al., 2021). Physical processes encompass absorption, membrane separation, and ion exchange. The chemical process involved ozonation and oxidation. Meanwhile, biologically utilize plant, aerobic–anaerobic bacteria, and mixed microbial consortia (Saleh, et al., 2020).

This chapter explores the advancement of COFs for adsorption and subsequent degradation of pollutants through photocatalysis during light irradiation. Removal using adsorptive and photocatalytic disintegration of organic pollutants has evolved as an energy-efficient and low-cost innovation. Both have gained significant attention for treating the world's wastewater. As human industrialization and civilization progress rapidly, excessive dependence on fossil fuel consumption has led to an energy crisis and adverse environmental effects (Su et al., 2021). These two challenges pose the most serious threats to the long-term growth of human society. Sunlight, a green and virtually limitless renewable energy source is considered the best alternative to fossil fuels to address the energy crisis and mitigate environmental pollution (Yin, et al., 2021).

7.1.1 Structure of COFs

COFs are a new class of organic crystalline porous substances that are made using building blocks and reticular chemistry. COFs are porous polymers with organic and inorganic materials comprising light elements (C, N, O, and B) that have low density, crystalline structure, high porosity, easy operation, high thermal and mechanical stability, the ability to change the structure of holes, and high internal surface (Li et al., 2025). These elements were intricately linked by strong covalent bonds and extended into zero-dimensional (0D), one-dimensional (1D), two-dimensional (2D), and three-dimensional (3D) (Li et al., 2025). In contrast to 3D materials, 2D materials are considered more promising candidates for catalysis due to their high specific surface area and larger surface-to-volume ratio, which can give enhanced active sites for the target substrate. Furthermore, stacking of the single layers in 2D COFs and MOFs can substantially enhance the electron transfer (Qian et al., 2021). Because of their large specific surface area, excellent chemical stability, tuneable functionality, and numerous functional active sites, COFs have attracted a lot of attention.

7.1.2 Building Blocks and Linkages

COFs were synthesized using a range of functionalized building blocks of varying sizes and symmetries. We have summarized the most commonly utilized functional building units in Figure 7.2. Based on their reactive sites, they can be broadly categorized into amines, aldehydes, catechols, and boronic acids.

These building blocks typically consist of π-conjugated aromatic compounds or heterocycles, including triphenylene, pyrene, thiophene, hexabenzocoronene, naphthalene diimides, porphyrin, phthalocyanine, trioxazatrianguline, and tetrathiafulvalene. Notably, many of these functional groups exhibit photochemical, electrical, or redox properties that contribute significantly to the semiconductor behaviour of COFs. The complex structure of these building blocks enables the creation of sophisticated but orderly porous environments, which allows precise functional tuning. Additionally, binding is a crucial factor to consider when forming COFs. Theoretically, the majority of observed covalent groups can be used for the development of semiconducting COFs. The boroxine, imide, phenazine, imine, borate ester, and sp^2-C bonds have been utilized extensively in developing semiconductive COFs (Figure 7.2). This binding generally results in various stabilities, crystallinity, and photoelectric characteristics. The linkages utilized in preparing COFs are typically reversible or partially reversible to confirm self-healing and optimal crystal growth of defects. However, crystalline COFs have also been effectively constructed using several irreversible linkages. The imine linkage has excellent availability of monomers, the borate ester bonding produces high crystallinity, the phenazine can produce excellent conjugation and conductivity, and the sp^2–C bond exhibits robust conjugation and stability. Other semiconducting COFs were created and used in other related sectors by combining their advantages with the creation of building blocks and topological diagrams. Several semiconducting COFs have been created and used in various related fields by leveraging these advantages and incorporating them into the construction of topological diagrams and building blocks (Gendy, et al., 2021).

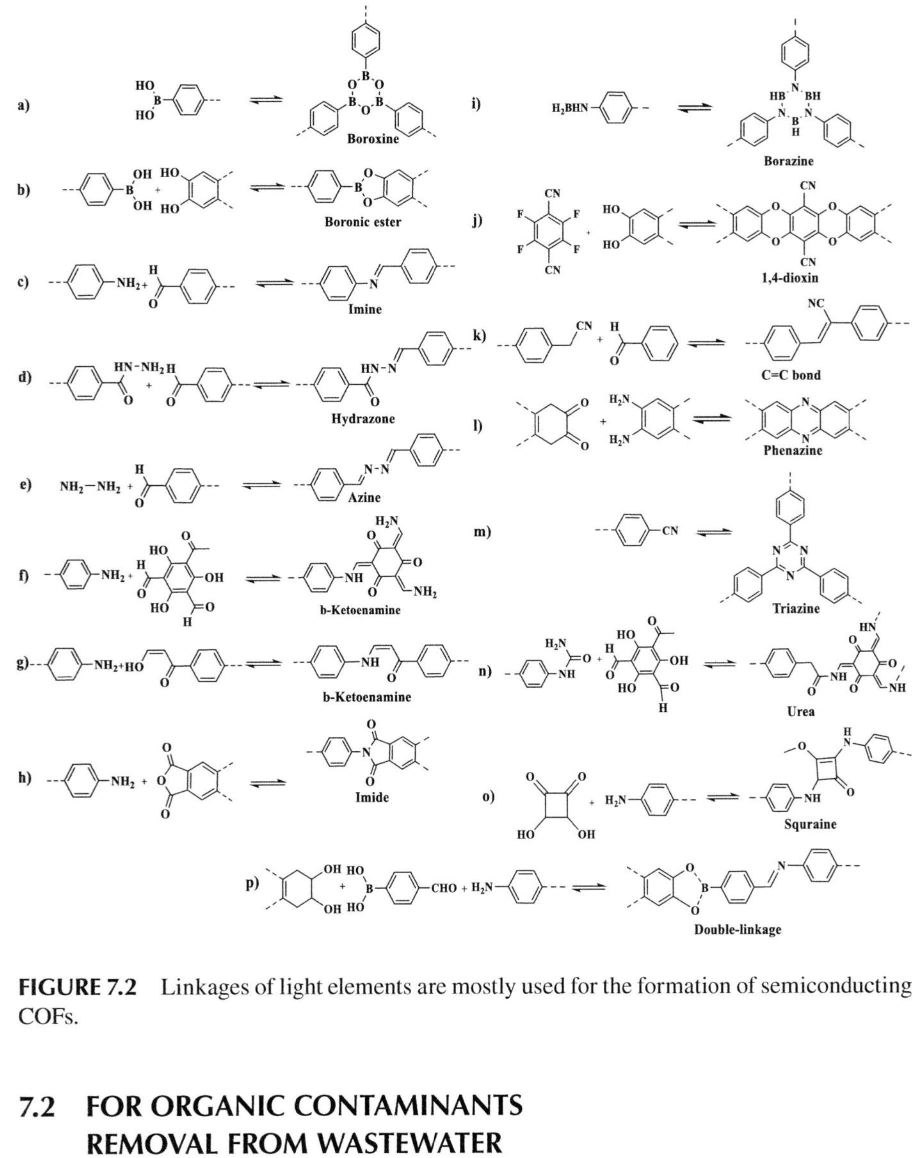

FIGURE 7.2 Linkages of light elements are mostly used for the formation of semiconducting COFs.

7.2 FOR ORGANIC CONTAMINANTS REMOVAL FROM WASTEWATER

COFs are now the best possible candidate materials for advanced uses in the removal of organic contaminants from polluted water due to their well-ordered structure, stability, good porosity, adjustable surface properties, and stability. These properties make COFs attractive for energy storage, photocatalytic degradation, absorption, cancer diagnosis and treatment, and pollution reduction. Many treatment technologies, like chemical, physical, and biological approaches, have been advanced for removing various emerging contaminants (Wang et al., 2020, He et al., 2021).

Physical processes encompass absorption, membrane separation, and ion exchange. The chemical process involved ozonation and oxidation. Meanwhile, biologically utilize plants, aerobic–anaerobic bacteria, and mixed microbial consortia (Saleh et al., 2020).

In general, conventional wastewater treatment processes, such as flocculation, adsorption, membrane filtration, and electrocoagulation, have been intensively developed to overcome their inadequate effectiveness for several organic pollutants (Jin et al., 2020, Swaminathan et al., 2024). Organic pollutants offer serious environmental risks due to their chemically stable constituents resisting degradation. The treatment of organic pollutants faces several technological challenges, such as material stability, low photocatalytic efficiency, slow reaction rates, limited visible light activity, hazardous intermediate production, scaling issues, high energy consumption, reusability, photocatalyst recovery, and catalyst deactivation. To address these issues, researchers are working on producing photocatalysts that can absorb a broader range of light, improving electron–hole separation to increase efficiency, and simple methods for regular use. As a result, technically reliable, highly efficient, cost-effective, and ecologically safe organic pollutant removal is favoured. Among these approaches, photocatalysis is the most promising, with the ability to mineralize and remove organic contaminants (Reyes et al., 2021). In recent decades, considerable progress has been made in photocatalytic techniques for organic pollution bioremediation. Modification techniques are required to improve photocatalyst efficiency, especially when integrating photocatalytic nanomaterials for environmental applications, such as organic pollutant degradation. This chapter explores the advancement of COFs in the adsorption and subsequent degradation of pollutants through photocatalysis during light irradiation. Adsorptive removal and photocatalytic disintegration of organic pollutants have evolved as energy-efficient and low-cost innovations. Both have gained significant attention for treating the world's wastewater (Al-Nuaim et al., 2023).

7.2.1 Adsorption

Over the past few decades, enormous research about natural and synthetic adsorbents, including clays, activated carbons (ACs), organic resins, carbon nanotubes, ordered mesoporous carbon, carbonaceous materials, graphene oxide, metal oxides, layered double hydroxides, and layered metal sulphides, has been applied in pollution clean-up (Liu et al., 2021). Generally, COF-derived materials are categorized into four groups: triazine-derived COFs, imine-based COFs, boron-containing COFs, and other COFs (Raghav et al., 2025). Functionalized COFs need to be tailored for particular applications. The absorption features of COFs are influenced by their surface area, distribution of pore size, highly organized structures, hydrophilicity and functionality of COFs, etc. (Rasheed, 2022).

The utilization of absorption technology as a substitute for polluted water treatment has been widely explored. When photons interact with the material surface, they excite electrons (e-) from the valence band (VB) to the conduction band (CB) and leave holes in the VB, forming electron–hole pairs that are responsible for redox reactions. Electrons and holes are generally known as the current carriers of a material (Pi et al., 2018, Arora et al., 2022).

It attempts to address the difficult problem of insufficient contaminant removal during wastewater treatment. Because of their low molecular weight and hydrophobicity, organic contaminants are typically more resistant to various forms of water treatment. For the adsorption process, the pollutant molecules are attracted to the surfaces of the adsorbent materials through a diffusion process from the bulk of the solution to the active pores of the adsorbent (Siipola et al., 2020, Marangi et al., 2021). Typically, this mechanism occurs via intermolecular attractive forces, such as chemisorption (e.g., ionic interactions) and physisorption (e.g., π–π interactions and van der Waals) (Kumar et al., 2019, Lv et al., 2019). Adsorption is highlighted by the novel adsorbent characteristic of materials, like high specific Brunner–Emmett–Teller surface area, high porosity, thermal stability, moisture, good selectivity for particular contaminants, low cost and availability, easy handling and regenerable, etc. One of the desirable properties of ideal adsorbent materials is their physical state as powder, cake, or beads (Fu et al., 2019, Zhan et al., 2020, Deng et al., 2025). In addition, the adsorption time is quite slow, and there is a problem in regenerating the used adsorbent. Advances in materials science have resulted in the introduction of a new generation of adsorbents with unusually high surface areas and porosity (Kaur et al., 2015, Xin et al., 2017, Peres et al., 2018). Various adsorbent materials with targeted pollutants are given in Table 7.1.

The absorption of photons:

$$\text{Photocatalyst} + h\nu\,(\text{photon energy}) \rightarrow e^-\,(\text{conduction band}) + h^+\,(\text{Valence band})$$

TABLE 7.1
Adsorption of Several Radionuclides, Heavy Metal Ions, and Organic Pollutants by COF-Based Materials and Other Promising Materials

Materials	Target	Adsorption Mechanism	References
Fe_3O_4@SiO_2@COF-DMSA	Au (III)	H-bonding	Fan et al. (2025)
Graphitic-C_3N_4	Ciprofloxacin	Heterogeneous adsorption	Ali et al. (2025)
TAPB-BPTA-S-β-CD and TAPB-BPTA-DTT	Pd(II)	Monolayer chemisorption	Liu et al. (2025)
Graphitic-C_3N_4	Co(II) Ni(II)	Inner-sphere complexation	Liao et al. (2018)
COF-LZU8	Hg(II)	Electrostatic interaction, coordination of Hg and S atoms (R-S-R)	Ding et al. (2016)
TAPB-DMTTPA-COF	Hg(II)	Coordination of Hg and S atoms (R-S-R group)	Huang et al. (2017)
ACOF	Hg(II)	Interaction between Hg and C=O groups	Li et al. (2019)
ACOF	U(VI)	Interaction between Hg and C=O groups	Li et al. (2019)
COF-SH	Pb(II)	Electrostatic interaction, coordination of Pb and S atoms	Cao et al. (2020)

(Continued)

TABLE 7.1 (*Continued*)
Adsorption of Several Radionuclides, Heavy Metal Ions, and Organic Pollutants by COF-Based Materials and Other Promising Materials

Materials	Target	Adsorption Mechanism	References
N-enriched COF	Cd(II)	Electrostatic interaction, coordination of Cd and N atoms	Dinari et al. (2019)
COF-BTA-DHBZ	Cr(VI)	Electrostatic interaction, H-bonding	Cui et al. (2019)
TpPa-1 Fe_3O_4@TpPa-1	Cr(VI)	Electrostatic interaction, H-bonding	Zhong et al. (2020)
COF1 & 2	Cr(VI)	Electrostatic interaction, H-bonding	Zhu et al. (2020)
Magnetic graphene oxide (GO)	As(V) Cd(II)	Electrostatic interaction; H-bonding, coordination of Cd and O atoms	Huang et al. (2019)
Ni-Fe_2O_3 Fe_2O_3 Co-Fe_2O_3	Pb(II)	Coordination of Pb and O atoms (-OH groups)	Chen et al. (2019)
TPT-TAPB-COF	Rhodamine B	Homogeneous adsorption	Li et al. (2018)
Fe_3O_4@TpBD	Bisphenol A	π–π interaction; H-bonding	Li et al. (2017)
TPB-DMTP-COF	Sulfamerazine	Electrostatic attraction; C–H···π interaction	Zhuang et al. (2020)
Magnetic-TPB-DMTP-COF	Diclofenac	π-π interaction; C–H···π interaction	Zhuang et al. (2020)
TpBD-Me_2-COF	Okadaic acid	Heterogeneous adsorption	Salonen et al. (2017)
TPT-DMBD-COF	Methylene blue	π–π interaction; electrostatic attraction	Huo et al. (2019)

7.2.2 Photocatalysis

Photocatalysis is a multifaceted method utilized in physics, chemistry, and chemical sciences. The photocatalysis process has numerous benefits and uses. It is economical and user-friendly for energy production, mitigating pollution and decreasing CO_2 emissions. The photocatalytic reaction is initiated by the absorption of solar light on the surface of solid materials, resulting in the generation of electron–hole pairs. The photocatalytic reaction consists of the interaction of semiconductors by themselves or in mixtures with metallic, organic, or organometallic organizers through the mechanism of photoreaction (Zango et al., 2020). In a photocatalytic process, there are two kinds of reactions: oxidation and reduction. Both of them are useful for the degradation of pollution (Tahir et al., 2020).

Adsorption-based approaches have been preferred mainly due to their simple operating procedures as well as high efficiency (Li et al., 2019, Wang et al., 2020). Thus, they are widely used in the remediation of heavy metals (Xu et al., 2020, Xiao et al., 2021), pharmaceutical pollutants (Wen et al., 2018, Jin et al., 2022), dyes, and other toxic compounds (Huang et al., 2018).

The adsorption-based and membrane-based technologies employing conventional porous materials such as MOFs (Liu et al., 2018), zeolites (Xu et al., 2020), ACs (Li et al., 2019), cyclodextrins and their derivatives (Huang et al., 2019), POPs (Sun et al., 2020), polymeric carbon nitride (Neale et al., 2020), and COF (Schwarzenbach et al., 2006, Kar et al., 2020), have received increasingly attention due to the excellent potential for wastewater treatment in the industrial process. Although membrane and adsorption separation techniques are different, the physicochemical factors of porous materials, including adsorption and desorption behaviour, diffusion coefficients, and selectivity, are important factors in the design of materials for these technologies (Da et al., 2023).

Photocatalysis began with transporting contaminants to the photocatalyst surface, which resulted in the adsorption of contaminants on the surface. When visible or UV light with energy equivalent to or higher than the photocatalytic material's bandgap is applied, it forms an electron–hole pair, where the electrons (e^-) transition to the CB and the holes (h^+) remain in the valence band (VB). The adsorbed molecules on the surface of photocatalytic materials undergo reduction and oxidation as a result of photonic activation (e^- and h^+). The reaction products are then absorbed and removed from the photocatalyst's surface.

The following photocatalysts are typically employed in photocatalytic oxidation: TiO_2 (Akpan et al., 2009), ZnO, Fe_2O_3 (Lee et al., 2013), CdS (Zhao et al., 2020), g-C_3N_4 (Du et al., 2019), etc. The most researched photocatalyst for organic pollutant degradation so far is TiO_2, which has been applied in the photocatalytic degradation of dyes (Pi et al., 2021) (Jiang et al., 2018), antibiotics (Zhang et al., 2020) (Chen et al., 2020), and other substances (Li et al., 2021). The earliest reported example of photocatalysis dates back to 1911, when Ebner used ZnO as a photocatalyst to bleach Prussian blue dye under light irradiation (Baur et al., 1921). Most research focuses on ZnO (bandgap of 3.3 eV) and TiO_2 (bandgap of 3.2 eV). Although these nanomaterials do not cause secondary pollution and are non-toxic, their wide band gaps limit their usefulness for UV light, which accounts for only 4% of solar energy. Although CdS has a narrow band gap of 2.4 eV, it has poor stability. Meanwhile, g-C_3N_4, a non-metallic semiconductor made up mainly of abundant carbon and nitrogen, has a band gap of 2.7 eV and relatively good stability. The stability is relatively good, but the photocatalytic performance still fails to meet the requirements of practical applications (Hu et al., 2022).

Photocatalysis is accelerating a photoreaction in the presence of a light-absorbing catalyst that produces reactive species. Top of FormBottom of FormIn accordance iIn accordance with the catalyst phase, the advanced oxidation mechanism can be classified as heterogeneous and homogeneous photocatalysis.

7.2.2.1 Homogeneous Photocatalysis

In homogeneous photocatalysis, both the reactant and the photocatalyst reside in a similar state (i.e., both can be in a gaseous state) and contain Fenton's reagent (Fe^+ and Fe^+/H_2O_2) and ozone, a combination of hydrogen peroxide (H_2O_2) and Fe^{+2} salts, to generate highly reactive hydroxyl radicals (•OH) under UV lights. These reactive species, including radicals and excited states, subsequently trigger chemical reactions with organic molecules in solution. The main advantage of homogeneous photocatalysis is the direct and rapid interaction between catalysts and reactants within the same step.

$$Fe^{2+} + H_2O_2 \rightarrow Fe^{3+} + HO^{\cdot} + OH^{-}$$

$$Fe^{3+} + H_2O_2^{\circ} \rightarrow Fe^{2+} + HO^{\cdot} + OH^{-}$$

$$Fe^{2+} + HO^{\cdot} \rightarrow Fe^{+3} + OH^{-}$$

The photo-Fenton synthesis is utilized to enhance the photoreduction of Fe^{2+} to Fe^{3+}, which is facilitated by UV-visible irradiation. This process leads to additional development of •OH, which further boosts the photocatalytic activity.

$$Fe^{3+} + H_2O + hv \rightarrow Fe^{2+} + HO^{\cdot} + H^{+}$$

$$Fe^{3+} + H_2O_2 + hv \rightarrow Fe^{2+} + HO^{\cdot} + H^{+}$$

$$Fe^{2+} + H_2O_2 \rightarrow Fe^{3+} + HO^{\cdot} + H^{-}$$

The photo-Fenton synthesis method is utilized to remediate organic pollutants in wastewater. The photo-Fenton process reactions require enhanced concentrations of ferrous catalyst, which also helps optimize sludge production and catalyst concentration. The concentration of dissolved ions in polluted water significantly affects the photo-Fenton. The photo-Fenton is environmentally friendly because of the low toxicity of Fe and high abundance in the environment (Gaur et al., 2022).

7.2.2.2 Heterogeneous Photocatalysis

Heterogeneous photocatalytic processes involve numerous phases, with the photocatalyst, such as titanium dioxide (TiO_2), being driven by ultraviolet or visible radiation from the solar spectrum in the presence of O_2. TiO_2 has gained popularity as a photocatalyst because of its low cost, safety, and excellent photocatalytic efficacy. It is essential in advanced oxidation methods for water purification (Zhao et al., 2020). Heterogeneous photocatalysis is more affected than homogeneous photocatalysis in more cases due to easy separation, greater stability, and recyclability of photocatalysts.

When visible or UV light with energy equivalent to or higher than the photocatalytic material's bandgap is applied, it forms an electron–hole pair, where the electrons (e^-) transition to the CB and the holes (h^+) remain in the VB. The adsorbed molecules on the surface of photocatalytic materials undergo reduction and oxidation as a result of photonic activation (e^- and h^+). The reaction products are then absorbed and removed from the photocatalyst's surface.

TiO_2 stimulate e^- transition to the CB, and h^+ remains in the VB. The adsorbed molecules on the surface of photocatalytic materials undergo reduction and oxidation due to photonic activation (e^- and h^+). The holes (h^+) oxidize the surface and absorb

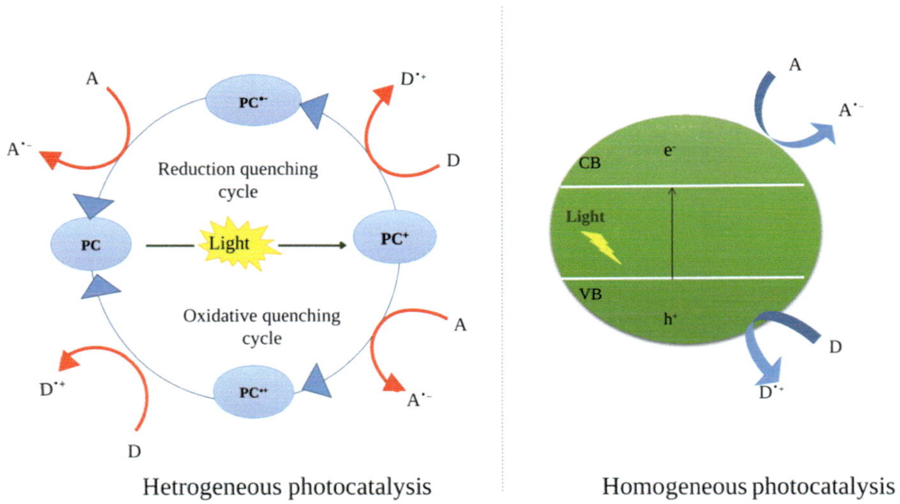

FIGURE 7.3 Different mechanisms of heterogeneous and homogeneous photocatalysis.

H_2O or OH^- to produce ·OH. The photoexcited electrons reduce O_2 and produce HO_2· (hydroperoxyl radicals) or $•O_2^-$ (superoxide radicals). These reactive O_2 species (ROS) and free electrons or holes react with impurities (e.g., organic and inorganic materials) absorbed on the surface and converted into eco-friendly products. Figure 7.3 depicts the mechanism of photocatalytic breakdown of organic contaminants using TiO_2 photocatalyst.

$$TiO_2 + h\nu \rightarrow e^-$$

$$O_2 + e^- \rightarrow {}^{\bullet}O^{2-}$$

$${}^{\bullet}O^{2-} + H_2O \rightarrow HO$$

$$H_2O/OH + h^+ \rightarrow {}^{\bullet}OH$$

$${}^{\bullet}O^{2-}/HO_2^{\bullet}/{}^{\bullet}OH + OP \rightarrow CO_2 + H_2O$$

TiO_2 is a naturally found mineral characterized by superior pigment characteristics, significant UV absorption, and remarkable durability, rendering it appropriate for photocatalytic applications. TiO_2 serves as a highly efficient catalyst for the photocatalytic degradation of various chemicals and hazardous chemicals, including dyes, herbicides, phenolic compounds, pesticides, and inorganic substances, while also facilitating the inactivation of microorganisms like yeasts, bacteria, and moulds in water and air. TiO_2 can be re-used on various solid surfaces, such as glass, stainless steel, and fibres (Tahir et al., 2020).

Heterogeneous photocatalysts have various benefits. Compared to homogeneous photocatalysis, the primary advantages of heterogeneous photocatalysis

TABLE 7.2
Various Features of Homogeneous Photocatalysts and Heterogeneous Photocatalysts

Features	Homogeneous Photocatalysts	Heterogeneous Photocatalysts
Catalyst phase	Same as reactants (liquid/gas)	Different from reactants (solid–liquid/gas)
Common catalysts	Fe^{3+}, H_2O_2, O_3	TiO_2, CdS, ZnO, g-C_3N_4
Reaction medium	Mostly aqueous or gaseous	Solid–liquid or solid–gas
Mechanism	Direct molecular interactions in solution	Absorption, formation of electron–hole pair
Stability	Can degrade or decompose over time	More stable and reusable
Efficiency	High quantum efficiency	Limited by charge recombination
Applications	Water treatment, oxidation process	Pollutant degradation, hydrogen evolution

involve chemical stability in liquid media, non-toxicity, cost-effectiveness, and non-selectivity of the catalytic reaction. Table 7.2 shows the different features of homogeneous photocatalysts and heterogeneous photocatalysts.

Photocatalysis is a promising green and sustainable technique that converts solar energy into chemical energy. A clear advantage is the ability of photocatalytic breakdown of organic pollutants to fully mineralize them into non-toxic inorganic chemicals, such as CO_2 and H_2O, in mild scenarios without producing secondary pollution. This technology saves much energy as it uses solar energy for the degradation process (Yang et al., 2022).

7.3 CURRENT CHALLENGES AND FUTURE PERSPECTIVES

Adsorption and photocatalytic degradation confront limited capacity, poor selectivity, low visible light utilization, and scaling problems. High material prices and challenges with catalyst recovery further limit viable usage. To increase efficiency, future advancements will focus on developing visible-light-responsive photocatalysts, 2D materials, and hybrid systems. Doping, heterojunction generation, and bio-based materials are all feasible, sustainable alternatives. Integrating solar energy with AI-driven optimization can raise efficiency even further, making these technologies more viable for large-scale wastewater treatment and environmental remediation.

7.4 CONCLUSION

The application of COF nanomaterials has broad prospects in absorption and photocatalysis degradation. COFs are emerging as versatile materials in the fight against environmental pollution, particularly through their absorption and photocatalytic degradation capabilities. COFs have a high specific surface area, tuneable pore structures, and great thermal and solvent stability, making them effective at capturing various pollutants, including heavy metal ions, endocrine-disrupting compounds, pharmaceutical toxic pollutants, and dye. Absorption and photocatalytic degradation of organic contaminants have developed as cost-effective and energy-efficient methods. Their ability to facilitate photocatalysis reactions under solar light results in the degradation of organic pollutants into eco-friendly compounds. COF materials give an efficient and sustainable material for removing pollutants by combining absorption and photocatalysis, making them clean water and air. COFs and COF-based materials will possess significant potential in managing environmental contamination.

REFERENCES

Akpan, Uduak G., and Bassim H. Hameed. "Parameters affecting the photocatalytic degradation of dyes using TiO_2-based photocatalysts: a review." *Journal of Hazardous Materials* 170, no. 2–3 (2009): 520–529.

Ali, Fatima D., Saad H. Ammar, Alaa I. Elaibi, Zaid H. Jabbar, and Mohammed D. Salman. "$CeFeO_3$/phosphorus-doped graphitic-C_3N_4 Z-scheme heterostructures for photocatalytic degradation of ciprofloxacin." *Case Studies in Chemical and Environmental Engineering* 11 (2025): 101156.

Al-Nuaim, Marwah A., Asawer A. Alwasiti, and Zainab Y. Shnain. "The photocatalytic process in the treatment of polluted water." *Chemical Papers* 77, no. 2 (2023): 677–701.

Arora, Isha, Harshita Chawla, Amrish Chandra, Suresh Sagadevan, and Seema Garg. "Advances in the strategies for enhancing the photocatalytic activity of TiO_2: conversion from UV-light active to visible-light active photocatalyst." *Inorganic Chemistry Communications* 143 (2022): 109700.

Asayesh-Ardakani, Elham, Maryam Rahmani, Akram Hosseinian, Seyed-Behnam Ghaffari, and Mohammad-Hossein Sarrafzadeh. "Improvement strategies on application of covalent organic frameworks in adsorption, photocatalytic, and membrane processes for organic pollution removal from water." *Coordination Chemistry Reviews* 518 (2024): 216087.

Baur, Emil, and Rebmann A. "Über versuchezurPhotolyse des Wassers." *Helvetica ChimicaActa* 4, no. 1 (1921): 256–262.

Bulgariu, Laura, Leticia Belén Escudero, Olugbenga Solomon Bello, Munawar Iqbal, Jan Nisar, Kayode Adesina Adegoke, Fadi Alakhras, Michael Kornaros, and IoannisAnastopoulos. "The utilization of leaf-based adsorbents for dyes removal: a review." *Journal of Molecular Liquids* 276 (2019): 728–747.

Cao, Ying, Xue Hu, Changqian Zhu, Shuangxi Zhou, Rui Li, Huilan Shi, Shiyu Miao, Mohammadtaghi Vakili, Wei Wang, and Delin Qi. "Sulfhydryl functionalized covalent organic framework as an efficient adsorbent for selective Pb (II) removal." *Colloids and Surfaces A: Physicochemical and Engineering Aspects* 600 (2020): 125004.

Chen, Tongfan, Bin Li, Wenbo Huang, Chunhui Lin, Guangshe Li, HaoRen, Yue Wu, Shuhui Chen, Wenxiang Zhang, and Heping Ma. "Highly crystalline ionic covalent organic framework membrane for nanofiltration and charge-controlled organic pollutants removal." *Separation and Purification Technology* 256 (2021): 117787.

Chen, Weiqiang, Zhanhui Lu, Binghuan Xiao, Pengcheng Gu, Wen Yao, Jinlu Xing, Abdullah M. Asiri, Khalid A. Alamry, Xiangke Wang, and Suhua Wang. "Enhanced removal of lead ions from aqueous solution by iron oxide nanomaterials with cobalt and nickel doping." *Journal of Cleaner Production* 211 (2019): 1250–1258.

Chen, Xiangyan, Xin Peng, Longbo Jiang, Xingzhong Yuan, Hanbo Yu, Hou Wang, Jin Zhang, and Qi Xia. "Recent advances in titanium metal–organic frameworks and their derived materials: features, fabrication, and photocatalytic applications." *Chemical Engineering Journal* 395 (2020): 125080.

Cui, Fu-Zhi, Rong-Ran Liang, Qiao-Yan Qi, Guo-Fang Jiang, and Xin Zhao. "Efficient removal of Cr (VI) from aqueous solutions by a dual-pore covalent organic framework." *Advanced Sustainable Systems* 3, no. 4 (2019): 1800150.

Da Silva, V. D., K. Zalewska, Z. Petrovski, C. D. Buarque, L. C. Branco, and P. M. Esteves. "Covalent organic frameworks as promising materials for the removal of metal and organic pollutants from water." *Materials Today Sustainability* 21 (2023): 100279.

Deng, Hao, Yanling Wu, Leyi Li, Peng Wang, Kui Fang, Jun Li, Derek Hao, Huayue Zhu, Qi Wang, and Qiang Li. "Synergistic mechanisms for efficient and safe antibiotic removal: effective adsorption and photocatalytic degradation using aerogels." *Separation and Purification Technology* 354 (2025): 129455.

Dinari, Mohammad, and Mohammad Hatami. "Novel N-riched crystalline covalent organic framework as a highly porous adsorbent for effective cadmium removal." *Journal of Environmental Chemical Engineering* 7, no. 1 (2019): 102907.

Ding, San-Yuan, Ming Dong, Ya-Wen Wang, Yan-Tao Chen, Huai-Zhen Wang, Cheng-Yong Su, and Wei Wang. "Thioether-based fluorescent covalent organic framework for selective detection and facile removal of mercury (II)." *Journal of the American Chemical Society* 138, no. 9 (2016): 3031–3037.

Du, Xuedong, Xiaohong Yi, Peng Wang, Jiguang Deng, and Chong-chen Wang. "Enhanced photocatalytic Cr (VI) reduction and diclofenac sodium degradation under simulated sunlight irradiation over MIL-100 (Fe)/g-C_3N_4 heterojunctions." *Chinese Journal of Catalysis* 40, no. 1 (2019): 70–79.

Fan, Yinghui, Wenqing Wang, Abdul Haleem, Jianming Pan, and Hao Li. "Construction of magnetic covalent organic frameworks based on the monomer-mediated in situ layer-by-layer growth strategy for efficient gold recovery." *Separation and Purification Technology* 353 (2025): 128582.

Fu, Likang, Shixing Wang, Guo Lin, Libo Zhang, Qiming Liu, Ju Fang, Chenhuinan Wei, and Gang Liu. "Post-functionalization of UiO-66-NH2 by 2, 5-Dimercapto-1, 3, 4-thiadiazole for the high efficient removal of Hg (II) in water." *Journal of Hazardous Materials* 368 (2019): 42–51.

Gaur, Nisha, Dhiraj Dutta, Ayushi Singh, Rama Dubey, and DevVratKamboj. "Recent advances in the elimination of persistent organic pollutants by photocatalysis." *Frontiers in Environmental Science* 10 (2022): 872514.

Gendy, Eman A., JeroshaIfthikar, Jawad Ali, Daniel T. Oyekunle, ZouhairElkhlifia, IrshadIbranShahib, Ahmed I. Khodair, and Zhuqi Chen. "Removal of heavy metals by covalent organic frameworks (COFs): a review on its mechanism and adsorption properties." *Journal of Environmental Chemical Engineering* 9, no. 4 (2021): 105687.

He, Yasan, and Xiaogeng Lin. "Fabricating compact covalent organic framework membranes with superior performance in dye separation." *Journal of Membrane Science* 637 (2021): 119667.

Hu, Si-Yu, Yue-Na Sun, Zhi-Wen Feng, Fei-Ou Wang, and Yun-kai Lv. "Design and construction strategies to improve covalent organic frameworks photocatalyst's performance for degradation of organic pollutants." *Chemosphere* 286 (2022): 131646.

Huang, Dan, Jizi Wu, Lu Wang, Xingmei Liu, Jun Meng, Xianjin Tang, Caixian Tang, and Jianming Xu. "Novel insight into adsorption and co-adsorption of heavy metal ions and an organic pollutant by magnetic graphene nanomaterials in water." *Chemical Engineering Journal* 358 (2019): 1399–1409.

Huang, Lijin, Qin Shuai, and Shenghong Hu. "Tannin-based magnetic porous organic polymers as robust scavengers for methylene blue and lead ions." *Journal of Cleaner Production* 215 (2019): 280–289.

Huang, Ning, Lipeng Zhai, Hong Xu, and Donglin Jiang. "Stable covalent organic frameworks for exceptional mercury removal from aqueous solutions." *Journal of the American Chemical Society* 139, no. 6 (2017): 2428–2434.

Huang, Yan, Xiaofei Zeng, LinglingGuo, JianhuiLan, Liangliang Zhang, and Dapeng Cao. "Heavy metal ion removal of wastewater by zeolite-imidazolate frameworks." *Separation and Purification Technology* 194 (2018): 462–469.

Huo, J., Luo, B., & Chen, Y. "Crystalline Covalent Organic Frameworks from Triazine Nodes as Porous Adsorbents for Dye Pollutants." *ACS Omega*, 4, no. 27 (2019): 22504–22513. https://doi.org/10.1021/acsomega.9b03176

Jiang, Danni, PiaoXu, Han Wang, Guangming Zeng, Danlian Huang, Ming Chen, Cui Lai, Chen Zhang, Jia Wan, and WenjingXue. "Strategies to improve metal organic frameworks photocatalyst's performance for degradation of organic pollutants." *Coordination Chemistry Reviews* 376 (2018): 449–466.

Jin, Eunji, Soochan Lee, Eunyoung Kang, Yeongjin Kim, and WonyoungChoe. "Metal-organic frameworks as advanced adsorbents for pharmaceutical and personal care products." *Coordination Chemistry Reviews* 425 (2020): 213526.

Jin, Wei-Liang, Wei Li, Hai-Xian Wang, Xiao-Wei Liu, Hong-Xin Jiang, Li-Na Zhu, and De-Ming Kong. "Sponge-supported monolithic materials of porphyrin covalent organic frameworks for selective recognition, convenient removal and extraction of Cd^{2+}." *Journal of Environmental Chemical Engineering* 10, no. 3 (2022): 107662.

Kar, Supratik, Hans Sanderson, Kunal Roy, Emilio Benfenati, and Jerzy Leszczynski. "Ecotoxicological assessment of pharmaceuticals and personal care products using predictive toxicology approaches." *Green Chemistry* 22, no. 5 (2020): 1458–1516.

Kaur, Sumanjit, Seema Rani, R. K. Mahajan, M. Asif, and Vinod Kumar Gupta. "Synthesis and adsorption properties of mesoporous material for the removal of dye safranin: kinetics, equilibrium, and thermodynamics." *Journal of Industrial and Engineering Chemistry* 22 (2015): 19–27.

Kumar, Vanish, Suresh Kumar, Ki-Hyun Kim, Daniel CW Tsang, and Sang-Soo Lee. "Metal organic frameworks as potent treatment media for odorants and volatiles in air." *Environmental Research* 168 (2019): 336–356.

Lee, Seul-Yi, and Soo-Jin Park. "TiO2 photocatalyst for water treatment applications." *Journal of Industrial and Engineering Chemistry* 19, no. 6 (2013): 1761–1769.

Li, Y., Chen, W., Hao, W., Li, Y., & Chen, L. "Covalent organic frameworks constructed from flexible building blocks with high adsorption capacity for pollutants". *ACS Applied Nano Materials*, 1, no. 9 (2018): 4756–4761. https://doi.org/10.1021/acsanm.8b00983

Li, Y., Yang, C.-X., & Yan, X.-P. "Controllable preparation of core–shell magnetic covalent-organic framework nanospheres for efficient adsorption and removal of bisphenols in aqueous solution". *Chemical Communications*, 53, no. 16 (2017): 2511–2514. https://doi.org/10.1039/c6cc10188g

Li, Qing, Yuanyuan Zhu, Tao Pan, Guangxun Zhang, and Huan Pang. "Covalent organic framework nanomaterials: syntheses, architectures, and applications." *Advances in Colloid and Interface Science* 339 (2025): 103427.

Li, Ruidong, Jie Li, Qianwen Liu, Tingxi Li, Di Lan, and Yong Ma. "Recent progress on covalent organic frameworks and their composites as electrode materials for supercapacitors." *Advanced Composites and Hybrid Materials* 8, no. 1 (2025): 86.

Li, Shufeng, Yinhua Yang, Houchao Shan, Jing Zhao, Ze Wang, Di Cai, Peiyong Qin, Jan Baeyens, and Tianwei Tan. "Ultrafast and ultrahigh adsorption of furfural from aqueous solution via covalent organic framework-300." *Separation and Purification Technology 220* (2019): 283–292.

Li, Sumei, Saisai Shan, Sha Chen, Hanbing Li, Ziyi Li, Yixuan Liang, JiayingFei, LinhuaXie, and Jianrong Li. "Photocatalytic degradation of hazardous organic pollutants in water by Fe-MOFs and their composites: a review." *Journal of Environmental Chemical Engineering* 9, no. 5 (2021): 105967.

Li, Xing, Yue Qi, Guozong Yue, Qianxun Wu, Yang Li, Meicheng Zhang, Xinghua Guo, Xiaofeng Li, Lijian Ma, and Shoujian Li. "Solvent-and catalyst-free synthesis of an azine-linked covalent organic framework and the induced tautomerization in the adsorption of U (VI) and Hg (II)." *Green Chemistry* 21, no. 3 (2019): 649–657.

Liao, Qing, Wang Pan, Dongsheng Zou, Runpu Shen, Guodong Sheng, Xue Li, Yuling Zhu et al. "Using of g-C_3N_4 nanosheets for the highly efficient scavenging of heavy metals at environmental relevant concentrations." *Journal of Molecular Liquids* 261 (2018): 32–40.

Liu, Fenglei, Zhihao Li, Fang Xia, and Baowei Hu. "Efficient recovery of Pd (II) from nuclear wastewater using a functionalized covalent organic frameworks with uniform spherical structure: size control, performance and mechanism insight." *Separation and Purification Technology* 359 (2025): 130357.

Liu, Jing-Min, Xing-Zhi Wang, Chao-Yue Zhao, Jia-Li Hao, Guo-Zhen Fang, and Shuo Wang. "Fabrication of porous covalent organic frameworks as selective and advanced adsorbents for the on-line preconcentration of trace elements against the complex sample matrix." *Journal of Hazardous Materials* 344 (2018): 220–229.

Liu, Xiaolu, Hongwei Pang, Xuewei Liu, Qian Li, Ning Zhang, Liang Mao, MuqingQiu, Baowei Hu, Hui Yang, and Xiangke Wang. "Orderly porous covalent organic frameworks-based materials: superior adsorbents for pollutants removal from aqueous solutions." *The Innovation* 2, no. 1 (2021): 100076.

Lv, Shi-Wen, Jing-Min Liu, Hui Ma, Zhi-Hao Wang, Chun-Yang Li, Ning Zhao, and Shuo Wang. "Simultaneous adsorption of methyl orange and methylene blue from aqueous solution using amino functionalized Zr-based MOFs." *Microporous and Mesoporous Materials* 282 (2019): 179–187.

Ma, Zixuan, Lin Fang, Lijie Liu, Baowei Hu, Suhua Wang, Shujun Yu, and Xiangke Wang. "Efficient decontamination of organic pollutants from wastewater by covalent organic framework-based materials." *Science of the Total Environment* 901 (2023): 166453.

Marangi, Fabio, Matteo Lombardo, Andrea Villa, and Francesco Scotognella. "New strategies for solar cells beyond the visible spectral range." *Optical Materials*: X 11 (2021): 100083.

Neale, Peta A., Cedric Feliers, Lisa Glauch, Maria König, Caroline Lecarpentier, Rita Schlichting, Sylvie Thibert, and Beate I. Escher. "Application of in vitro bioassays for water quality monitoring in three drinking water treatment plants using different treatment processes including biological treatment, nanofiltration and ozonation coupled with disinfection." *Environmental Science: Water Research & Technology* 6, no. 9 (2020): 2444–2453.

Peres, Enrique C., Jenifer C. Slaviero, Anaelise M. Cunha, Ahmad Hosseini–Bandegharaei, and Guilherme L. Dotto. "Microwave synthesis of silica nanoparticles and its application for methylene blue adsorption." *Journal of Environmental Chemical Engineering* 6, no. 1 (2018): 649–659.

Pi, Yunhong, Xiyi Li, Qibin Xia, Junliang Wu, Yingwei Li, Jing Xiao, and Zhong Li. "Adsorptive and photocatalytic removal of Persistent Organic Pollutants (POPs) in water by metal-organic frameworks (MOFs)." *Chemical Engineering Journal* 337 (2018): 351–371.

Qian, Yuhang, and Dongge Ma. "Covalent organic frameworks: new materials platform for photocatalytic degradation of aqueous pollutants." *Materials* 14, no. 19 (2021): 5600.

Raghav, Sapna, Jyoti Raghav, Ma'aruf Abdulmumin Muhammad, Madhur Babu Singh, Pallavi Jain, Dinesh Kumar. "Recent advancements in synthesis and applications of covalent organic frameworks." *Applied Materials Today* 44 (2025): 102694.

Rasheed, Tahir. "Covalent organic frameworks as promising adsorbent paradigm for environmental pollutants from aqueous matrices: perspective and challenges." *Science of the Total Environment* 833 (2022): 155279.

Reyes, Nash Jett DG, Franz Kevin F. Geronimo, Kimberly Ann V. Yano, Heidi B. Guerra, and Lee-Hyung Kim. "Pharmaceutical and personal care products in different matrices: occurrence, pathways, and treatment processes." *Water* 13, no. 9 (2021): 1159.

Ritchie, Lyndsey K., Abbie Trewin, Aida Reguera-Galan, Tom Hasell, and Andrew I. Cooper. "Synthesis of COF-5 using microwave irradiation and conventional solvothermal routes." *Microporous and Mesoporous Materials* 132, no. 1–2 (2010): 132–136.

Saleh, Iman A., Nabil Zouari, and Mohammad A. Al-Ghouti. "Removal of pesticides from water and wastewater: chemical, physical and biological treatment approaches." *Environmental Technology & Innovation* 19 (2020): 101026.

Schwarzenbach, René P., Beate I. Escher, Kathrin Fenner, Thomas B. Hofstetter, C. Annette Johnson, Urs Von Gunten, and Bernhard Wehrli. "The challenge of micropollutants in aquatic systems." *Science* 313, no. 5790 (2006): 1072–1077.

Salonen, L. M., Pinela, S. R., Fernandes, S. P. S., Louçano, J., Carbó-Argibay, E., Sarriá, M. P., Rodríguez-Abreu, C., Peixoto, J., & Espiña, B. Adsorption of marine phycotoxin okadaic acid on a covalent organic framework. *Journal of Chromatography A*, 1525 (2017): 17–22. https://doi.org/10.1016/j.chroma.2017.10.017

Siipola, Virpi, Stephan Pflugmacher, Henrik Romar, Laura Wendling, and PerttiKoukkari. "Low-cost biochar adsorbents for water purification including microplastics removal." *Applied Sciences* 10, no. 3 (2020): 788.

Su, Yi-Yi, Xi Yan, Yan Chen, Xiao-Jing Guo, Xing-Fan Chen, and Wan-Zhong Lang. "Facile fabrication of COF-LZU1/PES composite membrane via interfacial polymerization on microfiltration substrate for dye/salt separation." *Journal of Membrane Science* 618 (2021): 118706.

Sun, Qi, Briana Aguila, Yanpei Song, and Shengqian Ma. "Tailored porous organic polymers for task-specific water purification." *Accounts of Chemical Research* 53, no. 4 (2020): 812–821.

Swaminathan, Pavithra, Saravanan A., Yaashikaa. P. R., and Vickram A. S. . "Recent advances in photocatalytic degradation of persistent organic pollutants: mechanisms, challenges, and modification strategies." *Sustainable Chemistry for the Environment* (2024): 100171.

Tahir, Muhammad Bilal, Tahir Iqbal, Muhammad Rafique, Muhammad ShahidRafique, Tasmia Nawaz, and M. Sagir. "Nanomaterials for photocatalysis." In Muhammad Bilal Tahir, Muhammad Rafique, and Muhammad Shahid Rafique (eds) *Nanotechnology and Photocatalysis for Environmental Applications*, pp. 65–76. Elsevier, 2020. https://doi.org/10.1016/B978-0-12-821192-2.00005-X

Wang, Min, ManjieGao, Linlin Deng, Xun Kang, Kailian Zhang, Qifeng Fu, Zhining Xia, and Die Gao. "A sensitive and selective fluorescent sensor for 2, 4, 6-trinitrophenol detection based on the composite material of magnetic covalent organic frameworks, molecularly imprinted polymers and carbon dots." *Microchemical Journal* 154 (2020): 104590.

Wang, Shizhao, XiaoyiXu, Yan Yue, Kunshan Yu, QingjunShui, Ning Huang, and Hongzheng Chen. "Semiconductive covalent organic frameworks: structural design, synthesis, and application." *Small Structures* 1, no. 2 (2020): 2000021.

Wen, Rui, Yang Li, Meicheng Zhang, XinghuaGuo, Xing Li, Xiaofeng Li, Jun Han et al. "Graphene-synergized 2D covalent organic framework for adsorption: a mutual promotion strategy to achieve stabilization and functionalization simultaneously." *Journal of Hazardous Materials* 358 (2018): 273–285.

Xiao, Yuling, Chao Ma, Ziyu Jin, Junling Wang, Lingxin He, Xiaowei Mu, Lei Song, and Yuan Hu. "Functional covalent organic framework for exceptional Fe^{2+}, Co^{2+} and Ni^{2+} removal: an upcycling strategy to achieve water decontamination and reutilization as smoke suppressant and flame retardant simultaneously." *Chemical Engineering Journal* 421 (2021): 127837.

Xin, Shengchang, Na Yang, FeiGao, Jing Zhao, Liang Li, and Chao Teng. "Three-dimensional polypyrrole-derived carbon nanotube framework for dye adsorption and electrochemical supercapacitor." *Applied Surface Science* 414 (2017): 218–223.

Xu, Shi-Xian, Zhao-Quan Yao, and Ying-Hui Zhang. "A covalent organic framework exhibiting amphiphilic selective adsorption toward ionic organic dyes tuned by pH value." *European Polymer Journal* 133 (2020): 109764.

Yang, Zheng, Heng-ShenXie, Wei-Yuan Lin, Yi-Wu Chen, DaoguangTeng, and Xing-Shun Cong. "Enhanced adsorption–photocatalytic degradation of organic pollutants via a ZIF-67-derived Co–N codoped carbon matrix catalyst." *ACS Omega* 7, no. 45 (2022): 40882–40891.

Yin, Congcong, Siyu Fang, Xiansong Shi, Zhe Zhang, and Yong Wang. "Pressure-modulated synthesis of self-repairing covalent organic frameworks (COFs) for high-flux nanofiltration." *Journal of Membrane Science* 618 (2021): 118727.

Zango, ZakariyyaUba, KhairulazharJumbri, NonniSorayaSambudi, Anita Ramli, Noor Hana Hanif Abu Bakar, BahruddinSaad, Muhammad Nur'HafizRozaini et al. "A critical review on metal-organic frameworks and their composites as advanced materials for adsorption and photocatalytic degradation of emerging organic pollutants from wastewater." *Polymers* 12, no. 11 (2020): 2648.

Zhan, Xue-Qing, Ya Zhang, Lei Xie, Huan-Li Liu, Xin-Ke Zhang, Bo Ruan, Hui Ding et al. "Magnetically treated Zr-based UiO-type porous coordination polymers study on adsorption of azo dye." *Microporous and Mesoporous Materials* 306 (2020): 110291.

Zhang, Xi, Jing Wang, Xing-Xing Dong, and Yun-Kai Lv. "Functionalized metal-organic frameworks for photocatalytic degradation of organic pollutants in environment." *Chemosphere* 242 (2020): 125144.

Zhang, Yawen, Hanchen Ye, Dongyun Chen, Najun Li, QingfengXu, Hua Li, Jinghui He, and Jianmei Lu. "In situ assembly of a covalent organic framework composite membrane for dye separation." *Journal of Membrane Science* 628 (2021): 119216.

Zhao, Feiping, Yongpeng Liu, Samia Ben Hammouda, BhairaviDoshi, NéstorGuijarro, Xiaobo Min, Chong-Jian Tang, Mika Sillanpää, Kevin Sivula, and Shaobin Wang. "MIL-101 (Fe)/g-C_3N_4 for enhanced visible-light-driven photocatalysis toward simultaneous reduction of Cr (VI) and oxidation of bisphenolA in aqueous media." *Applied Catalysis B: Environmental* 272 (2020): 119033.

Zhong, Xin, Zhipeng Lu, Wen Liang, and Baowei Hu. "The magnetic covalent organic framework as a platform for high-performance extraction of Cr (VI) and bisphenol a from aqueous solution." *Journal of Hazardous Materials* 393 (2020): 122353.

Zhu, Donghai, Shuangxi Zhou, Ziming Zhou, Rui Li, Jia Ye, Xu Ziyu, Shenjie Lan, Yunhai Zhang, Shiyu Miao, and Wei Wang. "Highly efficient and selective removal of Cr (VI) by covalent organic frameworks: structure, performance and mechanism." *Colloids and Surfaces A: Physicochemical and Engineering Aspects* 600 (2020): 124910.

Zhuang, Shuting, Yong Liu, and Jianlong Wang. "Covalent organic frameworks as efficient adsorbent for sulfamerazine removal from aqueous solution." *Journal of Hazardous Materials* 383 (2020): 121126.

8 Environmental Applications of Two- and Three-Dimensional Covalent Organic Frameworks

Satyesh Raj Anand

8.1 INTRODUCTION

In a few years, covalent organic frameworks (COFs) have appeared as a hopeful class of materials (Feng, Ding, & Jiang, 2012). Environmental applications (J. Wang & Zhuang, 2019) such as several processes using water purification, biomedical science, adsorption, gas phase separation, electrochemical (X. Zhao, Pachfule, & Thomas, 2021), energy conservations more prospective (B. Yu et al., 2024), fuel cell photocatalytic reduction (Y. Chen & Jiang, 2024) of pollutant heavy metal, removal of dyes (Raghav et al., 2025; Zhu et al., 2017), removal of organic and inorganic contamination including water purification, air filtration, and soil remediation developed as an advance water treatment application (Z. Ma et al., 2023; J. Wang & Zhuang, 2019). Due to COFs having high porosity, crystalline materials are self-assured of organic building blocks that are covalently associated with forming an intervallic assembly (Souto, Strutyński, Melle-Franco, & Rocha, 2020; Xu, Ding, An, Wu, & Wang, 2016). The exclusive properties of COFs include their high surface areas, tunable pore sizes, easy functionalization and chemical stability, membrane fabrication, regeneration, reusability, high magnetic separation, and various characteristic physicochemical properties ideal for various types of possible environmental applications. In the specialization of COF, growths in COFs for ecological applications are rapidly expanding, and advancements have been constructed in their applications in recent years (J. Wang & Zhuang, 2019). This chapter briefly familiarizes the environmental applications of two-dimensional (2D) and three-dimensional (3D) COFs, as well as their chemical and electrical properties and various removal methods for water contamination (J. Wang & Zhuang, 2019). COFs are vital to environmental challenges and the requirement for sustainable approaches and solutions. COFs similarly offer numerous advantages compared to traditional materials used in environmental applications. Such types of, e.g., COFs, regularly exist in better sustainable approaches and are environmentally more sustainable than traditional materials, for example, zeolites and activated carbon

DOI: 10.1201/9781003605386-8

(Wang & Zhuang, 2019). They can also be premeditated and synthesized to have precise properties tailored to address specific environmental challenges. The world faces numerous conservational challenges, from climate change to ecological glitches of air pollution, water pollution, and soil inorganic and organic contaminations. These challenges impact human health, economic progress, and environmental sustainability. Water pollution is a significant environmental challenge for millions worldwide, with deficient access to harmless drinking water. Water contamination can seriously influence human well-being, with the feast of water-borne diseases and ingesting toxic chemicals. Air effluence is a major conservational challenge, with millions of people facing air pollution seriously influencing human well-being, with respiratory difficulties, cardiovascular disease, and cancer. Soil contamination is an alternative substantial environmental challenge due to erosion, salinization, and pollution. Soil contamination can seriously affect ecosystem sustainability, with condensed biodiversity, reduced agricultural productivity, heavy metals, pesticides, industrial chemicals, and enlargement of greenhouse gas (GHG) emissions. COFs are a capable new class of materials for environmental applications (Figure 8.1).

They are outstanding for their unique characteristic properties, such as high surface zones, tuning pore sizes, and high chemical stability. COFs are self-possessed of organic building blocks covalently associated to form a periodic structure. The synthesis of COFs characteristically includes the condensation reaction of organic building blocks, such as boronic acids, aldehydes, and amines. The resultant COF materials would have high surface zones, oscillating from 100 to 1,000 $m^2 g^{-1}$. Tuning pore sizes are oscillating from 1 to 10 nm. This chapter delivers a comprehensive

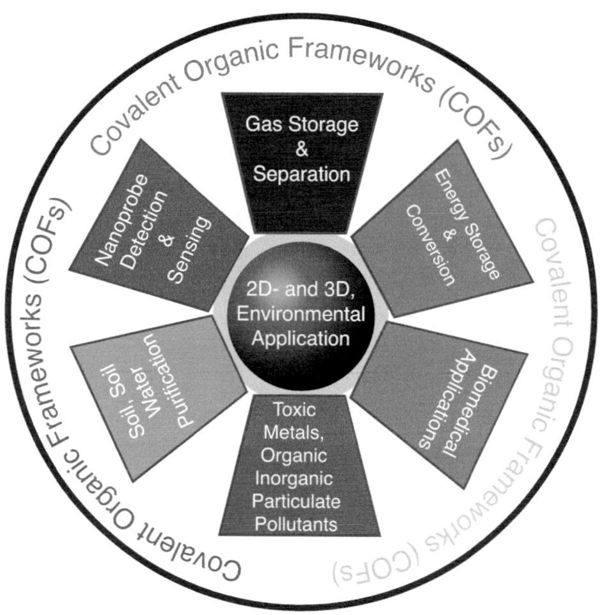

FIGURE 8.1 Experimental applications of covalent organic framework.

impression of the environmental applications of 2D and 3D COFs, including their scalability, stability, and cost-effectiveness. This chapter aims to stimulate a new research era and innovation in environmental science and technology.

8.2 STRUCTURE AND PHYSICOCHEMICAL PROPERTIES OF COFs

COFs are crystalline, porous materials, self-meticulous, completely bright elements. There are several elements present here, such as carbon (C), nitrogen (N), and oxygen (O). They are built over the covalent bonding of organic structure blocks, following the robust and highly stable framework building block structure. Their highly ordered, crystalline structure categorizes the COFs, which are self-possessed by repeating units of covalently bonded organic molecules. The structure of COFs should be chemically synthesized by choosing specific materials for the building blocks and various reaction conditions, permitting the construction of materials through precise high pore sizes, various shapes and sizes, characteristic functionalities, and physicochemical properties that provide these materials. COFs exhibit different physicochemical properties that make them striking for several applications. Most key factors show physicochemical properties and types of COFs material as shown in Figure 8.2.

8.2.1 High Surface Area

COFs occur in high surface areas, frequently beyond 1,000 m^2g^{-1}, which permits them to cooperate with and adsorb an extensive range of molecules.

8.2.2 Tunable Pore Size

The pore size of COFs should be precisely controlled over the assortment of building blocks and depend on chemical reaction conditions, permitting the construction of materials through specific pore sizes and shapes.

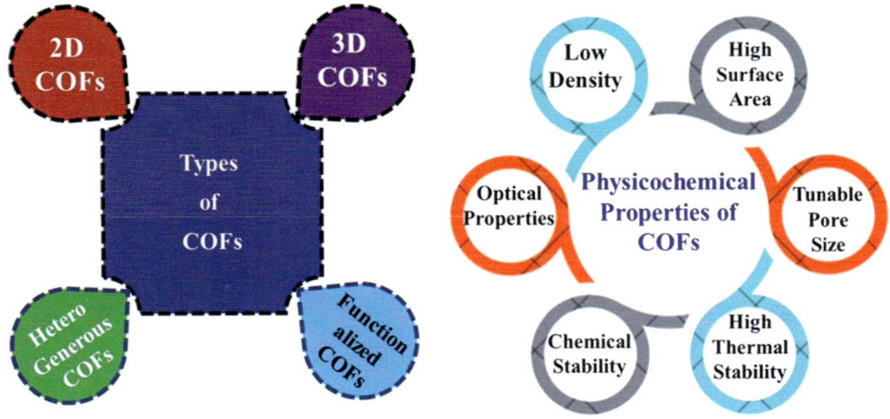

FIGURE 8.2 Schematic representation of physicochemical properties of COFs.

8.2.3 High Thermal Stability

COFs are highly stable thermally through several materials outstanding integral up to temperatures of 500°C or more.

8.2.4 Chemical Stability

COFs are correspondingly highly stable chemicals, resisting degradation in the presence of acids, bases, and other corrosive substances.

8.2.5 Low Density

COFs are typically low-density materials, which makes them attractive for applications where weight is a concern.

8.2.6 Optical Properties

COFs can show different optical properties, including phosphorescence, fluorescence, phosphorescence stability, and photoconductivity.

8.2.7 Electrical Properties of COFs

COFs exhibit unique electrical properties due to their ordered structure and tunable composition. They should be developed to be semiconducting, conducting, or protecting, and constructing appropriately for various applications. Such as, COF-5, a 2D COF comprised of boronate ester-linked phenyl rings, delivers a high electrical conductivity of 10^{-2} S cm^{-1}, making it a promising material for organic electronics and energy storage devices. The electrical effects of COF materials would alternate by changing their chemical structure compositional effects and encompass surface functionalization.

8.3 CLASSIFICATION OF COFs

The COF components would be categorized based on their chemical compositional behaviors, elementary composition, chemicals, and structural modification in the research fields, and they would obtain high physical and chemical characteristic properties. Some essential categories of COFs, like the subsequent parts, exist in the COFs developed with a single coating of covalently connected molecules.

8.3.1 2D and 3D Material of the COFs, Additionally More Advantages and Environmental Applications

The unique physicochemical characteristic properties of 2D and 3D COFs help with several environmental applications like water purification and removing different toxic dyes, heavy metals, volatile organic compounds (VOCs), and pesticides. The advantage based on the principle of COFs is that they are highly toxic to less toxic in nature and have material characteristic properties like surface area and better porosity materials. These properties enable them to absorb, adsorb, and remove waste contaminants effectively in environmental applications using COF s materials. Also, the tunability of the porous material and the shape of COFs permit highly selective adsorption of precise pollutants, making them extremely well-organized for environmental remediation. The COF materials are more chemically stable and have high electrical resistance to the contaminated component degradation. They effectively work as a COF material and are suitable composite materials for utilizing the effects of disciplinary environmental backgrounds. Additionally, the inconsequential and elastic nature of COFs empowers them to be effortlessly integrated into various environmental systems, such as water treatment membranes and air purification strainers. The 2D and 3D COF materials, in individual, have shown visible structural belongings to chemical composition effects toward excessive potential significance for environmental applications of materials due to their high porosity, stability, high surface passivation, flexibility compounds, and prosperity of surface functionalization. 3D COFs have established outstanding stability and mechanical strength, making them appropriate for high-temperature and high-pressure environmental applications. In general, the exclusive grouping of properties in 2D and 3D COFs is highly beneficial for environmental applications, with water purification, soil remediation, and air filtration. Their potential to efficiently remove pollutants and contaminants from the environment while also existing sustainable and ecofriendly variety makes them attractive for addressing numerous environmental challenges.

8.3.2 Materials of the 2D COF Materials

Advanced 2D COF polymeric materials are more challenging and offer vital opportunities for a class of materials self-possessed or covalently associated with organic building polymer blocks chain (Shuck, Xiao, & Wang, 2024). They exhibit chemical and physical morphological properties for tunable pore sizes, high surface areas, and outstanding chemical stability. Examples of 2D COFs include COFs, COF-1, a 2D COF self-possessed of phenylene and boronate ester associations, which is used for gas storage and separation. Another example is COF-5, a 2D COF self-possessed of phenylene and imine associations, which is used for environmental assessments for water purification (Kandambeth, Kale, Shekhah, Alshareef, & Eddaoudi, 2022) and photocatalysis and catalysis (Khalil, Das, & Thomas, 2024).

8.3.3 Materials of the 3D COFs

3D COFs are a class of materials self-possessed or covalently associated with organic building blocks that are arranged like connected units, such as 4, 6, and 8, associated with building blocks or nodes or chains. These are COFs collected from multiple layers of covalently bonded molecules like tetrahedral, triangular, prismatic, and cubic symmetry. A 3D network or polymer chain is connected with covalent bonds. They show quick structure tunable pore sizes, rigid structure high surface areas, and admirable chemical, physical, and mechanical stability. Examples of 3D COFs, such as COF-300, a 3D COF composed of tetrahedral and octahedral building blocks, which

TABLE 8.1
Comparative Studies of the 2D and 3D in the COFs Material Using Various Applications

S. No.	Material	2D & 3D COFs	Synthetic Methods	Applications	Ref.
1	Tp, Bpy, 1,4-dioxane, mesitylene, acetic acid, THF	COF	Heated methods	Hg^{2+} sensing	Cui et al. (2019)
2	Zn2(Bim)3, nanosheets, methanol, n-propanol, asymmetric D-Al_2O_3	2D COF	Ultrasonic treatment	Gas separation	Yuan Peng, Li, Ban, & Yang (2017)
3	BDBA, mesitylene, dioxane, COF-1, BDBA, HHTP, COF-5	2D COF	Thermal heating	Gas adsorption	Côté et al. (2005)
4	Acetic acid, 1,3,5 trimethylbenzene, dioxane, 1,3,5-triformylphloroglucinol, paraphenylenediamine	2D–2D COF	Hydrothermal method	Photocatalytic degradation	Khaing et al. (2020)
5	5,10,15,20-tetrakis (4-aminophenyl) porphyrin, acetic anhydride, n-Valeric acid	2D COF–3D	Refluxed	H_2/CO_2 gas permeabilities	Das, Ben, Qiu, & Valtchev (2020)
6	DTBT, TAPB, 1,3,5-trimethylbenzene, benzyl alcohol, acetic acid	3D COF	Soxhlet extractor	Photocyclization	T.-Y. Yu et al. (2023)

is used for sensing, electrochemistry, gas storage, and catalysis. COF-309, a 3D COF collected from diamond-shaped building blocks, is used in vital applications for adsorption, water purification, separation, photocatalysis, catalysis, optoelectronic, energy storage devices, chemical sensing, and drug delivery. Table 8.1 shows comparative studies of materials like 2D and 3D in the COFs using various applications.

8.3.4 Heterogeneous Catalysis COF Materials

Heterogeneous catalysis COFs are a rapidly unindustrialized new class of materials that involve a transformed organic building polymer blocks chain to yield the product of a framework with several physicochemical characteristic properties (J. Guo & Jiang, 2020). These COFs' organic and inorganic materials determine exclusive single characteristic properties, such as structural tunability, penetrability, pore size, synthetic versatility, ultralow density, crystalline properties, high surface areas, and multifunctionality and diversity. Heterogeneous COF material chemistry would be designed to assimilate diverse functional groups and interlinkage chemistry growth mechanisms necessary for enhanced high performance in various advantageous photocatalytic applications for organic transformation (H. Chen, Jena, Feng, Leus, & Van Der Voort, 2022). These are COFs self-controlled by significant types of building block chain polymers (Kubo, Nishiyabu, & James, 2015). The heterogeneous catalytic applications of COF-derived materials are driven by their active backbones, nanocomposite nature, heterostructures, and edge-active sites. By integrating the high surface area of COF-1 and COF-2 with the chemical stability of COF-derived frameworks, these materials exhibit enhanced performance in catalytic reactions and gas storage (Levy). Similarly, the heterogeneous COF-300/COF-309 composite combines the electrical and mechanical strength of COF-300 with the chemical and biosensing capabilities of COF-309, thereby demonstrating strong potential for applications in chemical sensing, biosensing, and environmental monitoring (Levy) (Rejali, Dinari, & Wang, 2023; F. Zhao, Liu, Mathe, Dong, & Zhang, 2017).

8.3.5 Functionalized COFs

Functionalized COFs are a class of composite materials designed with specific functional groups to increase the number of functional groups and enhance their physical and chemical properties of the COFs and high performance like selectivity and low sensitivity. These functional groups can be familiarized through the synthesis of COFs or through the advanced post-synthetic process of modification. Functionalized COFs show an enhanced quality of the characteristics, for example, increased chemical stability, tunable porosity, pore size, and more significant catalytic activity. Examples of functionalized COFs include azine and amine-functionalized COF-1, which show better catalytic activity for the decrease in nitrobenzene. Amine and sulfur group functionalized COF-5, which illustrates a higher quality of storage and CO_2 capture capabilities. Hydroxyl-functionalization of COF-300 establishes high performance in the catalytic oxidation of alcohols. Sulfonic acid-functionalization COF-309 exhibits high protonic conductivity and potential applications in fuel cells. These functionalized COFs have prolonged the scope of COF-driven materials and unlocked new paths for their application in several fields. These COFs have been adapted with specific functional groups to improve their properties and performance.

8.4 APPLICATIONS OF COFs

These are classified in the field of COFs and have a broad range of probable applications as follows:

8.4.1 Gas Storage and Separation

COFs should be utilized to store and distinguish different phases of gases, such as hydrogen (H_2) (Alabdulhadi et al., 2025), methane (CH_4), and carbon dioxide (CO_2) (Agwupuye et al., 2025). COFs have appeared as hopeful gas storage and dissociation materials due to their high surface areas, pore sizes, and low solidities. COFs can be premeditated to selectively adsorb specific gases, such as H_2, CH_4, and CO_2, from gas mixtures (Aksu & Keskin, 2025). For example, COF-1 and COF-5 have shown high CO_2 uptake gas capacities and selectivities, making them candidates for carbon capture and storage applications. Moreover, COFs can be functionalized to increase their gas storage and distinguished properties.

8.4.2 Catalysis

COFs can be utilized as catalysts for various chemical reactions. COFs have established first-rate responsibility for heterogeneous materials used as a catalyst. Because of their high surface passivation effects, highly porous material, variable sizes obtained, and highly functionalized compounds, COFs would be an advanced material in catalytic performances such as chemical reactions, redox reactions, and condensation (K. Wang, Qiao, Ren, Chen, & Zhang, 2025). For standard, e.g., COF-1, COF, and COF-5 for assembled organic frameworks and metal frameworks results in investigation to catalyze the redox reactions of alcohols and the reduction of nitrobenzene, individually. The high catalytic activity, high sensitivity for vital application, and selectivity of COFs make them notable substitutions to obsolete catalysts, with capable material for applications in acceptable energy storage for electrochemical, pharmaceuticals like drugs, and chemicals using pollutant contaminations (F. Chen et al., 2025).

8.4.3 Electrochemical Application Using Energy Storage and Advancement of Energy Conversion

COF materials are electrochemical energy conversions for energy storage, batteries, and water-splitting, and electrochemical sensing is in the energy conversion process toward supercapacitors, energy conversions, and biosensing for biomedical (Gao et al., 2020; X. Zhao et al., 2021). COFs have materialized as competent materials for energy storage advancements for energy conversion applications because of their increased surface passivation areas, porosities, pore sizes, and surface functionalization structures (Yifeng Zhang et al., 2022). COFs have remained learned as electrodes for fuel cells, energy radiant, and bioenergy, demonstrating high capacitance, power density analysis, electrical energy, and electric and energy density, and chemical energy investigations as electrochemical applications. COFs are the most helpful material for photocatalysts, wastewater splitting photoreduction, and CO_2 reduction (Yin, Zhang, & Lu, 2023), describing highly efficient materials and more stability (Yifeng Zhang et al., 2022). The increased porosity and the functionality of COFs distinguish COFs and designate the material design with energy storage and optimization of conversion possession. Examples such as COF-1, COF, and COF-5 represent an accomplished podium for emerging advanced technology in energy storage and energy conversion materials (Feng et al., 2012).

8.4.4 Sensing, Nanoprobe, and Detection

COFs can be utilized to sense a wide range of molecules and ions. COFs have been discovered as sensing, nanoprobe, and detection materials because of their high surface passivation areas, tunable nanolevels of the metal, biomolecules, and other compounds, and functionalized structures. COFs would be an architect for highly sensitive and selective detections and sense various analytes, gases, vapors, toxic metals, heavy metals, and biomolecules. The high porosity and functionality of COFs permit the incorporation of functional sites and signing units, permitting highly sensitive and selective sensing. COFs have been used for the nanoprobe of explosives, toxic metals, toxic gases, and biomarkers, indicating high sensitivity and selectivity. Also, COFs have been combined into various sensing platforms, with optical, electrical, fluorescence, phosphorescence, and electrochemical sensors, and calorimetric sensing, increasing their potential applications in fields, for example, environmental monitoring, paper-based sensing, healthcare, and homeland security. Inclusive COFs offer a hopeful platform for developing advanced sensing and nanoprobe materials.

8.4.5 Biomedical Applications

COFs can be used in biomedical applications, like drug delivery and tissue culture engineering. COFs have shown great potential in biomedical applications because of low energy molecules, high fluorescent materials, low toxicity in nature, and biocompatibility. COFs can be premeditated to selectively bring therapeutic agents, like drugs, proteins, and genes, to specific cells or tissues. COFs have been seen as bioimaging and fluorescent imaging biosensors for detecting biomarkers and various diseases. This material checks toxic behaviors of the material and then uses it in biological applications, including cancer, diabetes, therapies, and infectious diseases. COFs have also been cast off as frameworks for tissue culture engineering, biomedical regenerative medicine, and cell growth and differentiation. It has also been shown that COFs have antibacterial and antiviral activity. It has been shown to create potential candidates for treating infectious diseases. COFs present a flexible platform for advancing biomedical applications. Their excellent physicochemical properties are attractive to biomedical challenges.

8.4.6 Water Purification

COFs have developed as capable materials for water purification due to their various characteristic of material because of surface passivation and chemical stability. COFs can be synthetic processes and produce multiple products to selectively remove contaminants, such as heavy metals, pesticides, and industrial pollutants, from water. Also, COFs have been demonstrated to show microbial activity and antibacterial and antiviral properties, making them influential against waterborne pathogens. The increased water stability and recyclability of COFs make them an attractive choice for water purification applications, contributing to a bearable and efficient explanation for clean water production.

8.4.7 Removal of Inorganic and Organic Hybrid Contaminants

COFs have established an exceptional ability to remove heavy metals, nanopesticides, and other inorganic and organic pollutants from water. Several characteristic feature possessions in highest passivation zones, high porosity, variable components size, and functionalization variable functional groups in the structures documentation of the COFs to choosing selective new components captures facilitated molecules like inorganic compounds such as hydrogen gas (H_2), carbon mono oxide (CO), hydrogen dioxide (H_2O_2), nitrogen dioxide (NO_2), sulfur dioxide (SO_2), carbon dioxide (CO_2), nitrogen trioxide (NO_3), and sulfur trioxide (SO_3), with removal of atmospheric toxic gases depending upon the different binding energy. Numerous heavy metals are used for water purification and the removal of environmental contamination for COF materials, such as mercury (Hg), chromium (Cr), manganese (Mn), cadmium (Cd), arsenic (As), lead (Pb), copper (Cu), and cyanide (CN). COFs would be glowing assumed to remove other organic pollutants way to show water, soil, and drinking water produce among the several types of organic pollution. But also several examples such as, phenol (C_6H_5OH), Nitrophenols ($C_6H_5O_3N$), chlorophenols (C_6H_5OCl), pesticides, toxic dyes, per fluorinated comps. (PFCs), biopolymers, plastics, bleaching plants effluent, voltaic organic compounds like personal care products, polymeric aromatic hydrocarbons), chloroanilines (C_6H_6NCl), BTEX, polychlorinated biphenyls, and non-aqueous phase liquid (NAPL). belong to some pharmaceutical and industrial waste organic pollutants. The highest affinity and selectivity in choosing COF materials for these contaminated pollutants make it more possible to gain materials for water treatment, contributing to highly effective outcomes for alternative modifications to water pollution. The COFs have been constructed outstandingly, with the potential to remove hybrid inorganic and organic components like contaminants from water. Other compounds like Smog forming compounds, particulate components, and GHGs produce atmospheric air pollution. These would be advanced developments toward the targeted removal of toxic pollutants. These materials with high-affinity captain's selective material like COF, COFs, and MOFs are advantageous for water purification, air purification, and removal of soil contaminations. COFs can be professionally reestablished and recycled catalysts and prepare a sustainable approach. COFs offer a potential analytical tool for extenuating water pollution and covering public health. The applications of the removal of inorganic and organic material contaminations are shown in Figure 8.3.

8.4.8 Elimination of Inorganic and Organic Particulate Particles by Air Purification in the Environments

Air filtration methods high performance is an essential protagonist in the preservative systems for air purification or filtration methods, as it eliminates particulate inorganic and organic hazardous contaminations. Organic particulate matter, with pollen, dust, and smoke, can have harmful health effects and produce many implications, for example, cardiovascular diseases, headaches, dizziness, and respiratory difficulties. Operative air filtration systems use several methods to capture gas pollutants. Mechanical sieves, like HEPA sieves, use a physical barricade to trap nanoparticles as small as 0.3 μm. Activated nanocarbon sieves absorb various gases and odors into VOCs. Electronic air cleaners' environmental problems with electrostatic negative and positive charges attract polluted particles. AFS can remove various pollutants such as organic particulate matter, for example, PM2.5, PM10, nano, and ultrafine particles (Robert & Nallathambi, 2020). Toxic gases,

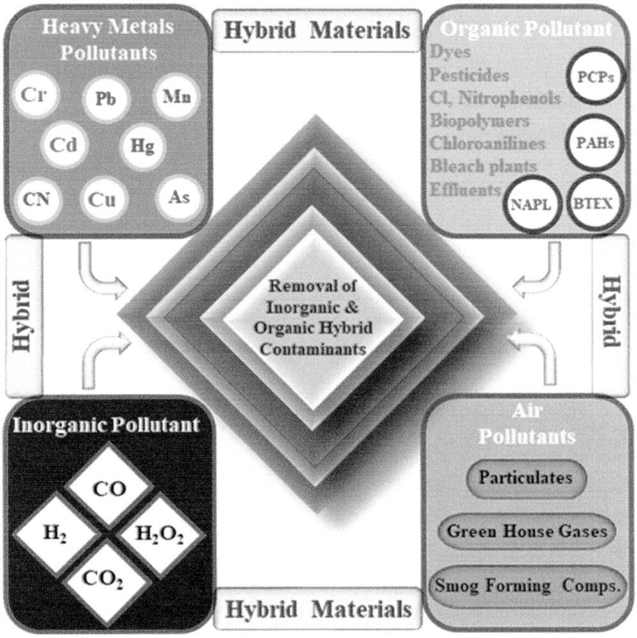

FIGURE 8.3 Schematic representation of the removal of inorganic and organic material contaminations.

for example, nitrogen dioxide (NO_2), ozone (O_3), sulfur dioxide (SO_2), sulfur trioxide (SO_3), carbon monoxide (CO), and VOCs are also recognized to be carcinogenic (Gaba & Iordache, 2011; Lipfert, 1994; Yu Zhang, He, Zhu, Wang, & Chen, 2021). Endotoxins and other toxic agents are involved in the treatment of AFS. The assistance of AFS is many. By removing organic pollutants, AFS can improve the quality of indoor air systems, decrease the health risk for respiratory system difficulties and other health issues, offer relief from antipathies and asthma, eliminate aromas and colognes, and increase productivity and cognitive function. AFS is vital for maintaining environmental clean air implications and improving public health. By considering the standing of AFS and implementing operative systems, entities can recover environmental pollution and decrease the risks associated with air pollution. Influential removal of VOCs and other gases requires technological filters using catalytic oxidation, which uses a catalyst to break down VOCs and other gases into harmless compounds. Photocatalytic oxidation for advanced technology uses light to activate a catalyst, which breaks down VOCs and other gases. Ionization for technology uses ions to break down VOCs and other gases. Remediation methods for organic pollutants include biodegradation, which means microorganisms break down organic pollutants into less toxic compounds. Phytoremediation: Plants absorb and degrade organic contaminants. Chemical oxidation: Oxidizing agents are used to break down organic pollutants. Thermal remediation: Heat is applied to vaporize or decompose organic pollutants.

8.4.9 Environmental Assessment of the Recycle Studies

Estimating COFs for several environmental potential applications requires a comprehensive understanding of their adsorption, chemical transformation and thermal properties. Many methods are used to examine the high-performance adsorption of COF compounds with adsorption isotherms and recycle studies. The fundamental thought of the adsorption performance of COFs, with the elevated and low quantity of gas adsorbed, is that additional binding energy is due to high surface zones. The thermodynamic and chemical kinetics provide evidence of the dynamic adsorption implementation of COFs via time-dependent studies of the desorption and adsorption of gases. Revival and recycling studies are active in evaluating the stability and reusability of COF compounds' separation applications.

8.4.10 Soil Remediation by Removal of Heavy Metal Pollutants

Soil remediation is required to remove heavy metals and other inorganic pollutants from contaminated soil. Toxic metals, e.g., lead (Pb), mercury (Hg), chromium (Cr), and arsenic (As), can pose environmental risks to human health and ecological issues. Diverse remediation technologies are utilized to remove heavy metals and inorganic pollutants from soil. These incorporate excavation and disposal of contaminated soil, which is scooped and disposed of in a landfill, agricultural practices, or waste disposal. In situ remediation strategies, such as biostimulation and bioaugmentation, can also effectively remove organic pollutants from soil. These techniques concern counting nutrients or microorganisms in the soil to enrich biodegradation. Soil washing: Contaminated soil is cleaned with a key to remove heavy metals. Phytoremediation of plants is used to absorb heavy metals from the soil. Chemical immobilization: Chemicals are added to the soil to immobilize heavy metals. In situ remediation technologies, such as absorbent reactive barricades and injection of modifications, are also effective in releasing heavy metals and other inorganic pollutants from soil. Soil remediation delivers multiple advantages, including a decreased risk of human exposure to heavy metals, protection of groundwater and surface water, conservation of ecosystem health, improved soil fertility, and agricultural productivity.

8.5 SYNTHESIS AND DESIGN OF COFs FOR ENVIRONMENTAL APPLICATIONS

The design and synthesis of COFs for environmental applications have been considered significantly due to their potential environmental challenges. COFs are a category of porous materials organized of organic building polymeric blocks linked by covalent bonds; their contribution to materials has various properties: high passivated surface areas, high porosity, variable sizes, and outstanding chemical stability. These inimitable properties prepare COFs as ideal applicants for environmental applications, including water purification, air filtration, and pollutant removal. COFs functionalized with amino or carboxyl groups have been shown to effectively remove pollutants from water with high efficiency. Another essential environmental application of COFs is in the air filtration of various gases, including nitrogen dioxide and methane, from air pollution by COFs. The design and synthesis of COFs for environmental applications requires careful consideration of several factors, including the type of pollutant and the properties of the COFs. COFs can be synthesized using various methods, including solvothermal, microwave-assisted, and mechanochemical synthesis.

8.5.1 Design Principles, Functionalization, Highly Porous, Variable Size and Shape of the COFs

The synthesis of COFs involves the development of a crystalline solid through a well-defined structure, surrounding a recapping preparation of organic building blocks linked by covalent bonds. The design principles of COFs are placed on various suitable building blocks, which would link building blocks together to form an anticipated structure (Huang et al., 2021). The building blocks are classically selected based on their ability to form strong covalent bonds and their architecture of shape and size, which can encourage the pore size and shape of the resulting COFs (Guan, Fang, Yan, & Qiu, 2022). The design principles of COFs are the tuning of pore size and shape, which are achieved by choosing building blocks through the precise various geometries. Using linear building blocks can result in COFs through one-dimensional pores. Various methods have functionalized COF surfaces, including postsynthetic transformation (PSM), linker exchange, and surface grafting. Postsynthetic transformation implicates the chemical transformation of existing COF linkages or building polymeric blocks, authorizing the introduction of new functional groups. Surface grafting consists of a branch of molecules or polymers to the COF surface, promoting the opening of specific properties, such as molecular warming abilities or catalytic activity. Using COFs for surface functionalization means connecting active functional sites (i.e., branching of one molecule to another) with more oriented molecules.. The fissile applications include catalysis activity, gas storage, energy storage, recycling, separation of catalysts, biosensing, and bioimaging cells, sensing and biosensing applications. Architect of the metal-catalytic activity, many functional sites simplified to have an efficient surface and group modification of chemical reactants into growth expansions (Fan et al., 2018; Pascanu, González Miera, Inge, & Martín-Matute, 2019). The surface passivation techniques of COF surfaces scheme a potential and efficient analytical process to quantify the compounds for alternate methods, and their chemical properties improve stability and durability for biomedical drugs using high performance and selection of their use in an essential scope of applications.

The design of COFs as nanomaterials uses solvothermal methods in the building blocks, such as polymeric materials dissolved in a solvent-free material. The heat promoted the establishment of covalent bonds. The outcome of COF materials is a way to isolate the process as a crystalline solid compound. Further characterization can be done using analytical techniques like NMR, powder X-ray diffraction (XRD), transmission electron microscopic (TEM), Raman spectroscopy, Single Crystal X-ray Diffraction (SCXRD), Scanning Electron Microscopy (SEM), and Fourier Transform

Infrared Spectroscopy (FTIR) analysis. The design and synthetic principles of COFs have also been applied throughout the high-performance quality functional material. Several functional groups associated with the polymeric building blocks chain COFs through functionality used a cost-effective postsynthetic modification.

8.5.2 Synthesis Methods of COFs

The synthetic process has also been used with conventional methods to design COFs material or compounds via solvothermal methods (Hu, Huang, & Liu, 2023). Mechanochemical design methods are a solvent-free process due to mechanical forces, like milling, or grinding to boost the chemical reaction among building blockchain. These processes would be used to prepare COFs through high-class characteristic properties and variable size and structures. Other synthetic processes include the synthesis of COFs using mechanochemical (Hamzehpour et al., 2024) and microwave-assisted methods (Rodríguez-Carríllo, Benítez, El Haskouri, Amorós, & Ros-Lis, 2023).

Generally, by synthesis of COF materials using such a solvothermal hydrothermal another technique used. The chemical reaction uses inorganic and organic components in the design of the building blockchain in a solvent and is solvent-free at high temperatures and pressures. These methods are recognized for the synthesis of COFs through excellent crystallinity material and high surface passivated areas (L. Chen et al., 2020). The microwave-assisted design process is a fast and furious well-systematization method that utilizes microwave-high radiation to accelerate the chemical reaction among building blockchains to produce polymer material. These processes can reduce the reaction time, fast reaction, and increase the growth in the product yield of COFs (X. Li et al., 2020; Yongwu Peng et al., 2016). Another synthetic methodology, such as sparking photochemical, catalytic process, and electrochemical syntheses, has been used to synthesize COF compounds (M. Ma, Lu, Guo, Wang, & Liang, 2022). Electrochemical combination refers to integrating electrochemical processes with other methods to increase the effectiveness and efficiency of various applications, especially in energy storage and environmental remediation. This can involve combining electrochemical methods with physical, chemical, or biological processes, or even integrating different electrochemical techniques. (Khojastehnezhad, Samie, Bisio, El-Kaderi, & Siaj, 2024). This chemical reaction process could have been contributing to high-performance quality and high chemical control over the synthesis methods and the subsequent COFs' different structures. Overall, the chemical and solvent optimal synthesis technique is contingent on the precise requirements of the COFs, through its shape, size, structure, and characteristic properties in the future scope of the potential applications (Li, Chen, Xing, Jiang, & Chen, 2020).

8.5.3 Solvothermal Synthetic Methods Using COF Materials

The solvothermal synthetic process is an approximately utilized method for making COFs. Solvothermal methods involve building polymer blocks in a solvent at high temperatures and pressures, classically in a wrapped autoclave or reactor (Cote et al.,

2005). The solvothermal synthesis of COFs consents for forming crystalline materials with high surface areas and porosity tunable sizes (Yuan et al., 2022).

The solvothermal synthesis of COFs classically contains the following steps: (a) selection of building blocks, for example, aldehydes, boronic acids, and amines; (b) termination of the building blocks in a solvent, for example, dioxane or mesitylene; (c) variable conditions of heating the resultant to a high temperature, classically between approximated 100°C and 200°C, under autogenous pressure; and (d) cooling conditions the resultant to permit for the initialization for crystallization of the COFs (Ding et al., 2011). The solvothermal synthesis process of COFs can prepare a wide range of materials through diverse structures and properties. Yaghi and coworkers used the solvothermal synthesis to prepare a COF with a hexagonal structure and a surface area of 4,210 $m^2 g^{-1}$ (El-Kaderi et al., 2007).

8.5.4 Mechanochemical, Another Synthesis Method

Mechanochemical synthesis is a solvent-free process that uses mechanical forces, such as grinding or milling, to create various reactions among building blocks and form COFs. These chemical reaction conditions can be acceptable structural organizing conditions for manufacturing methods developed for COF compounds routine chemical contributions to several practical and advantageous compounds. The synthesis process, for example, compacted the solvent method and improved the chemical reaction for effectiveness. These syntheses were developed by James and coworkers using a mechanochemical combination to synthesize methods of the COF compounds through variable size, porosity, and tunable characteristic properties. Another synthesis procedure involves transformations to synthetic COFs, such as electrochemical designs and routine synthesis practice.

8.6 CHARACTERIZATION OF COF COMPOUNDS

Several analytical techniques used to characterize COF compounds are an integral part of morphological characterization of analysis and their appropriate environmental submission process. COFs are of variable shapes and sizes, porosity sustained COF compounds in characteristic properties should be modified to homogeneous materials, accurate size, shape, and size suitable for environmental applications (Cote et al., 2005). To evaluate COF compounds for conservational applications, several characterization techniques are working through regulated NMR spectroscopy, XRD, TEM, FTIR, UV spectroscopy, Fl spectroscopy, and SEM for morphological characterizations (Trickett et al., 2017). These dealings with material evidence on the COFs well crystallinity materials, internal and external morphology, functional groups, absorbance, chemical arrangement, emission, and excitation properties. The essential aspect of understanding chemical arrangements is their high performance in environmental applications (Ding et al., 2011). There are many types of pollution, and environmental purification involves evaluating their performance in eliminating pollutants from soil, water, and air. These progressions should be accomplished with numerous methods, e.g., advanced approaches, adsorption isotherms, thermal stability, photochemical stability, and kinetic studies (Xia, Zhao, & Darling, 2021). COF is used to remove heavy metals like lead (Pb) and mercury (Hg) from water pollution (Kobielska, Howarth, Farha, & Nayak, 2018). VOCs can be removed from various sources using various methods, including physical, chemical, and biological treatments. These methods aim to either destroy the VOCs or recover them. Standard techniques include adsorption, absorption, biofiltration, membrane filtration, incineration, and combustion. Specific examples include using activated carbon for adsorption, bio trickling filters for biofiltration, and catalytic oxidation for VOC destruction (Dong et al., 2022; Xie, Liu, He, & Li, 2018).

8.7 FUTURE SCOPE AND CHALLENGES FOR SCALABILITY AND COST-EFFECTIVENESS STABILITY, SELECTIVITY, AND DURABILITY IN THE COF MATERIALS

COFs have exposed excessive capacity in several applications, with high separation, gas storage, and catalysis performance. However, some challenges must be addressed to understand their potential application completely. One of the most critical challenges is scalability and cost-effectiveness (Bagheri, Aramesh, Sher, & Bilal, 2021; Hegazy et al., 2023). Now, COFs are synthesized and lead to complex and time-incontrollable developments, which limit their scalability and increase their cost. Emerging, more well-organized, and cost-effective synthesis approaches are essential to make various COFs more available. The alternative challenge is better stability and additional durability. COFs are sensitive to degradation due to several factors, such as humidity, high temperatures, and chemical exposure. Enhancing their stability and durability is required for their practical applications (Kandambeth, Dey, & Banerjee, 2018; X. Zhao et al., 2021). Sensitive, selectivity, and particularity are also significant challenges. COFs often have nonexistent selectivity and specificity, decreasing performance and high productivity. Preparing COFs with customized pore sizes, several structures, and functional groups can assist in their selectivity and particularity progress. Research prospective measurements should focus on materializing more efficient synthesis procedures, improving durability and stability, and preparing COFs with tailored effects to overcome these challenges. Moreover, studying advanced applications and experiencing COFs through other materials can increase their potential.

8.7.1 Future Opportunity and Possible Environmental Applications to COFs

COFs are most vibrant for significant application in the future for environments driven by the expansion of several removals of contaminated material in the fields toward the examination of new environmental potential survival applications (L. Guo et al., 2021). The synthesis and design of the COFs approach have improved several characteristic properties like high stability, industrial level scalability, and processability of these compounds (He, Zhang, & Jiang, 2021; Y. Li et al., 2020). These would be proficient in the advanced level growth of new synthetic technology for the synthetic methods, such as automated synthesis procedures and flow chemistry (Kandambeth et al., 2018). Also, several functional groups and environmental moieties can combine COFs with characteristic chemical properties, such as high-performance catalytic activity, selectivity, and water stability. Determining a new technique for environmental application, e.g., water contamination, carbon dioxide capture, soil remediation, toxic metals, nanopesticides, and industrial influence contaminants, presents a hopeful explanation for sustainable environmental approaches in a hazardous chemical removal scope. COFs can be synthesized and designed to mark particular environmental contaminants.

8.8 CONCLUSION

This chapter of the book helps motivate new research candidates. COFs are highly well-organized, porous materials that have precise applications in various environments. COFs are characterization strategies for understanding their structure, morphology, and properties. 2D and 3D COF materials have developed as competent materials for several environmental applications. Their exclusive properties include tunable pore sizes, porosity, high passivated surface areas, high porosity, variable dimensions, outstanding chemical stability, high passivated surface areas, variable sizes, and exceptional chemical stability of the COF material. These materials are superlative for remits demanding environmental challenges and prospective challenges. Using the architecture of the material, these processes produce various vital characteristic properties. Three fundamental characterization techniques used to analyze COFs: XRD, TEM, and SEM, to identify internal and external morphology. The COFs of the material used in applications such as water treatment, energy storage & conversion, gas separation, biomedical applications, toxic metals, organic–inorganic particulate pollutants, soil water Purification, nanoprobe, detection & sensing, and catalysis have shown outstanding potential applications. Another way is their capacity to selectively capture heavy metals and terminate pollutants, and GHGs from water and air giving them practical environmental remediation tools. Similarly, the design and synthesis of COFs would be appropriate for environmental applications, qualifying the innovation of materials with improved properties. The flexibility of COFs also qualifies their incorporation with other materials and new technologies, making them attractive due to high performance and productivity. Although significant development has been made in the growth of COFs for environmental applications, additional research is desirable to overcome stability, scalability, and cost-effectiveness challenges. The potential of COFs to address environmental difficulties and their prospects are resumed, and development in the application is predicted to profoundly impact the field of environmental science and technology.

ACKNOWLEDGEMENTS

SRA thanks all facilities for providing institute startup grants, a Seed Money Project ID: JU/SMPS/2023-24/ET01 grant funded by Faculty of Engineering & Technology for financial support in the Jagannath University, Jaipur- Rajasthan, India.

REFERENCES

Agwupuye, J. A., Romanus, E. N., Ihedioha, J. N., Amodu, I. O., Ekpong, B. O., Charlie, D. E., Zeeshan, M. (2025). Engineering of covalent organic framework (COF) via mono-doping and co-doping for the detection of CO_2 gas pollutant. *Next Materials, 7*, 100460.

Aksu, G. O., & Keskin, S. (2025). The COF space: Materials features, gas adsorption, and separation performances assessed by machine learning. *ACS Materials Letters, 7*, 954–960.

Alabdulhadi, R. A., Khan, S., Khan, A., Alfuhaid, L. T., Khan, M. Y., Usman, M., … Helal, A. (2025). Potential use of reticular materials (MOFs, ZIFs, and COFs) for hydrogen storage. *ACS Applied Energy Materials,* 8, 1397–1413.

Bagheri, A. R., Aramesh, N., Sher, F., & Bilal, M. (2021). Covalent organic frameworks as robust materials for mitigation of environmental pollutants. *Chemosphere, 270*, 129523.

Chen, F., Zheng, H., Yusran, Y., Li, H., Qiu, S., & Fang, Q. (2025). Exploring high-connectivity three-dimensional covalent organic frameworks: topologies, structures, and emerging applications. *Chemical Society Reviews*, *54*, 484–514.

Chen, H., Jena, H. S., Feng, X., Leus, K., & Van Der Voort, P. (2022). Engineering covalent organic frameworks as heterogeneous photocatalysts for organic transformations. *Angewandte Chemie International Edition, 61*(47), e202204938.

Chen, L., Du, J., Zhou, W., Shen, H., Tan, L., Zhou, C., & Dong, L. (2020). Microwave-assisted solvothermal synthesis of covalent organic frameworks (COFs) with stable superhydrophobicity for oil/water separation. *Chemistry–An Asian Journal, 15*(21), 3421–3427.

Chen, Y., & Jiang, D. (2024). Photocatalysis with covalent organic frameworks. *Accounts of Chemical Research, 57*(21), 3182–3193.

Cote, A. P., Benin, A. I., Ockwig, N. W., O'Keeffe, M., Matzger, A. J., & Yaghi, O. M. (2005). Porous, crystalline, covalent organic frameworks. *Science, 310*(5751), 1166–1170.

Cui, W.-R., Zhang, C.-R., Jiang, W., Liang, R.-P., Wen, S.-H., Peng, D., & Qiu, J.-D. (2019). Covalent organic framework nanosheet-based ultrasensitive and selective colorimetric sensor for trace Hg^{2+} detection. *ACS Sustainable Chemistry & Engineering, 7*(10), 9408–9415. doi:10.1021/acssuschemeng.9b00613

Das, S., Ben, T., Qiu, S., & Valtchev, V. (2020). Two-dimensional COF–three-dimensional MOF dual-layer membranes with unprecedentedly high H_2/CO_2 selectivity and ultrahigh gas permeabilities. *ACS Applied Materials & Interfaces, 12*(47), 52899–52907. doi:10.1021/acsami.0c17794

Ding, S.-Y., Gao, J., Wang, Q., Zhang, Y., Song, W.-G., Su, C.-Y., & Wang, W. (2011). Construction of covalent organic framework for catalysis: Pd/COF-LZU1 in Suzuki–Miyaura coupling reaction. *Journal of the American Chemical Society, 133*(49), 19816–19822.

Dong, C., Yang, J.-J., Xie, L.-H., Cui, G., Fang, W.-H., & Li, J.-R. (2022). Catalytic ozone decomposition and adsorptive VOCs removal in bimetallic metal-organic frameworks. *Nature Communications, 13*(1), 4991.

El-Kaderi, H. M., Hunt, J. R., Mendoza-Cortés, J. L., Côté, A. P., Taylor, R. E., O'Keeffe, M., & Yaghi, O. M. (2007). Designed synthesis of 3D covalent organic frameworks. *Science, 316*(5822), 268–272.

Fan, W., Wang, X., Xu, B., Wang, Y., Liu, D., Zhang, M., … Sun, D. (2018). Amino-functionalized MOFs with high physicochemical stability for efficient gas storage/separation, dye adsorption and catalytic performance. *Journal of Materials Chemistry A, 6*(47), 24486–24495. doi:10.1039/C8TA07839D

Feng, X., Ding, X., & Jiang, D. (2012). Covalent organic frameworks. *Chemical Society Reviews, 41*(18), 6010–6022.

Gaba, A., & Iordache, S. F. (2011). Reduction of air pollution by combustion processes. In Mohamed K. Khallaf (ed) *The Impact of Air Pollution on Health, Economy, Environment and Agricultural Sources*. InTech: London, UK, 119–142.

Gao, X., Dong, Y., Li, S., Zhou, J., Wang, L., & Wang, B. (2020). MOFs and COFs for batteries and supercapacitors. *Electrochemical Energy Reviews, 3*, 81–126.

Guan, X., Fang, Q., Yan, Y., & Qiu, S. (2022). Functional regulation and stability engineering of three-dimensional covalent organic frameworks. *Accounts of Chemical Research, 55*(14), 1912–1927. doi:10.1021/acs.accounts.2c00200

Guo, J., & Jiang, D. (2020). Covalent organic frameworks for heterogeneous catalysis: principle, current status, and challenges. *ACS Central Science, 6*(6), 869–879.

Guo, L., Yang, L., Li, M., Kuang, L., Song, Y., & Wang, L. (2021). Covalent organic frameworks for fluorescent sensing: Recent developments and future challenges. *Coordination Chemistry Reviews, 440*, 213957.

Hamzehpoor, E., Effaty, F., Borchers, T. H., Stein, R. S., Wahrhaftig-Lewis, A., Ottenwaelder, X., ... Perepichka, D. F. (2024). Mechanochemical synthesis of boroxine-linked covalent organic frameworks. *Angewandte Chemie, 136*(51), e202404539.

He, G., Zhang, R., & Jiang, Z. (2021). Engineering covalent organic framework membranes. *Accounts of Materials Research, 2*(8), 630–643.

Hegazy, H., Sana, S. S., Ramachandran, T., Kumar, Y. A., Kulurumotlakatla, D. K., Abd-Rabboh, H. S., & Kim, S. C. (2023). Covalent organic frameworks in supercapacitors: Unraveling the pros and cons for energy storage. *Journal of Energy Storage, 74*, 109405.

Hu, J., Huang, Z., & Liu, Y. (2023). Beyond solvothermal: Alternative synthetic methods for covalent organic frameworks. *Angewandte Chemie International Edition, 62*(37), e202306999.

Huang, J., Liu, X., Zhang, W., Liu, Z., Zhong, H., Shao, B., ... He, Q. (2021). Functionalization of covalent organic frameworks by metal modification: Construction, properties and applications. *Chemical Engineering Journal, 404*, 127136.

Kandambeth, S., Dey, K., & Banerjee, R. (2018). Covalent organic frameworks: chemistry beyond the structure. *Journal of the American Chemical Society, 141*(5), 1807–1822.

Kandambeth, S., Kale, V. S., Shekhah, O., Alshareef, H. N., & Eddaoudi, M. (2022). 2D covalent-organic framework electrodes for supercapacitors and rechargeable metal-ion batteries. *Advanced Energy Materials, 12*(4), 2100177.

Khaing, K. K., Yin, D., Ouyang, Y., Xiao, S., Liu, B., Deng, L., ... Zhang, Y. (2020). Fabrication of 2D–2D heterojunction catalyst with covalent organic framework (COF) and MoS_2 for highly efficient photocatalytic degradation of organic pollutants. *Inorganic Chemistry, 59*(10), 6942–6952. doi:10.1021/acs.inorgchem.0c00422

Khalil, I. E., Das, P., & Thomas, A. (2024). Two-dimensional covalent organic frameworks: structural insights across different length scales and their impact on photocatalytic efficiency. *Accounts of Chemical Research, 57*(21), 3138–3150. doi:10.1021/acs.accounts.4c00491

Khojastehnezhad, A., Samie, A., Bisio, A., El-Kaderi, H. M., & Siaj, M. (2024). Impact of postsynthetic modification on the covalent organic framework (COF) structures. *ACS Applied Materials & Interfaces, 17*(8), 11415–11442.

Kobielska, P. A., Howarth, A. J., Farha, O. K., & Nayak, S. (2018). Metal–organic frameworks for heavy metal removal from water. *Coordination Chemistry Reviews, 358*, 92–107.

Kubo, Y., Nishiyabu, R., & James, T. D. (2015). Hierarchical supramolecules and organization using boronic acid building blocks. *Chemical Communications, 51*(11), 2005–2020.

Levy, D. The Political Dynamics of Sustainable Coffee.

Li, X., Yang, C., Sun, B., Cai, S., Chen, Z., Lv, Y., ... Liu, Y. (2020). Expeditious synthesis of covalent organic frameworks: a review. *Journal of Materials Chemistry A, 8*(32), 16045–16060.

Li, Y., Chen, W., Xing, G., Jiang, D., & Chen, L. (2020). New synthetic strategies toward covalent organic frameworks. *Chemical Society Reviews, 49*(10), 2852–2868.

Lipfert, F. W. (1994). Air pollution and community health: a critical review and data sourcebook.

Ma, M., Lu, X., Guo, Y., Wang, L., & Liang, X. (2022). Combination of metal-organic frameworks (MOFs) and covalent organic frameworks (COFs): Recent advances in synthesis and analytical applications of MOF/COF composites. *TrAC Trends in Analytical Chemistry, 157*, 116741.

Ma, Z., Fang, L., Liu, L., Hu, B., Wang, S., Yu, S., & Wang, X. (2023). Efficient decontamination of organic pollutants from wastewater by covalent organic framework-based materials. *Science of the Total Environment*, *901*, 166453.

Pascanu, V., González Miera, G., Inge, A. K., & Martín-Matute, B. (2019). Metal–organic frameworks as catalysts for organic synthesis: A critical perspective. *Journal of the American Chemical Society, 141*(18), 7223–7234. doi:10.1021/jacs.9b00733

Peng, Y., Li, Y., Ban, Y., & Yang, W. (2017). Two-dimensional metal–organic framework nanosheets for membrane-based gas separation. *Angewandte Chemie, 129*(33), 9889–9893.

Peng, Y., Wong, W. K., Hu, Z., Cheng, Y., Yuan, D., Khan, S. A., & Zhao, D. (2016). Room temperature batch and continuous flow synthesis of water-stable covalent organic frameworks (COFs). *Chemistry of Materials, 28*(14), 5095–5101.

Raghav, S., Raghav, J., Muhammad, M.A., Singh, M.B., Jain, P. & Kumar, D. (2025). Recent advancements in synthesis and applications of covalent organic frameworks. *Applied Materials Today*, 44(June): 102694.

Rejali, N. A., Dinari, M., & Wang, Y. (2023). Post-synthetic modifications to covalent organic frameworks (COFs) for diverse applications. *Chemical Communications*, *59*, 11631–11647.

Robert, B., & Nallathambi, G. (2020). A concise review on electrospun nanofibres/nanonets for filtration of gaseous and solid constituents (PM2. 5) from polluted air. *Colloid and Interface Science Communications, 37*, 100275.

Rodríguez-Carríllo, C., Benítez, M., El Haskouri, J., Amorós, P., & Ros-Lis, J. V. (2023). Novel microwave-assisted synthesis of COFs: 2020–2022. *Molecules, 28*(7), 3112.

Shuck, C. E., Xiao, X., & Wang, Z. (2024). Challenges and opportunities in 2D materials. *Accounts of Chemical Research, 57*(21), 3079–3080. doi:10.1021/acs.accounts.4c00625

Souto, M., Strutyński, K., Melle-Franco, M., & Rocha, J. (2020). Electroactive organic building blocks for the chemical design of functional porous frameworks (MOFs and COFs) in electronics. *Chemistry–A European Journal, 26*(48), 10912–10935.

Trickett, C. A., Helal, A., Al-Maythalony, B. A., Yamani, Z. H., Cordova, K. E., & Yaghi, O. M. (2017). The chemistry of metal–organic frameworks for CO_2 capture, regeneration and conversion. *Nature Reviews Materials, 2*(8), 1–16.

Wang, J., & Zhuang, S. (2019). Covalent organic frameworks (COFs) for environmental applications. *Coordination Chemistry Reviews, 400*, 213046.

Wang, K., Qiao, X., Ren, H., Chen, Y., & Zhang, Z. (2025). Industrialization of covalent organic frameworks. *Journal of the American Chemical Society*, *147*, 8063–8082.

Xia, Z., Zhao, Y., & Darling, S. B. (2021). Covalent organic frameworks for water treatment. *Advanced Materials Interfaces, 8*(1), 2001507.

Xie, L.-H., Liu, X.-M., He, T., & Li, J.-R. (2018). Metal-organic frameworks for the capture of trace aromatic volatile organic compounds. *Chem, 4*(8), 1911–1927.

Xu, H.-S., Ding, S.-Y., An, W.-K., Wu, H., & Wang, W. (2016). Constructing crystalline covalent organic frameworks from chiral building blocks. *Journal of the American Chemical Society, 138*(36), 11489–11492.

Yin, H.-Q., Zhang, Z.-M., & Lu, T.-B. (2023). Ordered integration and heterogenization of catalysts and photosensitizers in metal-/covalent-organic frameworks for boosting CO_2 photoreduction. *Accounts of Chemical Research, 56*(19), 2676–2687.

Yu, B., Lin, R.-B., Xu, G., Fu, Z.-H., Wu, H., Zhou, W., … Li, J.-H. (2024). Linkage conversions in single-crystalline covalent organic frameworks. *Nature Chemistry, 16*(1), 114–121.

Yu, T.-Y., Niu, Q., Chen, Y., Lu, M., Zhang, M., Shi, J.-W., ... Lan, Y.-Q. (2023). Interpenetrating 3D covalent organic framework for selective stilbene photoisomerization and photocyclization. *Journal of the American Chemical Society, 145*(16), 8860–8870. doi:10.1021/jacs.2c12313

Yuan, C., Jia, W., Yu, Z., Li, Y., Zi, M., Yuan, L.-M., & Cui, Y. (2022). Are highly stable covalent organic frameworks the key to universal chiral stationary phases for liquid and gas chromatographic separations? *Journal of the American Chemical Society, 144*(2), 891–900.

Zhang, Y., He, X., Zhu, Z., Wang, W.-N., & Chen, S.-C. (2021). Simultaneous removal of VOCs and PM2. 5 by metal-organic framework coated electret filter media. *Journal of Membrane Science, 618*, 118629.

Zhang, Y., Liu, H., Gao, F., Tan, X., Cai, Y., Hu, B., ... Wang, X. (2022). Application of MOFs and COFs for photocatalysis in CO_2 reduction, H_2 generation, and environmental treatment. *EnergyChem, 4*(4), 100078.

Zhao, F., Liu, H., Mathe, S. D., Dong, A., & Zhang, J. (2017). Covalent organic frameworks: from materials design to biomedical application. *Nanomaterials, 8*(1), 15.

Zhao, X., Pachfule, P., & Thomas, A. (2021). Covalent organic frameworks (COFs) for electrochemical applications. *Chemical Society Reviews, 50*(12), 6871–6913.

Zhu, X., An, S., Liu, Y., Hu, J., Liu, H., Tian, C., ... Abney, C. W. (2017). Efficient removal of organic dye pollutants using covalent organic frameworks. *AIChE Journal, 63*(8), 3470–3478.

9 Environmental Applications of Magnetic Covalent Organic Frameworks

Vaibhav Sharma, Shobhit Saini, and Piyush Gupta

9.1 INTRODUCTION

Many innovative materials are developed for reducing the environmental problem, but there is a need to maintain a balance between industrialization and maintaining environmental aesthetics (He et al., 2021). Covalent organic frameworks (COFs) are very stable and flexible substances due to their excellent thermal and mechanical properties (Raghav et al., 2025; Wang et al., 2020b). They have extended in-plane Π conjugation (Wang et al., 2022a). There are some challenges with magnetic nanoparticles (NPs), such as oxidation, degradation, and instability (González-Sálamo et al., 2020). Magnetic COFs (MCOFs) are synthesized by adding magnetic NPs in COFs, and MCOFs are able to address the limitation of magnetic NPs. MCOFs had good stability and a larger surface area and could be used in controlling the environmental pollution. Their efficacy could be improved by modifying their surface with a particular functional group. MCOFs generate a link between the magnetic organic frame and COF and help in improving stability and tunability (Dong et al., 2020). MCOFs exhibit an excellent removal of toxic chemicals, emerging contaminants, and heavy metals from wastewater and help in the remediation of environmental pollution. Magnetic separation and recyclability are two main factors for their cost-effective and sustainable solutions (Mínguez Espallargas and Coronado, 2018). MCOFs show different environmental and biological applications due to their unique morphology and functional properties. There are two main synthesis methods: the first method is postsynthetic modification in which NPs are added after the preparation of COFs and the other is the in situ method, in which magnetic NPs are doped during the synthesis of COFs. These synthesis methods offer strong integration while preserving porosity and functionality, which depends on the solvothermal, mechanical, or hydrothermal process. But further research should be done on the long-term stability and toxicity of MCOFs. For versatile utilization of the potential of MCOFs in mitigating environmental pollution, a detailed study on the structure and their synthesis methods are crucial (Sargazi and Kaykhaii, 2024). This chapter highlights the synthesis methods of MCOFs and different environmental applications of MCOFs.

 DOI: 10.1201/9781003605386-9

9.2 SYNTHESIS METHODS OF MCOFs

Covalent triazine-based framework (CTF)/Fe_2O_3 composite was manufactured using a microwave-enhanced high-temperature ionothermal technique. CTF/Fe_2O_3 composite was prepared within 60 minutes with varying γ-Fe_2O_3 concentrations (Zhang et al., 2011). Microwave-assisted synthesis method was used to produce magnetic Fe_3O_4-coated melamine-based COFs (M-COFs) with high surface areas (344–600 $m^2\ g^{-1}$) and efficient magnetic separation (0.75–3.59 emu g^{-1}) (Ge et al., 2016). Magnetic porous nanocomposite was prepared through the grafting of COF, TpPa-1 onto Fe_3O_4 NPs (Fe_3O_4 NPs) (their surface was modified). 1,3,5-triformylphloroglucinol (Tp) and p-phenylenediamine (Pa-1) were used to prepare TpPa-1 MCOFs (He et al., 2017). CTF-1 works as a host for the in situ preparation of Fe_2O_3 NPs. γ-Fe_2O_3@CTF-1 was effectively used to eliminate heavy metals from both lab-prepared samples and water (surface water, groundwater, and wastewater) (Leus et al., 2018). mTpBD-Me_2 was prepared by using amino-functionalized magnetic NPs, and these NPs were synthesized by in situ dopamine functionalization of Fe_3O_4 (Romero et al., 2019). CTC-COF@MCNT was prepared by photochemical methods, and these COFs could be used to eliminate heterocyclic aromatic amines (Liang et al., 2020). The Schiff base reaction between melamine and terephthalaldehyde was performed to prepare Fe_3O_4@COF and in this reaction, no catalyst was used (You et al., 2020). Solvothermal process was used to produce Fe_3O_4@TpND (Wei et al., 2020). Fe_3O_4@TpPa-F4 was prepared by using the in situ growth method and monomer acts as a media for this method (Zhang et al., 2020). Hg^{2+} ions were removed from water by utilizing thiol-modified M-COF-SH, prepared by the reaction of disulfide derivatives (Huang et al., 2020b). Fe_3O_4@COF(TpPa-1), MCOF featuring a β-ketoenamine linkage, was synthesized by the hydrothermal method and used to remove contaminants from aqueous medium (Sargazi and Kaykhaii, 2024). Fe_3O_4@COF(TpDA) with 180.2 $m^2\ g^{-1}$ surface area and 62.3 emu g^{-1} saturation magnetization was employed as an adsorptive material for enrichment and detection of plant growth regulators in fruits and vegetables (Li et al., 2020). The ball-milling technique was employed to synthesize CTF/Fe_3O_4 (Wang et al., 2022b). Fe_3O_4@COF-Tp tetra(p-aminophenyl)-methane were the nanospheres of Fe_3O_4@3D COFs, which were prepared using the solvothermal process (Lu et al., 2021). M-TpDAB was prepared by using 3,3′-diaminobenzidine (DAB) and Tp (Guo et al., 2021). Mechanochemical process was used to prepare MCNTs@TpPa-1, and it could be used to eliminate microcystins (MCs) (Liu et al., 2021). NH_2–Fe_3O_4@COF displayed an excellent adsorption efficacy within 10 minutes for six benzoylurea (BU) insecticides (Xin et al., 2021). $MnFe_2O_4$@TpPa-1 (MFTP) was prepared by using the one-pot solvothermal process (Zhong et al., 2021). In situ method was used to prepare TpPa-1 COFs, in which nanoscale zero valent iron and dopamine as a connector were added. It was used to eliminate Cr(VI) ions from aqueous solution by the synergistic effect of adsorption and reduction process (Shen et al., 2022). Ethidium bromide (EB) and Tp were the constructing units of EB-COF@Fe_3O_4. This MCOF was used to eliminate five BU insecticides in water and food samples (Li et al., 2022). Co-precipitation method was used to generate Fe_3O_4/COF-DQTp and Benzo[a]pyrene could be removed by using this COF (Pang et al., 2022). A single-step technique was produced to generate a crystalline MCOF composite without prior functionalization of the substrate. The resultant composite preserves crystallinity and surface area similar to the bulk material (Romero et al., 2023). $Ni_{0.6}Fe_{2.4}O_4$-HT-COF was prepared by the condensation of $(NPCl_2)_3$, $C_3H_3N_3S_3$, and $Ni_{0.6}Fe_{2.4}O_4$. At 298 K and pH 5.0, it could adsorb Pb(II) (Wang et al., 2023). In situ growth method was utilized to prepare Fe_3O_4@TpPDA. Tp and 1,4-phenylenediamine (PDA) worked as precursors. Congo red dyes were removed by using Fe_3O_4@TpPDA (Lu et al., 2023).

9.3 MAGNETIC PROPERTIES OF MCOFs

MCOF composites exhibit numerous notable characteristics that are advantageous from the perspectives of green chemistry and sustainability, including facile and cost-effective recovery. The assessment of the material's magnetic characteristics is critically important. The magnetic characteristics of MCOFs are examined via vibrating sample magnetometry at ambient temperature (Sargazi and Kaykhaii, 2024). In the hysteresis loop, saturation magnetization is employed to analyze magnetism. Furthermore, MCOF is superparamagnetic in nature if hysteresis loop, coercivity, or remanence is not found in the curve (Kurmoo, 2009). The modification of magnetic NPs with a COF layer results in a decrease in the saturation magnetization value of the MCOF composite compared to the initial magnetic NPs. Despite the reduction in saturation magnetization of MCOFs, it could be readily extracted by using an external magnet within a few seconds. The primary attributes of magnetic composites are their reusability and recyclability, which govern the economic viability and efficiency of industrial processes. Despite the significant advantages of utilizing COFs as adsorbents, their widespread application remains hindered by high costs and filtration and centrifugation processes required for composite recovery. Magnetic NP-supported COFs have been employed as effective adsorbents for environmental clean-up. The incorporation of magnetism in COFs addresses the separation issue and facilitates ease of separation (Li et al., 2006).

9.4 ENVIRONMENTAL APPLICATIONS OF MCOFs

MCOFs are exceptionally adaptable materials that hold considerable promise for environmental applications, owing to their extensive surface area, adjustable porosity, and ease of magnetic recovery. These materials are efficiently employed in the treatment of wastewater, providing improved reusability, selectivity, and sustainable environmental clean-up along with advancements in green chemistry. Table 9.1 tabulates the different applications of MCOFs.

9.4.1 Adsorption of Insecticides from Eatables

MCOF (Fe_3O_4@COF-$(NO_2)_2$, was prepared for magnetic solid-phase extraction (MSPE) of six neonicotinoid insecticides (NNIs) in vegetables. 254.72 m^2 g^{-1}, 0.19 cm^3 g^{-1}, and 27.7 emu g^{-1} were surface area, pore volume, and magnetism, respectively. It demonstrated exceptional linearity (0.1–30 ng mL^{-1}), minimal detection limits (0.02–0.05 ng mL^{-1}), and substantial enrichment (170–250) (Lu et al., 2020). Six BU

TABLE 9.1
The Different Applications of MCOFs

MCOFs	Important Features	References
CTC-COF@MCNT	LOD ranging from 0.0058 to 0.025 ng g^{-1} for heterocyclic aromatic amines	Liang et al. (2020)
Fe_3O_4@TpPa-F4	Detected PFCs at concentrations between 10.36 and 729.34 ng L^{-1}.	Zhang et al. (2020)
Fe_3O_4@COF-$(NO_2)_2$	Low LOD (S/N = 3) between 0.02 and 0.05 ng mL^{-1} for six NNI pesticides	Lu et al. (2020)
NH_2–Fe_3O_4@COF	Minimal LOD (0.06–1.65 ng·L^{-1}) for six BU insecticides	Xin et al. (2021)
EB-COF@Fe_3O_4	Detection limits were 0.01–0.09, 0.03–0.3 ng mL^{-1}, and 0.3–2.0 ng g^{-1} for water, juice, and tomato, respectively	Li et al. (2022)
Fe_3O_4/M-COFs	Adsorbed Hg^{2+}, adsorption capacity 97.65 mg g^{-1}	Ge et al. (2016)
γ-Fe_2O_3@CTF-1	Adsorption capacities of 198.0 mg g^{-1} for AsO_3^{3-}, 102.3 mg g^{-1} for AsO_4^{3-}, and 165.8 mg g^{-1} for Hg^{2+}	Leus et al. (2018)
Fe_3O_4@TpPa-1	245.45 and 1,220.97 mg g^{-1} adsorption efficiency for Cr(VI) and BPA	Zhong et al. (2020)
M-COF-SH	Adsorbed Hg^{2+}, adsorption capacity 383 mg g^{-1}	Huang et al. (2020a, 2020b)
Fe^0/TpPa-1@DOPA	Adsorbed Cr(VI), maximum removal capacity of 516 mg g^{-1}	Shen et al. (2022)
MCOF composite	Removed okadaic acid from synthetic seawater	Romero et al. (2023)
$Ni_{0.6}Fe_{2.4}O_4$-HT-COF	Removed Pb(II), maximum adsorption capacity of 411.80 mg g^{-1}	Wang et al. (2023)
Fe_3O_4@COF	Removed uranium, adsorption capacity of 311 mg g^{-1}	Li et al. (2024)

insecticides could be eliminated by using NH_2–Fe_3O_4@COF, which had linearity and minimum limit of detection (LOD) of 10–1,000 ng·L^{-1} and 0.06–1.65 ng·L^{-1}, respectively (Xin et al., 2021). EB-COF@Fe_3O_4 could be used to remove BU insecticides. EB-COF@Fe_3O_4 showed linearity of 0.03–100.0 ng mL^{-1} for water and juice while 0.9–500 ng g^{-1} for tomatoes. 0.01–0.09, 0.03–0.3 ng mL^{-1}, and 0.3–2.0 ng g^{-1} were minimum LOD for water, juice, and tomatoes, respectively. Moreover, this COF could be reused up to 25 cycles (Li et al., 2022).

9.4.2 Adsorption of Heavy Metals from Wastewater

Fe_3O_4/M-COFs could adsorb about 97.65 mg g^{-1} of Hg^{2+} (Ge et al., 2016). Magnetic TpPa-1 was used for MSPE to separate polycyclic aromatic hydrocarbons (PAHs) from samples prior HPLC analysis (He et al., 2017). γ-Fe_2O_3@CTF-1 exhibited adsorption capacities of 198.0, 102.3, and 165.8 mg g^{-1} for AsO_3^{3-}, AsO_4^{3-}, and Hg^{2+}, respectively (Leus et al., 2018). mTpBD-Me_2 showed 812 mg g^{-1} and 830 mg g^{-1} elimination rate for okadaic acid and dinophysistoxin-1 (Romero et al., 2019). Fe_3O_4@TpPa-1 demonstrated 245.45 mg g^{-1} and 1,220.97 mg g^{-1} adsorption efficiency for Cr(VI) and bisphenol A (BPA) (Zhong et al., 2020). M-COF-SH exhibited 383 mg g^{-1} removal efficiency for Hg^{2+} (Huang et al., 2020a, 2020b). Fe_3O_4@TAPB-COF@ZIF-8 was utilized for the MSPE of BPA (Jiang et al., 2021). MFTP had a maximal adsorption efficiency of 1,235.01 mg g^{-1} for UO_2^{2+} ion, which is more than that of $MnFe_2O_4$ (212.1 mg g^{-1}) (Zhong et al., 2021). Fe0/TpPa-1@DOPA showed 516 mg g^{-1} removal efficacy for Cr(VI) (Shen et al., 2022). The MCOFs composite effectively removed okadaic acid from synthetic seawater, demonstrating an adsorption efficacy more than other COF. Furthermore, the material was recyclable for a minimum of three successive cycles of adsorption/desorption utilizing 2-propanol as desorption solvent, exhibiting only negligible reductions in adsorption capability (Romero et al., 2023). $Ni_{0\cdot6}Fe_{2\cdot4}O_4$-HT-COF attained a removal efficiency of 95.64% in 10 minutes, exhibiting the highest adsorption ability of 411.80 mg g^{-1} and helps in extracting Pb(II) from multi-ion wastewater (Wang et al., 2023). Fe_3O_4@COF exhibited a removal efficiency of 311 mg g^{-1} for uranium (Li et al., 2024).

9.4.3 Adsorption of Organic Pollutants and Dyes from Wastewater

CTF/Fe_2O_3 composites exhibited a surface area of 930–1,149 m^2 g^{-1}. 1.1 to 5.9 emu g^{-1} was saturation magnetizations at 300 K. They were employed to eliminate methyl orange from an aqueous solution. An adsorption capacity of 291 mg g^{-1} (0.889 mmol g^{-1}) and quick adsorption kinetics (kads = 4.31 m^2 mg^{-1} min^{-1}) were observed (Zhang et al., 2011). In situ growth technique with monomer mediation was proposed for the precise fabrication of porous nanospheres including a magnetic core and a flexible COF shell. The composite demonstrated significant stability and superior efficacy in eliminating a common category of endocrine-disrupting pollutants, specifically BPA from aqueous solutions. The adsorption rate of BPA was 160.6 mg g^{-1} (Li et al., 2017). Fe_3O_4@COFs could eliminate triclosan (TCS) and triclocarban (TCC) from water and serum samples. Fe_3O_4@COFs had 82.3% to 95.4% removal efficiency and 92.9% to 109.5% were the recovery rates of TCS and TCC from bovine serum, along with reusability for a minimum of ten cycles. Fe_3O_4@COFs highlighted greater stability and excellent efficacy in removing TCS and TCC from aqueous and biological samples (Li et al., 2019). MCOFs were synthesized by using an impregnation method. The prepared MCOFs had a saturation magnetization of 5.2 emu g^{-1}. 109 mg g^{-1} and 113.2 mg g^{-1} were the elimination rates for diclofenac and sulfamethazine, respectively (Z Huang et al., 2020a, 2020b). A Fe(III)-doped MCOF showed an adsorption efficiency of 60.04 mg g^{-1} at 25°C for phenylbutazone (Zheng et al., 2020). Carboxylate-functionalized Fe_3O_4@COF–COOH MCOF composites were used for the removal of tetracyclines (TCs), PAHs, and triphenylmethane dyes (TDs) by using the mixed-mode solid-phase extraction process. High sensitivity was attained with LODs varying from 0.003 to 0.0081 for PAHs, 0.02 to 0.06 for TCs, and 0.006 to 0.008 µg L^{-1} for TDs (Hu et al., 2020). MCOF-2 could remove about 565 mg g^{-1} of diclofenac sodium (Huang et al., 2020a). Fe_3O_4@COF (TpDA) had an LOD of 4.68–7.51 µg L^{-1} and 83.0% to 105.0% was the recovery rate. This MCOF was effectively utilized for the detection of plant development regulators in foodstuffs (Li et al., 2020). Fe_3O_4@TpPa-1 highlighted the outstanding adsorption capabilities, i.e. 245.45 mg g^{-1} for Cr(VI) and 1,220.97 mg g^{-1} for BPA (Zhong et al., 2020). The Fe_3O_4@COF had 335.2 m^2 g^{-1} surface area and 49.6 emu g^{-1} saturation magnetization, facilitating quick magnetic separation and effective adsorption during the adsorption process. The synthesized adsorbent demonstrated an exceptional adsorption capacity of 140 and 290.4 mg g^{-1} for BPA and Bisphenol AF (BPAF), respectively (You et al., 2020). Fe_3O_4@TpND had 114.97 mg L^{-1} removal efficiency at 298 K and pH 6.0 for the elimination of BPA (Wei et al., 2020). MCOFs@MIPs@CDs exhibited excellent fluorescence characteristics and stability. The composite material had excellent sensitivity for 2,4,6-trinitrophenol, achieving a 100 pM limit of detection, an 8.4 imprinting factor, and less than 1 minute detection time (Wang et al., 2020a). CTF2/Fe_3O_4 demonstrated a superior adsorption performance. It achieved adsorption capacities of 1.18 mmol g^{-1} for sodium p-perfluorous nonenoxybenzene sulfonate and 1.02 mmol g^{-1} for hexafluoropropylene oxide trimer acid, with adsorption equilibrium attained within 3 hours (Wang et al., 2022b). MCOF effectively eliminated 20 mg L^{-1} of dye solution within 2 minutes. It showed the highest adsorption rate of 228.07 mg g^{-1} for Amido Black 10B, attributed to a high surface area of 120 m^2 g^{-1} (Kamalabadi et al., 2022). Under optimum adsorption circumstances, the adsorption efficiency of Fe_3O_4/COF-DQTp reached 99% for Benzo[a]pyrene and could remove 19 mg g^{-1} within 10 minutes (Pang et al., 2022). The highest adsorption efficiency of Fe_3O_4@TpPDA was 179.4 mg g^{-1} for Congo red and could be reused up to seven times (Lu et al., 2023). COF@CS could increase the adsorption efficiency 2.8–6.2 mmol g^{-1} h^{-1} at pH 5 for PFOA, which surpasses COF's 5.9 mmol g^{-1} h^{-1} and CS's 3.4 mmol g^{-1} h^{-1}. COF@CS demonstrated an exceptional removal efficacy for ten additional PFAS variants. Furthermore, COF@CS has effectively maintained the stable reusability for up to five cycles (Zhang et al., 2024).

9.4.4 Sensitive Detection of Lethal Contamination in Aqueous Medium

Fe_3O_4@TpPa-F4 exhibited consistent porosity (0.81 nm), substantial surface area (188.04 m^2 g^{-1}), fluorine concentration (4.85%), and robust magnetism (50 emu g^{-1}), rendering it optimal for perfluorinated compound (PFC) extraction (Zhang et al., 2020). Fe_3O_4@TpPa-F4 showed 0.005–0.05 ng L^{-1} LOD under optimum conditions. When applied to milk samples, it identified PFCs (10.36–729.34 ng L^{-1}), validating its efficacy as an adsorbent for sample pretreatment (Liu et al., 2021). M-TpDAB had a linearity within the range of 0.15–100 and 1.0–100.0 ng mL^{-1} for water and tea samples. The detection limits (in ng mL^{-1}) and recoveries for the phenylurea herbicides were 0.05-0.15 and 84.6% to 105% in water samples and 0.30–0.50 and 80.3% to 102% in tea samples (Guo et al., 2021). The utilization of Fe_3O_4@3D COF nanospheres resulted in a superior adsorption capacity for BPA, reaching 209.9 mg g^{-1}, surpassing that of conventional 2D-COFs (Lu et al., 2021). The adsorption efficiencies for BPA of Fe_3O_4/TpPa-2, Fe_3O_4/TpBD, and Fe_3O_4/TFPB-BD were 44.85 mg g^{-1}, 24.14 mg g^{-1}, and 2.61 mg g^{-1}, respectively (Wang et al., 2022c). Figure 9.1 illustrates the utilization of MCOFs in the remediation of different contaminants.

9.5 CHALLENGES

MCOF composites have numerous problems that must be resolved in order to fully utilize their properties as versatile adsorbents. This encompasses the necessity for enhanced extraction of microorganisms along with a more profound comprehension of the adsorption–desorption mechanism to eliminate the reliance on expensive solvents.

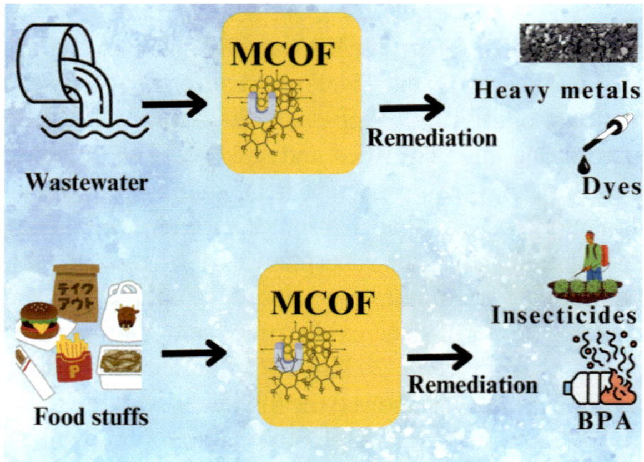

FIGURE 9.1 Role of MCOFs in environmental remediation applications.

The elevated synthetic expense of COF materials, mostly attributed to costly monomers, necessitates ways to integrate cheap functional groups for feasible large-scale production. The scalability of magnetic separation techniques for various pollutants, including insecticides, BPA, heavy metals, dyes, and organic pollutants from wastewater under standard circumstances, presents a considerable challenge. Understanding the toxicity of these NPs is essential for their safe long-term use. Furthermore, although MCOFs have shown encouraging outcomes in extraction applications, their application in catalysis and other advanced fields remains nascent, necessitating additional investigation and postsynthetic modifications to optimize photocatalytic reactions and incorporate specific active sites for enhanced catalytic efficacy.

9.6 FUTURE PROSPECTS

Future investigations into MCOFs should prioritize the improvement of their adsorption performance, selectivity, and recyclability for expanded environmental and analytical applications. Creating MCOFs with adjustable porous architectures, modified surfaces, and enhanced magnetic responsiveness may enhance their adsorption efficacy for microplastics, pharmaceutical residues, and persistent organic pollutants. Incorporating MCOFs with sophisticated detection methodologies will enhance their utilization in environmental applications. Enhancing synthesis strategies while ensuring cost-effectiveness and stability is essential for practical applications. Furthermore, investigating biodegradable or environmentally friendly MCOFs may mitigate secondary pollutants, hence fostering sustainability. There is a need of more research in the advancement of next-generation MCOFs for enhanced efficiency and sustainability in applications.

9.7 CONCLUSION

MCOFs represent the considerable potential in diverse applications such as pesticide and insecticide detection. MCOFs are utilized in the removal of heavy metals, dyes, and organic pollutants from wastewater. Their unique characteristics, such as large surface area, porosity, thermal and chemical durability, and robust magnetic properties, facilitate the effective extraction and elimination of pollutants from wastewater. MCOFs have been utilized in environmental samples for the detection of heterocyclic aromatic amines, perfluorinated chemicals, NNI pesticides, BU insecticides, and MCs, demonstrating exceptional sensitivity and selectivity. Furthermore, their adsorption properties were excellent for Hg^{2+}, Pb^{2+}, Cr(VI), UO_2^{2+}, BPA, dyes, PAHs, and pharmaceutical residues. MCOFs can be regenerated and reutilized, hence cost-efficient adsorbent. Thus, this chapter summarizes various applications of MCOFs in analytical and environmental sciences, rendering them viable materials to reduce the concentration of pollutants in the ecosystem.

REFERENCES

Dong, J., X. Han, Y. Liu, H. Li, and Y. Cui. 2020. Metal–covalent organic frameworks (MCOFs): A bridge between metal–organic frameworks and covalent organic frameworks. *Angewandte Chemie International Edition*. 59(33):13722–13733. https://doi.org/10.1002/ANIE.202004796.

Ge, J., J. Xiao, L. Liu, L. Qiu, and X. Jiang. 2016. Facile microwave-assisted production of Fe_3O_4 decorated porous melamine-based covalent organic framework for highly selective removal of Hg^{2+}. *Journal of Porous Materials*. 23(3):91–800. https://doi.org/10.1007/S10934-016-0134-Y/FIGURES/12.

González-Sálamo, J., G. Jiménez-Skrzypek, C. Ortega-Zamora, M.A. González-Curbelo, and J. Hernández-Borges. 2020. Covalent organic frameworks in sample preparation. *Molecules*. 25(14):3288. https://doi.org/10.3390/MOLECULES25143288.

Guo, L., J. Liu, J. Li, L. Hao, W. Liu, C. Wang, W. Wu and Z. Wang. 2021. A core-shell structured magnetic covalent organic framework as a magnetic solid-phase extraction adsorbent for phenylurea herbicides. *Journal of Chromatography A*. 1651:462301. https://doi.org/10.1016/J.CHROMA.2021.462301.

He, M., Q. Liang, L. Tang, Z. Liu, B. Shao, Q. He, T. Wu, S. Luo, Y. Pan, C. Zhao, C. Niu, and Y. Hu 2021. Advances of covalent organic frameworks based on magnetism: Classification, synthesis, properties, applications. *Coordination Chemistry Reviews*. 449:214219. https://doi.org/10.1016/J.CCR.2021.214219.

He, S., T. Zeng, S. Wang, H. Niu, and Y. Cai. 2017. Facile synthesis of magnetic covalent organic framework with three-dimensional bouquet-like structure for enhanced extraction of organic targets. *ACS Applied Materials and Interfaces*. 9(3):2959–2965. https://doi.org/10.1021/ACSAMI.6B13643/SUPPL_FILE/AM6B13643_SI_001.PDF.

Hu, K., J. Cheng, W. Zhang, T. Pang, X. Wu, Z. Zhang, Y. Huang, W. Zhao, and S. Zhang. 2020. Simultaneous extraction of diverse organic pollutants from environmental water using a magnetic covalent organic framework composite. *Analytica Chimica Acta*. 1140:132–144. https://doi.org/10.1016/J.ACA.2020.10.019.

Huang, L., N. Mao, Q. Yan, D. Zhang, and Q. Shuai. 2020a. Magnetic covalent organic frameworks for the removal of diclofenac sodium from water. *ACS Applied Nano Materials*. 3(1):319–326. https://doi.org/10.1021/ACSANM.9B01969/SUPPL_FILE/AN9B01969_SI_001.PDF.

Huang, L., R. Shen, R. Liu, and Q. Shuai. 2020b. Thiol-functionalized magnetic covalent organic frameworks by a cutting strategy for efficient removal of Hg^{2+} from water. *Journal of Hazardous Materials*. 392:122320. https://doi.org/10.1016/J.JHAZMAT.2020.122320.

Jiang, H.L., Fu, Q. Bin, M.L. Wang, J.M. Lin and R.S. Zhao. 2021. Determination of trace bisphenols in functional beverages through the magnetic solid-phase extraction with MOF-COF composite. *Food Chemistry*. 345:128841. https://doi.org/10.1016/J.FOODCHEM.2020.128841.

Kamalabadi, M., S. Khalili, T. Madrakian, and A. Afkhami. 2022. Facile synthesis of magnetic melamine-based covalent organic framework for removal of Amido Black 10B. *The European Physical Journal Plus*. 137(5):1–11. https://doi.org/10.1140/EPJP/S13360-022-02758-9.

Kurmoo, M. 2009. Magnetic metal–organic frameworks. *Chemical Society Reviews*. 38(5):1353–1379. https://doi.org/10.1039/B804757J.

Leus, K., K. Folens, N.R. Nicomel, J.P.H. Perez, M. Filippousi, M. Meledina, M.M Dîrtu, S. Turner, G. Van Tendeloo, Y. Garcia, G. Du Laing, and P. Van Der Voort. 2018. Removal of arsenic and mercury species from water by covalent triazine framework encapsulated γ-Fe_2O_3 nanoparticles. *Journal of Hazardous Materials*. 353:312–319. https://doi.org/10.1016/J.JHAZMAT.2018.04.027.

Li, N., D. Wu, X. Li, X. Zhou, G. Fan, G. Li, and Y. Wu. 2020. Effective enrichment and detection of plant growth regulators in fruits and vegetables using a novel magnetic covalent organic framework material as the adsorbents. *Food Chemistry*. 306:125455. https://doi.org/10.1016/J.FOODCHEM.2019.125455.

Li, S., W. Liu, Q. Wang, M. Xu, Y. An, L. Hao, C. Wang, Q. Wu, and Z. Wang. 2022. Constructing magnetic covalent organic framework EB-COF@Fe_3O_4 for sensitive determination of five benzoylurea insecticides. *Food Chemistry*. 382:132362. https://doi.org/10.1016/J.FOODCHEM.2022.132362.

Li, X.J., X.Y. Wang, S. Gao, and R. Cao. 2006. Two three-dimensional metal-organic frameworks containing one-dimensional hydroxyl/carboxylate mixed bridged metal chains: Syntheses, crystal structures, and magnetic properties. *Inorganic Chemistry*. 45(4):1508–1516. https://doi.org/10.1021/IC050999X/SUPPL_FILE/IC050999XSI20051014_111805.PDF.

Li, Y., C.X. Yang, and X.P. Yan. 2017. Controllable preparation of core–shell magnetic covalent-organic framework nanospheres for efficient adsorption and removal of bisphenols in aqueous solution. *Chemical Communications*. 53(16):2511–2514. https://doi.org/10.1039/C6CC10188G.

Li, Y., H. Zhang, Y. Chen, L. Huang, Z. Lin, and Z. Cai. 2019. Core-shell structured magnetic covalent organic framework nanocomposites for triclosan and triclocarban adsorption. *ACS Applied Materials and Interfaces*. 11(25):22492–22500. https://doi.org/10.1021/ACSAMI.9B06953/SUPPL_FILE/AM9B06953_SI_001.PDF.

Li, Y., J. Hu, C. Li, and X. Hou. 2024. Magnetic covalent organic framework for efficient solid-phase extraction of uranium for on-site determination by portable x-ray fluorescence spectrometry. *Analytical Chemistry*. 96(15):5757–5762. https://doi.org/10.1021/ACS.ANALCHEM.4C00063.

Liang, R., Y. Hu, and G. Li. 2020. Photochemical synthesis of magnetic covalent organic framework/carbon nanotube composite and its enrichment of heterocyclic aromatic amines in food samples. *Journal of Chromatography A*. 1618:460867. https://doi.org/10.1016/J.CHROMA.2020.460867.

Liu, G., H. Chen, W. Zhang, Q. Ding, J. Wang, and L. Zhang. 2021. Facile mechanochemistry synthesis of magnetic covalent organic framework composites for efficient extraction of microcystins in lake water samples. *Analytica Chimica Acta*. 1166:338539. https://doi.org/10.1016/J.ACA.2021.338539.

Lu, F., J. Lin, C. Lin, G. Qi, X. Lin, and Z. Xie. 2021. Heteroporous 3D covalent organic framework-based magnetic nanospheres for sensitive detection of bisphenol A. *Talanta*. 231:122343. https://doi.org/10.1016/J.TALANTA.2021.122343.

Lu, J., R. Wang, J. Luan, Y. Li, X. He, L. Chen, and Y. Zhang. 2020. A functionalized magnetic covalent organic framework for sensitive determination of trace neonicotinoid residues in vegetable samples. *Journal of Chromatography A*. 1618:460898. https://doi.org/10.1016/J.CHROMA.2020.460898.

Lu, S., Y. Wei, S. Long, Z. Chen, F. Chen, H. Lin, and J. Lu. 2023. Efficient adsorption and removal of congo red from aqueous solution using magnetic covalent organic framework nanocomposites. *ChemistrySelect*, 8(1):e202203621. https://doi.org/10.1002/SLCT.202203621.

Mínguez Espallargas, G. and E. Coronado. 2018. Magnetic functionalities in MOFs: From the framework to the pore. *Chemical Society Reviews*. 47(2):533–557. https://doi.org/10.1039/C7CS00653E.

Pang, Y.H., N.C. Yang, J.Y. Qiao, X.F Shen, and Y.Y Huang. 2022. Preparation and characterization of magnetic covalent organic framework and its application for efficient adsorption of Benzo[a]pyrene. *Journal of Porous Materials*. 29(1):169–179. https://doi.org/10.1007/S10934-021-01151-8/TABLES/3.

Raghav, S., J. Raghav, M.A. Muhammad, M. B. Singh, P.Jain, D. Kumar. 2025. Recent advancements in synthesis and applications of covalent organic frameworks. *Applied Materials Today*, 44(June): 102694. https://doi.org/10.1016/j.apmt.2025.102694

Romero, V., S.P.S. Fernandes, L. Rodriguez-Lorenzo, Y.V. Kolen'ko, B. Espiña, and L.M. Salonen. 2019. Recyclable magnetic covalent organic framework for the extraction of marine biotoxins. *Nanoscale*. 11(13):6072–6079. https://doi.org/10.1039/C9NR00388F.

Romero, V., S.P.S. Fernandes, L.P.L. Gonçalves, O. Oliveira, M. Meledina, K. Strutyński, M. Melle-Franco, Y.V. Kolen'ko, B. Espiña, and L.M. Salonen. 2023. One-step synthesis of magnetic covalent organic framework composite for the adsorption of marine toxin okadaic acid. *CrystEngComm*. 25(16):2456–2462. https://doi.org/10.1039/D3CE00178D.

Sargazi, M. and M. Kaykhaii. 2024. Magnetic covalent organic frameworks-fundamentals and applications in analytical chemistry. *Critical reviews in analytical chemistry*. 54(5):1200–1226. https://doi.org/10.1080/10408347.2022.2107872.

Shen, H., L. Chen, C. Zhou, J. Du, C. Lu, H. Yang, L. Tan, X. Zeng and L. Dong. 2022. Immobilizing Fe0 nanoparticles on covalent organic framework towards enhancement of Cr(VI) removal by adsorption and reduction synergistic effect. *Separation and Purification Technology*. 290:120883. https://doi.org/10.1016/J.SEPPUR.2022.120883.

Wang, C., Z. Zhang, Y. Zhu, C. Yang, J. Wu, and W. Hu. 2022a. 2D covalent organic frameworks: From synthetic strategies to advanced optical-electrical-magnetic functionalities. *Advanced Materials*. 34(17):2102290. https://doi.org/10.1002/ADMA.202102290.

Wang, M., M. Gao, L. Deng, X. Kang, K. Zhang, Q. Fu, Z. Xia, and D. Gao. 2020a. A sensitive and selective fluorescent sensor for 2,4,6-trinitrophenol detection based on the composite material of magnetic covalent organic frameworks, molecularly imprinted polymers and carbon dots. *Microchemical Journal*. 154:104590. https://doi.org/10.1016/J.MICROC.2019.104590.

Wang, Q., T. Gao, L. Hao, Y. Guo, W. Liu, L. Guo, C. Wang, Z. Wang, and Q. Wu. 2020b. Advances in magnetic porous organic frameworks for analysis and adsorption applications. *TrAC Trends in Analytical Chemistry*. 132:116048. https://doi.org/10.1016/J.TRAC.2020.116048.

Wang, S., H. Wang, S. Wang, L. Fu, and L. Zhang. 2023. Novel magnetic covalent organic framework for the selective and effective removal of hazardous metal Pb(II) from solution: Synthesis and adsorption characteristics. *Separation and Purification Technology*. 307:122783. https://doi.org/10.1016/J.SEPPUR.2022.122783.

Wang, W., H. Shao, C. Sun, X. Jiang, S. Zhou, G. Yu, and S. Deng. 2022b. Preparation of magnetic covalent triazine frameworks by ball milling for efficient removal of PFOS and PFOA substitutes from water. *Environmental Science: Nano*. 9(4):1466–1475. https://doi.org/10.1039/D1EN01158H.

Wang, Y., Y. Pang, Y. Yang, and X. Shen. 2022c. Facile synthesis of recyclable magnetic covalent organic frameworks for adsorption of bisphenol A from aqueous solution. *Journal of Porous Materials*. 29(5):1411–1421. https://doi.org/10.1007/S10934-022-01266-6/FIGURES/7.

Wei, D., J. Li, Z. Chen, L. Liang, J. Ma, M. Wei, Y. Ai, and X. Wang. 2020. Understanding bisphenol-A adsorption in magnetic modified covalent organic frameworks: Experiments coupled with DFT calculations. *Journal of Molecular Liquids*. 301:112431. https://doi.org/10.1016/J.MOLLIQ.2019.112431.

Xin, J., Y. Zhou, X. Wang, G. Xu, M. Xie, L. Liu, R. Zhao, Y. Wu, and M. Wang. 2021. Room-temperature synthesis of magnetic covalent organic frameworks for analyzing trace benzoylurea insecticide residue in tea beverages. *Food Chemistry*. 347:129075. https://doi.org/10.1016/J.FOODCHEM.2021.129075.

You, L., K. Xu, G. Ding, X. Shi, J. Li, S. Wang, and J. Wang. 2020. Facile synthesis of Fe_3O_4@COF covalent organic frameworks for the adsorption of bisphenols from aqueous solution. *Journal of Molecular Liquids*. 320:114456. https://doi.org/10.1016/J.MOLLIQ.2020.114456.

Zhang, M., J. Li, C. Zhang, Z. Wu, Y. Yang, J. Li, F. Fu, and Z. Lin. 2020. In-situ synthesis of fluorinated magnetic covalent organic frameworks for fluorinated magnetic solid-phase extraction of ultratrace perfluorinated compounds from milk. *Journal of Chromatography A*. 1615:460773. https://doi.org/10.1016/J.CHROMA.2019.460773.

Zhang, W., F. Liang, C. Li, L.G. Qiu, Y.P. Yuan, F.M Peng, X. Jiang, A.J. Xie, Y.H. Shen, and J.F. Zhu. 2011. Microwave-enhanced synthesis of magnetic porous covalent triazine-based framework composites for fast separation of organic dye from aqueous solution. *Journal of Hazardous Materials*. 186(2–3):984–990. https://doi.org/10.1016/J.JHAZMAT.2010.11.093.

Zhang, X., S. Wang, X. Zhu, D. Zhu, W. Wang, B. Wang, S. Deng, and G. Yu. 2024. Efficient removal of per/polyfluoroalkyl substances from water using recyclable chitosan-coated covalent organic frameworks: Experimental and theoretical methods. *Chemosphere*. 356:141942. https://doi.org/10.1016/J.CHEMOSPHERE.2024.141942.

Zheng, R., D. Feng, and Y. Xia. 2020. Fe(III)-functionalized magnetic covalent organic frameworks for fast adsorption and removal of phenylbutazone in aqueous solution. *ChemistrySelect*, 5(25):7497–7504. https://doi.org/10.1002/SLCT.202001671.

Zhong, X., Z. Lu, W. Liang, and B. Hu 2021. Incorporating bimetal oxide $MnFe_2O_4$ onto covalent organic frameworks for the removal of UO_2^{2+} ion from aqueous solution. *Applied Surface Science*. 556:149581. https://doi.org/10.1016/J.APSUSC.2021.149581.

Zhong, X., Z. Lu, W. Liang, and B. Hu. 2020. The magnetic covalent organic framework as a platform for high-performance extraction of Cr(VI) and bisphenol a from aqueous solution. *Journal of Hazardous Materials*, 393:122353. https://doi.org/10.1016/J.JHAZMAT.2020.122353.

Zhuang, S., R. Chen, Y. Liu, and J. Wang, 2020. Magnetic COFs for the adsorptive removal of diclofenac and sulfamethazine from aqueous solution: Adsorption kinetics, isotherms study and DFT calculation. *Journal of Hazardous Materials*. 385:121596. https://doi.org/10.1016/J.JHAZMAT.2019.121596.

10 Adsorption of Heavy Metal Ions by Covalent Organic Frameworks

Priyanshi Shukla, Manisha Rawat, and Rekha Sharma

10.1 INTRODUCTION

The discharge of heavy metals into natural aquatic ecosystems, including rivers, lakes, and streams, has been directly or indirectly induced by environmental toxification stemming from various factory-based processes, such as fertilizer production, metallic layering, tanning, battery manufacturing, mining, paper production, and pesticide application (Smith et al., 2015; Datta et al., 2025). Unlike organic pollutants, which can naturally degrade over time, heavy metals are nonbiodegradable, can be dissolved easily in water, and can be transported easily. As a result, organic pollutants break down easily while heavy metals remain in the water for a longer period of time (Lv et al., 2019). Biotoxins and food additives, also known as emerging pollutants, are some harmful compounds that are produced by natural contaminants (Sud et al., 2008). Different types of removal methods have been developed, which include electrolysis (Al-Shannag et al., 2015), precipitation (Peligro et al., 2016), adsorption (Xue et al., 2016; Maftouh et al., 2023), filtering (Lam et al., 2018), and photocatalytic reduction (Li et al., 2016; Chen et al., 2019; Wang et al., 2020a, b, c; Zhang et al., 2020). Due to its flexibility, low cost, effectiveness, and versatility, adsorption has stood out as one of the most remarkable methods over the past 30 years (Huang et al., 2017). Metal oxides (Lingamdinne et al., 2017), zeolites (Yang et al., 2020; Kumari et al., 2024), aluminosilicate minerals (Katsou et al., 2010; Zha et al., 2018), clays (Uddin, 2017), polymers (Zhao et al., 2018), and activated carbons (Nayak et al., 2017) are some traditional adsorbents that are widely used in the water treatment industry. Disordered surface structures that are not crystalline, small pore sizes, limited surface areas (SAs), and chemical bonds or chemical linkages that create hindrance in the adsorption process are some drawbacks of these adsorbents (Peng et al., 2018). Therefore, the investigation of the development of the new adsorbents, which have a wide SA, high porosity, and some well-defined adsorption sites, is essential for the effective removal of heavy metals. COFs are designed with large SAs (Raghav et al., 2025) and outstanding high porosity, which enhances their adsorption capacity and also overcomes these limitations (Gendy et al., 2021). COFs or covalent organic frameworks have several advantageous properties, which are due to their unique combination of crystallinity and organic functionality.

DOI: 10.1201/9781003605386-10

For instance, one of the most extensively studied properties of the COF is porosity. COFs possess permanent porosity, which has consistently been shown by the gas sorption isotherms measurements. Studies have shown that with highly tunable pore sizes, COFs can have both mesopores and micropores. When Yaghi and colleagues synthesized the first successful instances of COFs in 2005, they showed the effectiveness of the topological design principle by creating porous organic frameworks joined by covalent bonds (Cote et al., 2005; Sajid, 2024). The complex integration of organic building pieces into an organized framework with atomic accuracy is made possible by COFs, which are covalent porous crystalline polymers. COFs offer persistent porosity, relatively small densities, and excellent thermal stabilities due to their composition of lightweight materials linked by strong covalent bonds. 2D and 3D COFs can be distinguished based on the dimensions of the construction blocks. Only 2D sheets can have a covalently bonded framework in 2D COFs. These sheets stack to create an eclipsed layered arrangement with columns that are steadily coordinated (Zhang et al., 2023).

It is challenging to produce ordered π systems employing established noncovalent and/or covalent approaches, but this columnar stacking structure offers a special way to do it. Charge carrier movement in the stacking direction may be facilitated by the ordered columns in 2D COFs, suggesting that two-dimensional COFs may be used to create novel p-electronic and photo-functional materials for optoelectronics and photovoltaics. On the other hand, three-dimensional COFs, which extend this framework in three dimensions using a building block that contains a silane or sp (Xu et al., 2023) carbon, are distinguished by their low densities (not exceeding 0.17 g cm^{-3}), many unclosed sites, and enormous characterized SAs (in some cases, more than 4,000 $m^2 g^{-1}$). Owing to these features, three-dimensional COFs are the best options for gas storage. Based on reports, multilayer COFs with square channels have also been created utilizing imine and boron ester linkages and linkers produced from porphyrins or phthalocyanines (Waller, 2019; Sun et al., 2025). Also, these layers stack in an eclipsed manner, creating square pores with diagonals up to 44 Å, some of which exhibit thin film formation (Vardhan et al., 2023). COFs show possibilities for usage in gas storage applications due to their low densities and wide surfaces. Important gases, including carbon dioxide, hydrogen, and methane, may be stored in COFs with great capacity (Chen et al., 2025). Specifically, even at room temperature, COFs have shown themselves to be excellent methane storage materials. 3D organic covalent frameworks often exhibit the highest storage capacities due to their pore volumes and large SAs as compared to layered materials.

10.2 ADSORPTION

When ions or atoms from a liquid, solid, or gas accumulate on a surface, the process is known as adsorption. A layer of adsorbate, such as metal ions, accumulates on the surface of the adsorbent because of this process. Due to its exceptional ability to remove heavy metals and cost-effectiveness, this method is widely used. The conventional methods that are used to eliminate or remove heavy metals are effective, but they are limited and also have some drawbacks, which include a large amount of hazardous sludge, high cost, need for excessive chemical demand, and heavy energy consumption. Therefore, there is a growing need to develop a therapeutic strategy that is not only environmentally sustainable and effective but also easy to use and affordable (Chakraborty et al., 2021). Adsorption is considered one of the most preferred methods because of its cost-effectiveness and remarkable efficiency. However, this method also has its limitations, which include the high-cost preparation of the adsorbent, like activated carbon, the adsorptive potential with each regeneration cycle gradually declines, and the need for chemical regeneration (Bilal et al., 2013). Certain waste materials and some natural resources from agricultural and industrial or commercial activities offer an affordable alternative to remove heavy metals. Cost remains a crucial factor when comparing different adsorbent options. When adsorbent requires minimal processing, is readily available, or is derived as a byproduct from commercial or agricultural activities, then the adsorbent is often considered a "low-cost adsorbent" (Yagub et al., 2014). Numerous natural adsorbents such as sugar beet pulp (Pehlivan et al., 2008), coal fly ash (Patel et al., 2023), coir pith (Jatkar et al., 2024), cellulose extracted from rice husk (Kumar & Sharma, 2019), and castor tree leaf powder (Al Rmalli et al., 2008) have been documented in previous studies as effective wastewater treatment methods for eliminating harmful contaminants including colorants and heavy metals.

10.3 ADSORPTIVE REMOVAL MECHANISMS USING COFs

Research on the mechanisms of porous crystalline formations in order to determine the kind of polycondensation and crystallization of COFs is extremely important. To improve the effectiveness of heavy metal capture, it is essential to comprehend the adsorption mechanism. For pollutants to be effectively removed, metal ions must have a significant affinity for a certain functional group of COFs. Electrostatic interaction, hydrogen bonds, ion exchange, and coordination bond/chelation all contribute to the explanation of the absorption mechanism.

10.3.1 Electrostatic Interaction

pH has an impact on the electrostatic attraction process. Because the H^+ ions around the surface contend with metal ions for the adsorption sites on the COFs, the adsorption performance is poor at low pH. Consequently, the adsorption sites' effective adsorption affinity has dropped. The rate at which metal ions were removed was constant at high pH and reversed with other metals. To collect Hg (II), the S-functional group of COFs in their various forms (disulfide group, thioether group, thiol group, sulfur, and thioamide group) is essential. The sulfur-functionalized COF materials have a positively charged surface at low pH that obstructs the interaction with the mercury complex, which is cationic (such as S-Mn^+ or Mn^+) and stops further adsorption. According to reports, the O-functionalized group of COFs in COF–BTA–DHBZ removes Cr^{6+} by electrostatic attraction and hydrogen bonding. The elimination mechanism is regulated by the hydroxyl group (OH) on the COF backbone. Adsorption was enhanced at low pH by the electrostatic attraction and hydrogen bonding between anionic Cr (VI) species and (-OH) units. The largest amount of Cr (VI) adsorption occurs at pH = 1. The protons of the phenol's hydroxyl (OH) group are extracted at high pH values. This makes it difficult for Cr^{6+} to be absorbed since the remaining anionic O^- and anionic Cr (VI) species are hostile to one another (Cui et al., 2019). COFs, which possess nitrogen or N-functional groups, have been identified as the (=N^+, NH) groups in CTF-1, γ-Fe_2O_3@CTF-1, BT-DGC1, and COF(V). The elimination mechanism is shown in water via electrostatic interactions between the (=N^+, NH) groups of COFs and metal ions (Ghazi et al., 2018). Nitrogen and oxygen (N, O) were the functionalized groups, which were introduced in TpPa-1, Fe_3O_4@COF(TpPa1), EB-COF: Br, iCON, and TpODH as (–CONH, C=N–, C=O) and in COF (COF-SH) as (–NH, –SH), N, S-functionalized groups were reported.

10.3.2 Hydrogen Bond and Ion Exchange

The process responsible for capturing Se^{6+}, Cr^{6+} S^{6+}, As^{5+}, Mo^{6+}, and As^{5+} in an oxyanion and BT-DGCl has been explained by the hydrogen link that exists between a functional N-H group. Additionally, hydrogen bond is formed between the carbonyl (–C=O) group of EB-COF: Br and As (V) and p^{5+}. In $[NH_4]+\left[\text{COF-SO}_3^-\right]$, the cation interchange between $[NH_4]+$ and Th^{4+} ions was described as the ion exchange process. In addition to chemical structures, it may be inferred that the heteroatoms N, S, and O are crucial to the mechanism of COF elimination. Particular functional groups of COFs were shown to have strong heavy metal sensitivity and adsorption capability. For instance, S-functionalized groups like TTB-COF and TTB-COF have shown great capacity and sensitivity for mercury and gold.

10.3.3 The Effect of Chelation and Coordination Bonds

The two most resilient kinds of interactions are chelation and coordination bonds. These bonds are based on heteroatoms such as N, S, and O that may form stable metal complexes through heavy metal coordination. The S, O, and N atoms' unpaired electrons and the high electron density of COFs allow them to establish coordinated bonds and draw metal ions out of water. Prior research has shown that S-functionalized groups exhibit great sensitivity to Hg^{2+} and Au and strong selectivity for various heavy metals, including Pb^{2+}, Cu^{2+}, Hg^{2+}, UO_2^{2+}, and Au. The exceptional sensitivity and selectivity of COFs may be due to the effective charge that is transferred by the electron-rich thiols to the electron-deficient Mn^+ ions with a large adsorption capacity. For example, the (–SR) group of TAPB-BMTTPA-COF and COF-LZU8 was primarily responsible for the mercury removal process, which included the –SR group forming a chelation bond with Hg^{2+}. Moreover, the chelation connection between Hg^{2+} and Sn^{2+} and the (–SH) group of TPB-DMTP-COF-SH. A strong coordination between ethylenediamine tetraacetic acid (EDTA) in these COF and Cr (III) allowed for the observation of the O-functional group in the removal process of Cr (III) by TpPa-NH_2@EDTA. In addition to the coordination bond, the aromatic skeleton of COFs plays a significant role in improving the adsorption potential. For example, when a multi-coordination bond was formed between Pb^{2+} and COF-TP/COF-TE, COF-TE exhibited a larger capacity for lead adsorption than COF-TP due to its weak π-π stacking and less aromatic skeleton, which enables more internal diffusion of

Pb^{2+} adsorption. It became apparent that COF-616-CY and COF-4PE-3P captured copper with N-functionalized groups. When the (–NH, $=N^{+}$) group of COFs and Cu^{2+} forms a chelation bond, the adsorption processes may be described. COF-SO_3H, TpPa-NO_2, and TpPa-1 atoms S, N, and O are capable of forming a coordination link with Th^{4+}. Two (2) or three (3) heteroatoms can combine to form the coordination bond and chelation outcome. In COF (COF-SH), N, S-functionalized groups are obtained as (–NH, –SH), whereas TpPa-1 and TpPa-NO_2 have N, O-functionalized groups as amide bonds (–CONH) and COF-616-MTE and COF-616-DTT have O, S-functionalized groups as (–OR/SR, –OH, SH), respectively. $[NH_4]+[COF\text{-}SO_3]$ COF contains N, O, and S-functionalized groups as (–CO, NH, $-SO_3$) (Gendy et al., 2021). As seen in Table 10.1, there is an evaluation of the effectiveness of different COF materials in the adsorption of As (III, IV) and Cr (III, IV).

10.4 SYNTHESIS OF COFs

Understanding the right conditions for COF synthesis is essential if they are to be applied in practical settings. Numerous more techniques have been devised to synthesize COFs in addition to the widely utilized solvothermal methods (Li et al., 2020). Figure 10.1 shows the techniques that have been employed to create COFs.

10.4.1 Microwave Synthesis

A lot of research has been done on microwave irradiation as a substitute technique for creating COFs (Gawande et al., 2014). Nontoxic products, the ability to concurrently adjust temperature and pressure, faster reaction speed, and phase behavior monitoring are only a few of its numerous benefits over the solvothermal approach (Nüchter et al., 2004). Cooper and colleagues recently synthesized COF-5 using microwave irradiation combined with stirring at a temperature of 100°C for 20 minutes at a power setting of 200 W (Campbell et al., 2009; Ritchie et al., 2010). The following engineered method demonstrated a speed increase of 200 times compared to the previously described solvothermal synthesis, which required 72 hours (Cote et al., 2005). Furthermore, the removal of impurities through the microwave-assisted synthesis enhanced the SA of COF-5; its BET (Brunauer–Emmett–Teller) surface area surpassed that of the analogous material produced via the solvothermal method. Impurities were removed from this substance by heating the gray precipitate of isolated COF-5 that formed in the reaction vessel with dry acetone for 20 minutes at 200 W and 55°C. For the effective removal of the supernatant's purple hue, the procedure was conducted two times, which was a consequence of the formation of an oxidized form of hexahydroxy triphenylene (HHTP).

10.4.2 Mechanochemical (MC) Synthesis

Due to its simplicity, speed, lack of solvents, and environmental friendliness, the MC route—which involves hand grinding at room temperature using just a mortar and pestle—is attracting a lot of interest (Friščić, 2010; Li et al., 2024). Banerjee et al. suggested employing a manual grinding arrangement with a mortar and pestle to

TABLE 10.1
Evaluation of the Effectiveness of Different COF Materials in the Adsorption of As (III, IV) and Cr (III, IV)

COF Adsorbents	Optimum pH for Cr (III, IV)	Competitive Adsorption for Heavy Metals	Selectivity and Sensitivity for Heavy Metals	Max. Cr (III, IV) Removal Capacity ($mg\ g^{-1}$)	Equilibrium Time of the Cr (III, IV) Adsorption	Ref.
Fe_3O_4@COF (TpPa-1)	1	Cr^{6+}	Cr^{6+}	245.45 $mg\ g^{-1}$	60 minutes	Zhong et al. (2020)
COF–BTA–DHBZ	1	Cr^{3+}	Cr^{6+}	272 $mg\ g^{-1}$	12 hours	Cui et al. (2019)
EB-COF: Br	7	As^{5+}, P^{5+}	As^{5+}, p^{5+}	35.3 $mg\ g^{-1}$ and 53.1 $mg\ g^{-1}$	60 minutes	Yang et al. (2020)
TpPa-NH_2@EDTA	3	Cr^{3+}, Ag^{+}, Ni^{2+}, Cu^{2+}, Fe^{3+}, Pd^{2+}	all > 85%	232.55 $mg\ g^{-1}$	10 minutes	Jiang et al. (2019)
CTF-1	8	Ca^{2+}, As^{5+}, Mg^{2+}, Hg^{2+}, As^{3+} and natural organic matter	As^{5+}, Hg^{2+}	28.6 $mg\ g^{-1}$ and 32.8 $mg\ g^{-1}$	1,400 minutes	Leus et al. (2018)
BT-DGCl COF	2	Cr^{6+}, Mo^{6+}, S^{6+}, Se^{6+}, As^{5+}	Cr^{6+}	200 $mg\ g^{-1}$	<5 minutes	Jansone-Popova et al. (2018)
γ-Fe_2O_3@CTF-1	7	As^{3+}, As^{5+}, Ca^{2+}, Mg^{2+}, Hg^{2+}, and natural organic matter	As^{3+}, As^{5+}, and Hg^{2+}	198 $mg\ g^{-1}$, 101 mg g^{-1}, and 166 $mg\ g^{-1}$	10 minutes	Leus et al. (2018)
TpPa-1	1	Cr^{6+}	Cr^{6+}	310.8 $mg\ g^{-1}$	10 minutes	Zhong et al. (2020)

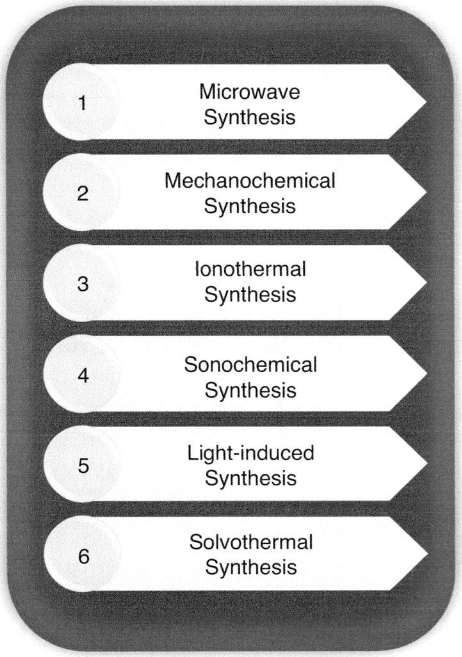

FIGURE 10.1 Approaches for covalent organic frameworks (COFs) synthesis.

quickly and solvent-free MC synthesize the COFs using Schiff base condensations at ambient temperature (Biswal et al., 2013). The color shift that happened while grinding was a sign that COFs were forming. Furthermore, unlike COFs made solvothermally, the COFs formed graphene-like layered structures as a result of the exfoliation of the COF layers during the MC synthesis. By condensing 4,4′-azodianiline and Tp, Wang et al. recently synthesized the COF TpAzo in 20 minutes using the same technique; this COF possessed a sheet-like shape, good thermal stability, and porosity (Wang et al., 2018).

10.4.3 Ionothermal Synthesis

The ionothermal method uses a molten salt or an ionic liquid as a solvent and a catalyst to create solids at high temperatures (about 400°C) and pressures (Kuhn et al., 2008). By cyclizing and trimerizing nitrile structure units in molten $ZnCl_2$ at 400°C, Thomas et al. created porous crystalline COFs using this technique; these CTFs showed excellent crystallinity as well as strong chemical and thermal stabilities

(Kuhn et al., 2008; Bojdys et al., 2010). However, the severe conditions needed for this process reduce its appeal since they restrict the range of building blocks that may be used and may also generate unfavorable side effects and breakdown, which restricts its application. To create 3D COFs, Qiu et al. have created an ionothermal method that uses ambient temperature and pressure. They employed 1-butyl-3-methylimidazolium bis((trifluoromethyl)sulfonyl) imide, a room-temperature liquid, as the catalyst and solvent for the synthesis of Schiff bases. Comparing these 3D-IL-COFs to the time-frames needed for conventional solvothermal procedures (3–7 days), the former developed in a substantially quicker reaction time (12 hours) (Guan et al., 2018).

10.4.4 Sonochemical Synthesis

Given its cost-effectiveness and rapidity, sonochemical synthesis might be favored compared to conventional methods (Bang & Suslick, 2010). This method employs ultrasonic waves to generate bubbles within the solvent, which subsequently expand and collapse in a phenomenon known as cavitation. This generates significantly elevated regional temperature and pressure, initiating and accelerating chemical processes. Ahn et al. utilized this method to synthesize COF-1 and COF-5 within a concise 1-hour reaction period (Yang et al., 2012). The old technique took 400 times as long to synthesize COF-1, while this sonochemical approach produced a high SA of 2,122 $m^2 g^{-1}$ while increasing the reaction batch size to 0.5 L. To generate core/shell structures, Shim et al. synthesized COF-5 using this gentle technique on the surfaces of graphene and carbon nanotubes (Yoo et al., 2015).

10.4.5 Light-Induced Synthesis

A novel highly conjugated and crystalline COF was developed under simulated sunshine irradiation at a wavelength ranging from 200 to 2,500 nm, facilitated by the presence of a co-catalyst. In a controlled environment, a mixture of 1,2,4,5-benzenetetramine tetrahydrochloride and hexa keto cyclohexane octahydrate was solubilized in a blended solvent. Acetic acid and water were introduced as co-catalysts to induce condensation. Without light, an amorphous form of hcc-COF was produced. The creation of an extended conjugated structure that improved charge transport is likely the reason for the crystalline hcc-COF's exceptional electrical conductivity (2.22×10^{-3} S m^{-1}) (Kim & Choi, 2019).

10.4.6 Solvothermal Synthesis

Solvothermal techniques have been used to generate several COFs, which are comparable to the standard autoclave synthesis of inorganic zeolites (Davis, 2002). The typical approach involves heating a blend of monomers, which have been combined in a range of meticulously chosen solvents, within a sealed Pyrex tube at temperatures between 80°C and 120°C for 3–7 days. Selecting the right solvent is essential to improving the monomers' solubility. Solvent combinations may expedite the nucleation of COFs by facilitating the diffusion of the monomers into the solution. Additionally, the enclosed environment guarantees the availability of water during the reaction, which enhances the COFs' crystalline development (Lu et al., 2015). Using this technique, Yaghi et al. published the first COFs (Cote et al., 2005). Through the self-condensation of 1,4-benzenediboronic acid, COF-1 was created; it had a multilayer hexagonal structure. Likewise, BDBA and HHTP were condensed to create COF-5, which featured a coplanar extended sheet structure. Their prospective uses may be limited since most COFs made with solvothermal techniques have been intractable and unprocessable powders. Dichtel and colleagues employed a solvothermal technique to create 2D COF films on single-layer graphene; this approach might be useful for creating organic electronic devices, where the production of thin films is crucial to improving optoelectronic characteristics (Colson et al., 2011). Several additional 2D COF films with special qualities have also been produced with this technique (Wang et al., 2019).

10.5 CHARACTERISTICS AND USES OF COFs

10.5.1 Porosity and Gas Storage

COFs have several beneficial qualities due to their unique blend of organic functionality and crystallinity. For instance, porosity is one of the most researched properties of COFs. Measurements of gas sorption isotherms have frequently demonstrated their persistent porosity.

COFs that have pores have been documented, and their pore size is also quite adjustable. In two-dimensional layered formations, channels that run parallel to the stacking direction are formed and give rise to porosity. A careful choice of linkers helps fine-tune the channel diameters. For instance, COF-1 has a 7 Å pore aperture due to its structure, which is structured on hexagonal layers that stack in a staggered manner. Frameworks with eclipsed layers and channels with pores that are hexagonal and their widths varying from 27 Å (COF-5) to 47 Å (HHTP-DPB COF) were created by isoreticular extension of this COF (Cote et al., 2005; Sajid, 2024). Similar to that, multilayer COFs that have specific square channels were reported; they are constructed using imine and boron ester linkers that come from phthalocyanines or porphyrins (Waller, 2019; Sun et al., 2025). Moreover, some of these layers evolved as thin films. These layers stack in an eclipsed form, resulting in square pores with diagonals as broad as 44 Å (Vardhan et al., 2023). Because of their low densities and large SAs, COFs have potential for use in gas storage technologies. Using COFs, high storage capacity for important gases such as carbon dioxide, hydrogen, and methane has been shown (Chen et al., 2025). COFs have shown themselves as an outstanding methane storage material, particularly at room temperature. 3D COFs generally showed higher potential to store because of having a bigger SA in comparison to layered materials. One example shows that at 298 K and 80 bar, the 3D non-interpenetrating COF-102 has a Langmuir SA of 4,650 m^2g^{-1} and a methane storage capacity of 25 wt.% (Li et al., 1999). With the capacity of 15 wt.% (Kang et al., 2023), the 9-fold interpenetrating COF-320 shows significant adsorption of methane under the same conditions. Due to their significance in renewable energy, materials with higher capacity to store hydrogen have been a main focus, and the evaluation of COFs has also been done for this purpose. Both COFs, i.e., COF-102 and COF-103, have a higher capacity to store hydrogen in comparison to all the COFs that have been reported. The hydrogen uptake value for COF-102 at 77 K and 85 bar is 72.4 mg g^{-1} and for COF-103 it is 70.5 mg g^{-1}. The mentioned values match those reported for the MOFs (metal-organic frameworks) with optimal performance (Chen et al., 2025). When metal sites are added to COFs, then the bond between the hydrogen molecules and the adsorbent strengthens, which results in the capacity to store at room temperature. Simulations show that metalating COF-301 with $PdCl_2$ yields COF-301-$PdCl_2$, a substance that would surpass the DOE (design of experiments) objective for room-temperature hydrogen storage (Mendoza-Cortes et al., 2012). The linkers that are tetrahedral in shape and are called DHTA (2,5-dihydroxyterephthalaldehyde) and TAM (Tris(4-aminophenyl) methane) undergo imine condensation to produce multiple interpenetrating dia frameworks, which is how this framework is created. The imine linkage has -OH (hydroxyl) group next to it, which establishes metal-bonding sites for metal centers, making this COF's distinctive property. Grand canonical Monte Carlo computations were performed to model H_2 adsorption isotherms, and hybrid DFT (density functional theory) functionals were employed to determine the interactions between H_2 molecules and the framework. At 100 bar and room temperature, COF-301-$PdCl_2$ might hold up to 60 g L^{-1}, which is indicated by the outcomes of these models. Moreover, to trap and then for the storage of hazardous gases, COFs can be used. Ammonia, for example, is a chemical that is used extensively but must be carried out properly since it is dangerous and corrosive. Engaging with Lewis acidic (LA) groups, such as the boron atoms found in boroxine or boronate ester-based COFs, may successfully capture ammonia, a Lewis base. By the condensation of the bi-phenyl di-boronic acid and hexa hydroxy triphenylene construction units, the COF-10 is created, which has an elevated ammonia storage potential of 15 mol kg^{-1} (Doonan et al., 2010). After three successive cycles of ammonia adsorption and desorption, it is noteworthy that neither capacity nor crystallinity is lost.

10.5.2 Optoelectronics

The optoelectronic characteristics of the COFs have also found use beyond the realm of catalysis. For example, semiconducting and luminous characteristics are shown by a TP-COF, which is made from triphenylene and pyrene linkers (Wan et al., 2008). Furthermore, two COFs, i.e., COF-366 and COF-66, which are based on porphyrin, showed that they have an extensive charge in the mobility of the charge carriers (Sun et al., 2025). As these materials have small interlayer distances and adopt an eclipsed structure with square channels, they are an ideal choice for this purpose. Hole conduction is enabled in both frameworks and displays values of charge mobility that are greater than those of typical conducting polymers, with COF-366 and COF-66 having 8.1 and 3.0 V/s/cm (Li et al., 1999), respectively. Additionally, charge carrier mobility has been shown for the highly conjugated CS-COF (Guo et al., 2013) made by condensing HATP with tBPyTO and for phthalocyanine-based COFs (Waller, 2019). In the latter instance, COF and C_{60} were combined to create a donor/acceptor composite material with superior photoconductive qualities. Lastly, the interaction of the redox-active anthraquinone building block, 2,6-diaminoanthraquinone (DAAQ), with 1,3,5-triformylphloroglucinol (TFP) produced a COF that could store energy

(DeBlase et al., 2013). Compared to monomeric 2,6-diaminoanthraquinone (DAAQ) and non-redox-active COF materials, this framework shows a capacitance that is much greater and remains constant even after 5,000 charge–discharge cycles.

10.5.3 Catalysis

Porous substances or materials, which are used as heterogeneous catalysts, show an exciting approach, combining the favorable points of established heterogeneous catalysis with enhanced access to active sites. Nevertheless, the application of COFs in catalysis remains inadequately investigated in comparison to other types of porous materials, even characterized by having exceptional adaptability and being highly porous. In various substrate-based Knoevenagel condensation processes, two base-functionalized COFs—BF-COF-1 and BF-COF-2—have been employed as catalysts not long ago (Fang et al., 2014). These materials are generated through the process of imine condensation between TAA and either TFB (1,3,5-Triformylbenzene) or TFP. Subsequently, they construct BF-COF-2 or BF-COF-1. The resulting compounds exhibit high activity when benzaldehyde is converted to 2-benzylidenemalononitrile, with BF-COF-1 and BF-COF-2 exhibiting 96% and 98% activity, respectively. However, the pore aperture diameters of these two COFs are crucially different, resulting in altered selectivities. Postsynthesis, the catalytic species can be introduced into the framework despite the inherent activity of BF-COF-1 and BF-COF-2. For example, postsynthetically adding Pd $(OAc)_2$ to an imine-linked COF led to the establishment of Pd/COF-LZU-1 (Ding et al., 2011). The Pd atom is coordinated with the nitrogen atoms of the framework and is bound between two adjacent COF layers, as indicated by data from X-ray photoelectron spectroscopy. In Suzuki–Miyaura couplings, the substance has displayed exceptional reactivity and significant recyclability. Following the synthesis of COF, a layered imine COF generated by the solution infiltration approach was also utilized to immobilize gold nanoparticles. When 4-nitrophenol is reduced to 4-aminophenol, the resultant compound exhibits a significant activity (Pachfule et al., 2014). Furthermore, a platinum-doped hydrazone-based COF has recently demonstrated catalytic activity in the photocatalytic generation of hydrogen (Stegbauer et al., 2014). Here, the COF serves as a photosensitizer to produce excitons, while platinum catalyzes proton reduction.

10.5.4 Thin-Film COFs

The goal of more studies is to enhance the applicability of COFs in optoelectronics. One ongoing objective in this field is the COF thin film construction, and two methods have been used to achieve this goal. The substance may first develop in situ on a surface and second before being discharged on a surface, the COF might develop first. On a single layer of graphene and an Ag (111), COFs have been developed using a basic approach (Colson et al., 2011; Vardhan et al., 2023). A layered COF as a powder is created in the second method, and the powder is then exfoliated to generate materials with a limited number of sheets that might be used on a surface. Sonication (Gaikwad & Han, 2024), mechanical delamination (Chandra et al., 2013), and immersion (Bunck & Dichtel, 2013) in suitable solvents are various types of

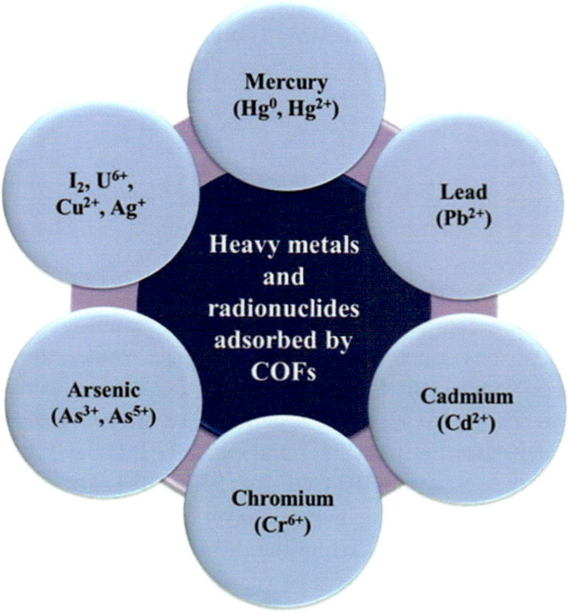

FIGURE 10.2 Illustration showing the types of heavy metals and radionuclides, which are adsorbed by COFs.

techniques that have been verified. Linkers that have inherent electrical conductivity, for example, thiophene derivatives, are under examination to improve and modify the electronic properties of COFs. Two COFs of this kind have been documented using TTDVA (4,4′-(Thiazolo[5,4-d]thiazole-2,5-diyl)divanillin) (Dogru et al., 2013) or BTDVA (4,4′-(Benzothiadiazole-4,7-diyl)divanillin) (Medina et al., 2014) combined with HHTP. A thin film of these COFs was developed on the surfaces that were polycrystalline in nature, so their electrical properties can be improved. Then, they were mixed with fullerene derivatives to produce merged donor–acceptor materials. An isoreticular sequence of COFs is synthesized when TDBA (2,4,6-triformyl pyrogallol), TTDBA (4,4′-(Thiazolo[5,4-d]thiazole-2,5-diyl)dibenzaldehyde), and BTDBA (4,4′-(Benzothiadiazole-4,7-diyl)dibenzaldehyde) condense with HHTP (Bertrand et al., 2013) (Figure 10.2).

10.6 HEAVY METAL REMOVAL VIA COFs

10.6.1 Removal of Lead (Pb)

In industries like battery production, mining, ceramic glass, smelting, and plating, lead ions (Pb^{2+}) are predominantly produced by human activities (Feng et al., 2018; Awual, 2019; Wu et al., 2019). The human body parts such as kidneys, reproductive system, brain, and liver are all harmed by the documented Pb^{2+} toxicity, which is lethal to living things (Awual, 2019; Wu et al., 2019). Increased occurrences of disorders like encephalopathy and hepatitis may result from higher levels of lead buildup in fresh water (Du et al., 2017; Awual, 2019; Zhu et al., 2019; Ma et al., 2021). The US Environmental Protection Agency sets the permissible limits at 0.05 mg L^{-1}, while the European Union restricts 0.01 mg L^{-1} (Awual, 2019).

To remove Pb^{2+} ions from aqueous solutions via an ion exchange technique, Gupta and colleagues (Gupta et al., 2018) created an ionic COF in 2018. According to the results, applications for water purification might find a feasible remedy in synthetic COF material. An innovative electrochemical sensor, a TAPB-DMTP-COF, which detected the lead (II) ions in an aqueous medium, was designed by Zhang et al. in 2018 (Zhang et al., 2018). At the surface of the TAPB-DMTP-modified COF, lead (II) ions have been coupled by using complexation with amine groups. Due to the outstanding structural characteristics of the COFs, the sensor, which was developed, showed excellent fast electron transfer and a high lead adsorption capacity. Because of having active sites and significantly large SA, the novel composite COF material showed a low detectable level, remarkable stability, enhanced sensitivity, extended linear range, and consistent performance. Because of having all of the above-mentioned characteristics, the researchers came to the conclusion that the sensors based on COF will be favorable for environmental remediation and clinical evaluation. Using a specific process, known as polymerization reaction between acyl chloride (R-COCl) and amino groups ($-NH_2$) via MC ball milling at room temperature, two amide-based COF materials were developed by Li et al. (Li et al., 2019) that functioned as Pb^{2+} adsorbents. Two diamine monomers, the first one COF-TP, which was derived from the aromatic diamine p-phenylenediamine, and the second one COF-TE, which was derived from the linear diamine ethylenediamine — which displayed different framework structures and functional groups—were selected for COFs' synthesis. A distinct lamellar structure of the synthesized COF materials showed that the amide groups that were present in the COF, worked as active sites for the adsorption of the lead (II) ions. COF-TE showed a higher adsorption potential for Pb^{2+} than COF-TP because the weak π-π stacking and smaller aromatic skeleton of COF-TE allowed the enhanced internal diffusion of Pb^{+2}. Moreover, COF-TE had a superior saturation adsorption capacity of 185.7 mg g^{-1}, which is due to its enhanced amide group concentration, than COF-TP, which has an adsorption capacity of 140 mg g^{-1}. Researchers also concluded that the COF, which is based on amide, has the potential to work as adsorbent materials for Pb^{+2} ions because it has an excellent ability to be recycled for Pb^{+2} adsorption. Modified with ultra-filtration membranes that are made of polyvinylidene fluoride, Xu and his team (Xu et al., 2020) introduced COFs as a simple approach for the fabrication of ultra-filtration membrane, which were meant to eliminate lead ions. The elimination efficiency of synthesized COF membranes was found to be 87.5% during the fourth reusability cycle, while it was equal to 92.4%. Hence, for the effective treatment of water that has Pb^{2+} ions and wastewater, this novel composite material plays an important role. A mild solvothermal solution suspension synthesis method was used by Cao et al. (2020) to synthesize a novel COF. The lead ions were effectively removed from the aqueous solutions using it. The synthesized COF material possessed a lamellar structure, and it was also shown that on the skeleton of COF the sulfhydryl groups were attached, which were able to work as an active adsorption site for the Pb^{+2} ions. The COF material that was sulfhydryl functionalized showed the exceptional adsorption potential of 239 mg g^{-1}.

This potential significantly surpasses the capacity of traditional adsorptive substances, and it was able to attain adsorption equilibrium in a period of 48 hours.

10.6.2 Removal of Cadmium (Cd)

Highly soluble in comparison with other heavy metals and which can easily penetrate the biosystem, cadmium (Cd) is known as one of the most hazardous heavy metals (Dinari & Hatami, 2019). Cadmium is identified as a carcinogenic or cancer-causing agent and categorized under group B1 classification according to the United States (US) environment protection agency (EPA) but according to the international agency for research on cancer, cadmium is carcinogenic (cancer causing) and falls under category 1 classification (al Sadat Shafiof & Nezamzadeh-Ejhieh, 2020; Nguyen et al., 2020). Cadmium causes hypertension, renal failure, testicular shrinkage, emphysema, and a well-known disease itai-itai in human beings because of its ability to build up in the human body in specific regions such as reproductive, renal, skeletal, and neurological systems (Liu et al., 2020; Nguyen et al., 2020). A unique heteropore COF was synthesized by Liu and colleagues in the year 2019 (Liu et al., 2020). The adsorption of the cadmium (II) ions from the aqueous medium can be handled using this framework. The outstanding high level of chemical stability toward water across a broad range of pH values was shown by this synthesized COF. Moreover, it has a broad variety of easily accessible metal ion chelating sites, potentially large SA of 1476.6 $m^2 g^{-1}$, increased crystallinity, and organized porosity. Liu and his colleagues synthesized a COF, having some outstanding characteristics that include rapid adsorption rates, outstanding absorptive potential of 116 mg g^{-1} to absorb Pb^{+2} ions, and elevated ability to be recycled. Moreover, crystalline COF that was reinforced with nitrogen was constructed by Dinari and Hatami in the year 2019 (Dinari & Hatami, 2019). The use of this framework was to work as a Cd^{+2} adsorbent and eliminate cadmium from the aqueous medium, and the synthesis of this framework included the condensation of 2,4,6-tris(hydrazine)-1,3,5-triazine and 2,4,6-tris (p-formyl phenoxy)-1,3,5-triazine. When it came to Cd^{2+} adsorption, the as-prepared COF performed well with an increased potential of 396 mg g^{-1}. The synthesized substance was also very efficient, economical, and environmentally benign.

10.6.3 Removal of Mercury (Hg)

When taking into account the amazing affinity toward binding that exists between mercury (II) and sulfur derivatives, a great number of sulfur-containing COFs have been developed and tested for their powerful adsorption capabilities and selectivity for Hg^{2+} ions (Wang & Zhuang, 2019). To occupy the unfilled 6s orbital, the ion Hg^{2+} tends to react with electron-donating elements to form stable compounds (like sulfur to form mercuric sulfide) and ligands that contain phosphorus, nitrogen, and oxygen. However, mercuric sulfide is unstable at high temperatures (greater than 400°C) and breaks down into its metallic form (Velempini & Pillay, 2019; Liu et al., 2023). Hard bases tend to form stable complexes with hard acids, according to the hard–soft acid and base theory. On the other hand, soft bases are more likely to form stable complexes with soft acids. Regarding mercury and sulfur, the latter is a soft base that forms a covalent bond with Hg, which is the soft acid. It is distinguished by its high polarizability, low electronegativity, and comparatively large ionic size (Ram & Chauhan, 2018; Velempini & Pillay, 2019). A well-designed COF with triazole and thiol groups in its cavities was investigated by Merí-Bofí and colleagues (Merí-Bofí et al., 2017) in 2017 as a possible Hg^{2+} sorbent material. Incorporation of thiol and triazole groups into the extremely porous organic polymer, which is known as TPB-DMTP-COF (triphenylbenzene; diethoxyterephth aldehyde), was done which ultimately resulted in positive output under mild circumstances. When it comes to adsorption of Hg^{+2} ions from both concentrated and dilute solutions, the novel synthesized material showed a remarkable efficiency in eliminating Hg^{2+} ions. The resulting COF that is imine based showed an exceptional affinity for Hg^{2+} ions with fast kinetics, and it potentially reduced the concentration of Hg^{2+} from 10 mg L^{-1} to 1.5 μg L^{-1}, and this limit is below the allowed limit of drinking water. The new perspectives are presented by this material on the usage of COFs in environment remediation and Hg^{2+} elimination. A 2D mesoporous COF, which is loaded with flexible and densely packed thiol and thioether chelating groups, was developed by Sun et al. (2017) in the year 2017. A strong affinity for Hg^{2+} with impressive distribution coefficient values over a broad pH range of 3–10, as well as extremely rapid kinetics for Hg^{2+} adsorption, was all shown by the as-synthesized COF. The mercury (0) and mercury (II) adsorption capacities were also very high, accounting for 863 and 1,350 mg g^{-1}, respectively. Their results validated the significant potential of COFs as an adsorptive substance for mercury (II). In 2018, Leus and associates developed a composite material using Fe_2O_3 nanoparticles (in the γ-Fe_2O_3 phase of maghemite) and a CTF-1 (called covalent triazine framework) (Leus et al., 2018). The adsorptive potential of γ-Fe_2O_3@CTF-1 for Hg^{2+} was 165.8 mg g^{-1}, much higher than that of pure CTFs and other materials based on iron. The composite material, which is as-prepared, may also be recycled for several cycles due to its great adsorption capabilities. Furthermore, a new technique for creating porous covalent poly triazines with high sulfur levels (13.2 wt.%) was introduced by Fu et al. (Fu et al., 2018). As they have enhanced chemical and physical stability, substantial adsorption potential, and Hg ion elimination potential, even at trace levels, the synthesized covalent triazine framework is ideal for water decontamination and effluent treatment applications. In 2019, Mondal and colleagues (Mondal et al., 2019) reported a remarkable adsorption capacity of 1,252 mg g^{-1} for mercury (II) and 813 mg g^{-1} for mercury (0), by using an ionothermal technique to synthesize a covalent triazine network functionalized with SCTN-1 or thioester groups. The elimination of hazardous Hg through the use of the organic porous material is facilitated by a simple operating procedure, high distribution coefficients, fast adsorption kinetics, and simplicity in adsorbent renewal, as demonstrated by these researchers. Because of the previously mentioned properties, these COFs show great adsorptive potential for Hg^{+}, so they are great adsorbent materials for the removal of mercury ions. A simple one-step infiltration synthesis to enable the in situ synthesis of the AgNPs (silver nanoparticles) was utilized by Wang et al. (Wang et al., 2020a, 2020b, 2020c). Due to its high adsorption capacity, outstanding Ag atom utilization, enhanced selectivity and stability, fast adsorption kinetics, and better reusability, the synthesized AgNPs@COF material has shown high potential as an Hg^{2+} ion adsorbent from acidic solutions. This creates new opportunities for utilizing COFs as carriers to stabilize metal NPs (nanoparticles), making them potential options for acidic water treatment. With the help of Suzuki polymerization of two monomers derived from TAA (triarylamine), He et al. (He et al., 2020) developed a novel COF polymer. They also designed a Schiff base-based COF fluorescent sensor using that polymer, which was potentially capable to detect and remove the mercury (II) ions. Tao and his colleagues (Tao et al., 2020) in 2020 developed a SO_3-anchored COF, which was porous, and was used for the adsorption of both types of mercury, i.e., mercury (0) and mercury (II). Exceptional selectivity, rapid adsorption kinetics, potential binding affinity, exceptionally high adsorption potential for both types of mercury, i.e., mercury Hg^0 and Hg^{2+}, and good sustainability, are some outstanding properties shown by the synthesized material. A revolutionary metal-free, fluorescent COF that functions as a chemosensor for the simultaneous detection and elimination of mercury (II) ions was designed by Panda et al. (Panda et al., 2020). The synthesized COF was characterized by its large pore dimensions as well as enhanced π-conjugation, which permitted it to be used as a highly effective sensor for the Hg^{2+} ions with the exceptional removal performance of 99.8%. This substance is among the few metal-free adsorbents and sensors applicable in various industrial effluent management contexts. A one-step approach was developed using a sulfhydryl (-SH) modified COF to eliminate the mercury (II) ions from the contaminated water in 2020 by Ma and his team. For Hg^{2+} ions, the synthesized COFs had an outstanding adsorption potential of 1,283 mg g^{-1}, overtaking that of the COF lacking -SH by almost 25 times. Having rapid adsorption kinetics, the material eliminated 95% of 1,000 μg L^{-1} Hg^{2+} in 30 minutes and more than 2 hours. Besides that, it demonstrated an incredible selectivity for the binding of the mercury (II) ions across several other metal cations, which led to a contaminant removal output above 99% throughout the broad pH range. Hence, in an environment of circumstances, the previously specified COF-SH material conveys the notable mercury (II) elimination potential from contaminated water samples. Yu et al. ultimately described (Yu et al., 2020) the gradual development of a reprocessable fluorescent COF, based on a specific approach, which is functionalized with hydroxyl (–OH) and allyl groups. This structure or framework has potential applications in the remediation of wastewater, which is potentially hazardous, and can also be utilized for the identification of mercury (II) ions and their effective removal.

10.6.4 Removal of Chromium (Cr)

Chromium (VI) is characterized by its mutagenic and carcinogenic properties, strong oxidizing activity, tendency to potentially harm living beings, and the exceeding concentration of 100 parts per billion (Jansone-Popova et al., 2018). In the year 2018, Jansone-Popova and team developed a novel COF using a one-step solvothermal polycondensation reaction to tackle the growing requirement for materials with functional properties that are easily effective in eliminating hazardous heavy metals from the ground and industrial wastewater. At the edges of the hexagonal pores that were formed within the 2D sheets, which were assembled from their fundamental elements, guanidinium groups were found. Because of its ability to generate robust ion pairs with oxo anions via electrostatic forces and non-covalent hydrogen bonding, guanidinium, a nitrogen-cation equivalent of urea (NH_2CONH_2), was selected for use. With capacities that varied from 90 to 200 mg g^{-1} across multiple solution pH levels, the as-prepared COF material went beyond typical adsorbents and implemented ion exchange materials in terms of its potential to absorb Cr^{+6} ions. For the effective removal of the Cr (VI) from the aqueous solutions, a novel dual-pore COF containing –OH groups was designed by Cui et al. (Cui et al., 2019). Having a peak adsorptive potential of 384 mg g^{-1} Cr^{+6}, the synthesized material had fast adsorption kinetics, attaining 272 mg g^{-1} Cr^{+6} within only 10 minutes. Among the most significant attributes observed for porous materials so far is this one. A magnetic COF having β-keto enamine linkage was developed by Zhong and his colleagues in the year 2020 (Zhong et al., 2020). It had a higher adsorptive potential of 245.45 mg g^{-1} for Cr^{+6}, was reusable, had a really great stability, and was very easy to synthesize. Because of these extraordinary characteristics, the material that is previously mentioned is an ideal choice for the application in environmental remediation. Two COFs (COF1 and COF2) that had distinct hydroxy group distribution were successfully synthesized by Zhu and his coworkers in the year 2020 (Zhu et al., 2020). By showing enhanced Cr^{+6} adsorption capacity and Cr^{+6} selectivity in mixed ion solutions throughout a wide pH range, both COFs, i.e., COF1 and COF2, elevated their potential for adsorption applications.

10.6.5 Removal of Arsenic (As)

Arsenic (As) is very well known to cause cancer in human beings and is one of the most hazardous and abundant metalloids found in the earth's crust (Borah et al., 2018; Mukherjee et al., 2021). Ecosystems can be contaminated with arsenic through various natural and human-caused causes, including sedimentary rocks, geothermal water, and environmental pollution. In geothermal and marine environments, arsenosulfides and arsenides induce the continuous accumulation of poisonous inorganic and organic arsenic species, which is a threat to aquatic and terrestrial plants and animals (Mukherjee et al., 2021). A biological oxidation/reduction process, which is part of a geochemical cycle involving organic substrates, facilitates the release of arsenic from hydro-geosphere environments, leading to its dispersion and speciation within groundwater (Kumar et al., 2020; Mukherjee et al., 2021). Depending on the nearby geochemical equilibrium systems, inorganic arsenic (III) (reducing) and arsenic (V) (oxidizing) might cause long-term harm to human health by affecting food chains (Kumar et al., 2020; Mukherjee et al., 2021). Arsenic is constantly detected in the groundwater exceeding the World Health Organization maximum contamination limit which is >10 μg L^{-1} and due to the adverse physiological effects of arsenic, it is necessary to develop new materials for the arsenic removal applications. Considering all of these factors, a composite material using @CTF-1 and γ-Fe_2O_3 NPs was synthesized by Leus et al. (Leus et al., 2018). Making use of the resulting γ-Fe_2O_3 @CTF-1, the research investigated the adsorption of As (III) and As (V) from synthetic solution as well as from real groundwater, surface water, and wastewater. The synthetic composite material showed an outstanding potential to remove both types of arsenic, i.e., arsenic (III) and arsenic (V), in comparison to typical adsorbents, as a result, bringing out the potential use of the synthetic composite material in water purification and wastewater treatment operations. A Fe^0/COFs that showed an outstanding adsorption of As^{3+}, attaining an upper limit of 135.78 mg g^{-1}, was developed by Liu and his team (Liu et al., 2020). An EB-COF:Br material, which was synthesized by Yang et al. (Yang et al., 2020), is a very good arsenate-binding material and has significant adsorptive potential. These synthetic materials or composites show great potential to completely remove As (arsenic) from wastewater.

10.7 CONCLUSION

COFs have various advantageous characteristics over conventional materials, which include exceptional adsorption potential, fast adsorption kinetics, increased preference, and reusing capability. The current issues encompass: Limited large-scale manufacturing; constrained synthesis of COFs using simple and ecologically benign technologies; the elevated expense of synthetic COF-based products, which obstructs their practical utilization in ecological restoration; the proper handling of COFs about environment and human well-being; the absence of consideration for regeneration for prolonged reutilization; and the recycling process must use cost-effective and eco-friendly solvents. The removal of heavy metals under environmental circumstances must be performed on real water samples. Thus, the improved synthetic method and varied functional groups are expected to significantly lower the price of COFs and enable large-scale manufacturing. The bond formation and chemical structure characteristics enhance the affinity and selectivity of COF for particular heavy metal ions. When the various functional groups are incorporated within the core units of COFs, the adsorption potential of the materials, which are based on COF and target heavy metals, is enhanced. The selectivity and sensitivity of COFs to target pollutants are influenced by the nature of bond formation and the presence of nitrogen, oxygen, and sulfur functional groups. The adsorption performance of COF-based materials exceeded that of isolated COFs due to the formation of coordination bonds, chelation, and electrostatic interactions. The oxygen, sulfur, and nitrogen functional groups demonstrated notable effectiveness. The adsorption efficacy is enhanced by the tautomerization of ketone-enol. Consequently, it is essential to analyze the effectiveness of COF-based functional materials in efficiently eliminating contaminants from the environment and the coordination between their performance and structure. There remains a significant knowledge gap that future studies must address to synthesize appropriate COF materials capable of fulfilling all the essentials for the capture of heavy metals from aqueous solutions.

REFERENCES

Al Rmalli, Shaban W., Abdella A. Dahmani, Mohamed M. Abuein, and Amar A. Gleza. "Biosorption of mercury from aqueous solutions by powdered leaves of castor tree (*Ricinus communis* L.)." *Journal of Hazardous Materials* 152, no. 3 (2008): 955–959.

al Sadat Shafiof, Maryam, and Alireza Nezamzadeh-Ejhieh. "A comprehensive study on the removal of Cd (II) from aqueous solution on a novel pentetic acid-clinoptilolite nanoparticles adsorbent: experimental design, kinetic and thermodynamic aspects." *Solid State Sciences* 99 (2020): 106071.

Al-Shannag, Mohammad, Zakaria Al-Qodah, Khalid Bani-Melhem, Mohammed Rasool Qtaishat, and Malek Alkasrawi. "Heavy metal ions removal from metal plating wastewater using electrocoagulation: kinetic study and process performance." *Chemical Engineering Journal* 260 (2015): 749–756.

Awual, Md Rabiul. "Mesoporous composite material for efficient lead (II) detection and removal from aqueous media." *Journal of Environmental Chemical Engineering* 7, no. 3 (2019): 103124.

Bang, Jin Ho, and Kenneth S. Suslick. "Applications of ultrasound to the synthesis of nanostructured materials." *Advanced Materials* 22, no. 10 (2010): 1039–1059.

Bertrand, Guillaume HV, Vladimir K. Michaelis, Ta-Chung Ong, Robert G. Griffin, and Mircea Dincă. "Thiophene-based covalent organic frameworks." *Proceedings of the National Academy of Sciences* 110, no. 13 (2013): 4923–4928.

Bilal, Muhammad, Jehanzeb Ali Shah, Tayyab Ashfaq, Syed Mubashar Hussain Gardazi, Adnan Ahmad Tahir, Arshid Pervez, Hajira Haroon, and Qaisar Mahmood. "Waste biomass adsorbents for copper removal from industrial wastewater—a review." *Journal of Hazardous Materials* 263 (2013): 322–333.

Biswal, Bishnu P., Suman Chandra, Sharath Kandambeth, Binit Lukose, Thomas Heine, and Rahul Banerjee. "Mechanochemical synthesis of chemically stable isoreticular covalent organic frameworks." *Journal of the American Chemical Society* 135, no. 14 (2013): 5328–5331.

Bojdys, Michael J., Jekaterina Jeromenok, Arne Thomas, and Markus Antonietti. "Rational extension of the family of layered, covalent, triazine-based frameworks with regular porosity." *Advanced Materials* 22, no. 19 (2010): 2202–2205.

Borah, Rinkumoni, Kaling Taki, Anindita Gogoi, Pallavi Das, and Manish Kumar. "Contemporary distribution and impending mobility of arsenic, copper and zinc in a tropical (Brahmaputra) river bed sediment, Assam, India." *Ecotoxicology and Environmental Safety* 161 (2018): 769–776.

Bunck, David N., and William R. Dichtel. "Bulk synthesis of exfoliated two-dimensional polymers using hydrazone-linked covalent organic frameworks." *Journal of the American Chemical Society* 135, no. 40 (2013): 14952–14955.

Campbell, Neil L., Rob Clowes, Lyndsey K. Ritchie, and Andrew I. Cooper. "Rapid microwave synthesis and purification of porous covalent organic frameworks." *Chemistry of Materials* 21, no. 2 (2009): 204–206.

Cao, Ying, Xue Hu, Changqian Zhu, Shuangxi Zhou, Rui Li, Huilan Shi, Shiyu Miao, Mohammadtaghi Vakili, Wei Wang, and Delin Qi. "Sulfhydryl functionalized covalent organic framework as an efficient adsorbent for selective Pb (II) removal." *Colloids and Surfaces A: Physicochemical and Engineering Aspects* 600 (2020): 125004.

Chakraborty, Rupa, Renu Verma, Anupama Asthana, S. Sree Vidya, and Ajaya Kumar Singh. "Adsorption of hazardous chromium (VI) ions from aqueous solutions using modified sawdust: kinetics, isotherm and thermodynamic modelling." *International Journal of Environmental Analytical Chemistry* 101, no. 7 (2021): 911–928.

Chandra, Suman, Sharath Kandambeth, Bishnu P. Biswal, Binit Lukose, Shrikant M. Kunjir, Minakshi Chaudhary, Ravichandar Babarao, Thomas Heine, and Rahul Banerjee. "Chemically stable multilayered covalent organic nanosheets from covalent organic frameworks via mechanical delamination." *Journal of the American Chemical Society* 135, no. 47 (2013): 17853–17861.

Chen, Fengqian, Haorui Zheng, Yusran Yusran, Hui Li, Shilun Qiu, and Qianrong Fang. "Exploring high-connectivity three-dimensional covalent organic frameworks: topologies, structures, and emerging applications." *Chemical Society Reviews* 54 (2025) 484–514.

Chen, Weiben, Zongfan Yang, Zhen Xie, Yusen Li, Xiang Yu, Fanli Lu, and Long Chen. "Benzothiadiazole functionalized D–A type covalent organic frameworks for effective photocatalytic reduction of aqueous chromium (VI)." *Journal of Materials Chemistry A* 7, no. 3 (2019): 998–1004.

Colson, John W., Arthur R. Woll, Arnab Mukherjee, Mark P. Levendorf, Eric L. Spitler, Virgil B. Shields, Michael G. Spencer, Jiwoong Park, and William R. Dichtel. "Oriented 2D covalent organic framework thin films on single-layer graphene." *Science* 332, no. 6026 (2011): 228–231.

Cote, Adrien P., Annabelle I. Benin, Nathan W. Ockwig, Michael O'Keeffe, Adam J. Matzger, and Omar M. Yaghi. "Porous, crystalline, covalent organic frameworks." *Science* 310, no. 5751 (2005): 1166–1170.

Cui, Fu-Zhi, Rong-Ran Liang, Qiao-Yan Qi, Guo-Fang Jiang, and Xin Zhao. "Efficient removal of Cr (VI) from aqueous solutions by a dual-pore covalent organic framework." *Advanced Sustainable Systems* 3, no. 4 (2019): 1800150.

KankanDatta, Swarnavo Chakraborty, and Aryadeep Roychoudhury. "Management of soil, waste and water in the context of global climate change." In *Environmental Nexus for Resource Management*, pp. 1–26. CRC Press, 2025.

Davis, Mark E. "Ordered porous materials for emerging applications." *Nature* 417, no. 6891 (2002): 813–821.

DeBlase, Catherine R., Katharine E. Silberstein, Thanh-Tam Truong, Héctor D. Abruña, and William R. Dichtel. "β-Ketoenamine-linked covalent organic frameworks capable of pseudocapacitive energy storage." *Journal of the American Chemical Society* 135, no. 45 (2013): 16821–16824.

Dinari, Mohammad, and Mohammad Hatami. "Novel N-riched crystalline covalent organic framework as a highly porous adsorbent for effective cadmium removal." *Journal of Environmental Chemical Engineering* 7, no. 1 (2019): 102907.

Ding, San-Yuan, Jia Gao, Qiong Wang, Yuan Zhang, Wei-Guo Song, Cheng-Yong Su, and Wei Wang. "Construction of covalent organic framework for catalysis: Pd/COF-LZU1 in Suzuki–Miyaura coupling reaction." Journal of the American Chemical Society 133, no. 49 (2011): 19816–19822.

Dogru, Mirjam, Matthias Handloser, Florian Auras, Thomas Kunz, Dana Medina, Achim Hartschuh, Paul Knochel, and Thomas Bein. "A photoconductive thienothiophene-based covalent organic framework showing charge transfer towards included fullerene." *Angewandte Chemie International Edition* 52, no. 10 (2013): 2920–2924.

Doonan, Christian J., David J. Tranchemontagne, T. Grant Glover, Joseph R. Hunt, and Omar M. Yaghi. "Exceptional ammonia uptake by a covalent organic framework." *Nature Chemistry* 2, no. 3 (2010): 235–238.

Du, Yi, Jian Wang, Yidong Zou, Wen Yao, Jing Hou, Liangshu Xia, Anguo Peng, Ahmed Alsaedi, Tasawar Hayat, and Xiangke Wang. "Synthesis of molybdenum disulfide/reduced graphene oxide composites for effective removal of Pb (II) from aqueous solutions." *Science Bulletin* 62, no. 13 (2017): 913–922.

Fang, Qianrong, Shuang Gu, Jie Zheng, Zhongbin Zhuang, Shilun Qiu, and Yushan Yan. "3D microporous base-functionalized covalent organic frameworks for size-selective catalysis." *Angewandte Chemie* 126, no. 11 (2014): 2922–2926.

Feng, Haopeng, Lin Tang, Jing Tang, Guangming Zeng, Haoran Dong, Yaocheng Deng, Longlu Wang, Yani Liu, Xiaoya Ren, and Yaoyu Zhou. "Cu-Doped Fe@ Fe_2O_3 core–shell nanoparticle shifted oxygen reduction pathway for high-efficiency arsenic removal in smelting wastewater." *Environmental Science: Nano* 5, no. 7 (2018): 1595–1607.

Friščić, Tomislav. "New opportunities for materials synthesis using mechanochemistry." *Journal of Materials Chemistry* 20, no. 36 (2010): 7599–7605.

Fu, Yu, Wenguang Yu, Weijie Zhang, Qiao Huang, Jun Yan, Chunyue Pan, and Guipeng Yu. "Sulfur-rich covalent triazine polymer nanospheres for environmental mercury removal and detection." *Polymer Chemistry* 9, no. 30 (2018): 4125–4131.

Gaikwad, Sanjit, and Sangil Han. "Microwave-and Ultrasound-Assisted Synthesis of Metal-Organic Frameworks (MOF) and Covalent Organic Frameworks (COF)." *Green Chemical Synthesis with Microwaves and Ultrasound* (2024): 249–282.

Gawande, Manoj B., Sharad N. Shelke, Radek Zboril, and Rajender S. Varma. "Microwave-assisted chemistry: synthetic applications for rapid assembly of nanomaterials and organics." *Accounts of Chemical Research* 47, no. 4 (2014): 1338–1348.

Gendy, Eman A., Jerosha Ifthikar, Jawad Ali, Daniel T. Oyekunle, Zouhair Elkhlifia, Irshad Ibran Shahib, Ahmed I. Khodair, and Zhuqi Chen. "Removal of heavy metals by covalent organic frameworks (COFs): a review on its mechanism and adsorption properties." *Journal of Environmental Chemical Engineering* 9, no. 4 (2021): 105687.

Ghazi, Zahid Ali, Abdul Muqsit Khattak, Rashid Iqbal, Rashid Ahmad, Adnan Ali Khan, Muhammad Usman, Faheem Nawaz et al. "Adsorptive removal of Cd^{2+} from aqueous solutions by a highly stable covalent triazine-based framework." *New Journal of Chemistry* 42, no. 12 (2018): 10234–10242.

Guan, Xinyu, Yunchao Ma, Hui Li, Yusran Yusran, Ming Xue, Qianrong Fang, Yushan Yan, Valentin Valtchev, and Shilun Qiu. "Fast, ambient temperature and pressure ionothermal synthesis of three-dimensional covalent organic frameworks." *Journal of the American Chemical Society* 140, no. 13 (2018): 4494–4498.

Guo, Jia, Yanhong Xu, Shangbin Jin, Long Chen, Toshihiko Kaji, Yoshihito Honsho, Matthew A. Addicoat et al. "Conjugated organic framework with three-dimensionally ordered stable structure and delocalized π clouds." *Nature Communications* 4, no. 1 (2013): 2736.

Gupta, Krishna M., Kang Zhang, and Jianwen Jiang. "Efficient removal of Pb^{2+} from aqueous solution by an ionic covalent–organic framework: molecular simulation study." *Industrial & Engineering Chemistry Research* 57, no. 18 (2018): 6477–6482.

He, Yulong, Xiaomin Wang, Kang Wang, and Limin Wang. "A triarylamine-based fluorescent covalent organic framework for efficient detection and removal of Mercury (II) ion." *Dyes and Pigments* 173 (2020): 107880.

Huang, Ning, Lipeng Zhai, Hong Xu, and Donglin Jiang. "Stable covalent organic frameworks for exceptional mercury removal from aqueous solutions." *Journal of the American Chemical Society* 139, no. 6 (2017): 2428–2434.

Jansone-Popova, Santa, Anthonin Moinel, Jennifer A. Schott, Shannon M. Mahurin, Ilja Popovs, Gabriel M. Veith, and Bruce A. Moyer. "Guanidinium-based ionic covalent organic framework for rapid and selective removal of toxic Cr (VI) oxoanions from water." *Environmental Science & Technology* 53, no. 2 (2018): 878–883.

Jatkar, Chandala, Rohant Dhabbe, Kalyanrao Garadkar, Ravindra Kupwade, Jian Shen, Rajendra Kumbhar, Jaykumar Patil, and Sandip Sabale. "A sustainable approach for tailoring coir-fibre based bio-adsorbent and its composite for industrial applications." *Biomass Conversion and Biorefinery* (2024): 1–13.

Jiang, Yunzhe, Chuanyao Liu, and Aisheng Huang. "EDTA-functionalized covalent organic framework for the removal of heavy-metal ions." *ACS Applied Materials & Interfaces* 11, no. 35 (2019): 32186–32191.

Kang, Chengjun, Zhaoqiang Zhang, Shinpei Kusaka, Kohei Negita, Adam K. Usadi, David C. Calabro, Lisa Saunders Baugh et al. "Covalent organic framework atropisomers with multiple gas-triggered structural flexibilities." *Nature Materials* 22, no. 5 (2023): 636–643.

Katsou, Evina, Simos Malamis, Katherine Joanne Haralambous, and Maria Loizidou. "Use of ultrafiltration membranes and aluminosilicate minerals for nickel removal from industrial wastewater." *Journal of Membrane Science* 360, no. 1–2 (2010): 234–249.

Kim, Soyoung, and Hee Cheul Choi. "Light-promoted synthesis of highly-conjugated crystalline covalent organic framework." *Communications Chemistry* 2, no. 1 (2019): 60.

Kuhn, Pierre, Markus Antonietti, and Arne Thomas. "Porous, covalent triazine-based frameworks prepared by ionothermal synthesis." *Angewandte Chemie International Edition* 47, no. 18 (2008): 3450–3453.

Kumar, Manish, Ritusmita Goswami, Arbind Kumar Patel, Medhavi Srivastava, and Nilotpal Das. "Scenario, perspectives and mechanism of arsenic and fluoride co-occurrence in the groundwater: a review." *Chemosphere* 249 (2020): 126126.

Kumar, Rajesh, and Rajeev Kr Sharma. "Synthesis and characterization of cellulose-based adsorbents for removal of Ni (II), Cu (II) and Pb (II) ions from aqueous solutions." *Reactive and Functional Polymers* 140 (2019): 82–92.

Kumari, Sheetal, Jyoti Chowdhry, Manish Kumar, and Manoj Chandra Garg. "Zeolites in wastewater treatment: a comprehensive review on scientometric analysis, adsorption mechanisms, and future prospects." *Environmental Research* (2024): 119782.

Lam, Boukary, Sébastien Déon, Nadia Morin-Crini, Gregorio Crini, and Patrick Fievet. "Polymer-enhanced ultrafiltration for heavy metal removal: influence of chitosan and carboxymethyl cellulose on filtration performances." *Journal of Cleaner Production* 171 (2018): 927–933.

Leus, Karen, Karel Folens, Nina Ricci Nicomel, Jeffrey Paulo H. Perez, Maria Filippousi, Maria Meledina, Marinela M. Dîrtu et al. "Removal of arsenic and mercury species from water by covalent triazine framework encapsulated γ-Fe_2O_3 nanoparticles." *Journal of Hazardous Materials* 353 (2018): 312–319.

Li, Guiliang, Jianrong Ye, Qile Fang, and Fu Liu. "Amide-based covalent organic frameworks materials for efficient and recyclable removal of heavy metal lead (II)." *Chemical Engineering Journal* 370 (2019): 822–830.

Li, Hailian, Mohamed Eddaoudi, Michael O'Keeffe, and Omar M. Yaghi. "Design and synthesis of an exceptionally stable and highly porous metal-organic framework." *Nature* 402, no. 6759 (1999): 276–279.

Li, Shaoxin, Jiajin Liu, Zhong Lin Wang, and Di Wei. "Mechano-driven chemical reactions." *Green Energy & Environment* 10, no. 5 (2024): 937–966.

Li, Y., Cui, W., Liu, L., Zong, R., Yao, W., Liang, Y. and Zhu, Y., 2016. Removal of Cr (VI) by 3D TiO_2-graphene hydrogel via adsorption enriched with photocatalytic reduction. *Applied Catalysis B: Environmental*, 199, pp.412–423.

Li, Yusen, Weiben Chen, Guolong Xing, Donglin Jiang, and Long Chen. "New synthetic strategies toward covalent organic frameworks." *Chemical Society Reviews* 49, no. 10 (2020): 2852–2868.

Lingamdinne, Lakshmi Prasanna, Yoon-Young Chang, Jae-Kyu Yang, Jiwan Singh, Eun-Ha Choi, Masaharu Shiratani, Janardhan Reddy Koduru, and Pankaj Attri. "Biogenic reductive preparation of magnetic inverse spinel iron oxide nanoparticles for the adsorption removal of heavy metals." *Chemical Engineering Journal* 307 (2017): 74–84.

Liu, Na, Liangfeng Shi, Xianghao Han, Qiao-Yan Qi, Zong-Quan Wu, and Xin Zhao. "A heteropore covalent organic framework for adsorptive removal of Cd (II) from aqueous solutions with high efficiency." *Chinese Chemical Letters* 31, no. 2 (2020): 386–390.

Liu, Ting, Zhuo Xiong, Peng Ni, Zizhen Ma, Yan Tan, Zishun Li, Shengnan Deng, Yincui Li, Qirong Yang, and Huawei Zhang. "Review on adsorbents in elemental mercury removal in coal combustion flue gas, smelting flue gas and natural gas." *Chemical Engineering Journal* 454 (2023): 140095.

Liu, Xiaoshuang, Haomiao Xu, Longlong Wang, Zan Qu, and Naiqiang Yan. "Surface nano-traps of Fe^0/COFs for arsenic (III) depth removal from wastewater in non-ferrous smelting industry." *Chemical Engineering Journal* 381 (2020): 122559.

Lu, Hui, Chang Wang, Juanjuan Chen, Rile Ge, Wenguang Leng, Bin Dong, Jun Huang, and Yanan Gao. "A novel 3D covalent organic framework membrane grown on a porous α-Al_2O_3 substrate under solvothermal conditions." *Chemical Communications* 51, no. 85 (2015): 15562–15565.

Lv, Shi-Wen, Jing-Min Liu, Zhi-Hao Wang, Hui Ma, Chun-Yang Li, Ning Zhao, and Shuo Wang. "Recent advances on porous organic frameworks for the adsorptive removal of hazardous materials." *Journal of Environmental Sciences* 80 (2019): 169–185.

Ma, Zhiyao, Fuyang Liu, Nengsheng Liu, Wen Liu, and Meiping Tong. "Facile synthesis of sulfhydryl modified covalent organic frameworks for highly efficient Hg (II) removal from water." *Journal of Hazardous Materials* 405 (2021): 124190.

Maftouh, Abderrahim, Omkaltoume El Fatni, Souad El Hajjaji, Mohammad Wadah Jawish, and Mika Sillanpää. "Comparative review of different adsorption techniques used in heavy metals removal in water." *Biointerface Research in Applied Chemistry* 13, no. 4 (2023).

Medina, Dana D., Veronika Werner, Florian Auras, Raphael Tautz, Mirjam Dogru, Jorg Schuster, Stephanie Linke et al. "Oriented thin films of a benzodithiophene covalent organic framework." *ACS Nano* 8, no. 4 (2014): 4042–4052.

Mendoza-Cortes, Jose L., William A. Goddard III, Hiroyasu Furukawa, and Omar M. Yaghi. "A covalent organic framework that exceeds the DOE 2015 volumetric target for H_2 uptake at 298 K." *The Journal of Physical Chemistry Letters* 3, no. 18 (2012): 2671–2675.

Merí-Bofí, Laura, Sergio Royuela, Félix Zamora, M. Luisa Ruiz-González, José L. Segura, Riansares Muñoz-Olivas, and María José Mancheño. "Thiol grafted imine-based covalent organic frameworks for water remediation through selective removal of Hg (II)." *Journal of Materials Chemistry A* 5, no. 34 (2017): 17973–17981.

Mondal, Sujan, Sauvik Chatterjee, Saptarsi Mondal, and Asim Bhaumik. "Thioether-functionalized covalent triazine nanospheres: a robust adsorbent for mercury removal." *ACS Sustainable Chemistry & Engineering* 7, no. 7 (2019): 7353–7361.

Mukherjee, Santanu, Alok Kumar Thakur, Ritusmita Goswami, Payal Mazumder, Kaling Taki, Meththika Vithanage, and Manish Kumar. "Efficacy of agricultural waste derived biochar for arsenic removal: tackling water quality in the Indo-Gangetic plain." *Journal of Environmental Management* 281 (2021): 111814.

Nayak, Arunima, Brij Bhushan, Vartika Gupta, and Poonam Sharma. "Chemically activated carbon from lignocellulosic wastes for heavy metal wastewater remediation: effect of activation conditions." *Journal of Colloid and Interface Science* 493 (2017): 228–240.

Nguyen, Vinh D., Tho T. Pham, Valerie Vranova, Hoa T.H. Nguyen, Linh T.N. Nguyen, Xuan T. Vuong, and Quy M. Bui. "Removal of cadmium from aqueous solution using sonochemically modified clinoptilolite: optimization and modeling." *Environmental Technology & Innovation* 20 (2020): 101166.

Matthias Nüchter,, Bernd Ondruschka,Werner Bonrath, andAmber Gum. "Microwave-assisted synthesis–a critical technology overview." *Green Chemistry* 6, no. 3 (2004): 128–141.

Pachfule, Pradip, Sharath Kandambeth, David Díaz Díaz, and Rahul Banerjee. "Highly stable covalent organic framework–Au nanoparticles hybrids for enhanced activity for nitrophenol reduction." *Chemical Communications* 50, no. 24 (2014): 3169–3172.

Panda, Atanu, Yanqin Yang, Sada Venkateswarlu, Younghu Son, Tae-Hyun Bae, and Minyoung Yoon. "Highly durable covalent organic framework for the simultaneous ultrasensitive detection and removal of noxious Hg^{2+}." *Microporous and Mesoporous Materials* 306 (2020): 110399.

Patel, Manav G., Parth G. Marakana, Anirban Dey, Bharti Saini, and Himanshu Chokshi. "Coal fly ash derived adsorbent for enhancing waste water treatment." *Materials Today: Proceedings* 77 (2023): 163–167.

Erol Pehlivan, B. H. Yanık, Gülnare Ahmetli, and Mustafa Pehlivan. "Equilibrium isotherm studies for the uptake of cadmium and lead ions onto sugar beet pulp." *Bioresource Technology* 99, no. 9 (2008): 3520–3527.

Peligro, Francisco R., Ivana Pavlovic, Ricardo Rojas, and Cristobalina Barriga. "Removal of heavy metals from simulated wastewater by in situ formation of layered double hydroxides." *Chemical Engineering Journal* 306 (2016): 1035–1040.

Peng, Yaguang, Hongliang Huang, Yuxi Zhang, Chufan Kang, Shuangming Chen, Li Song, Dahuan Liu, and Chongli Zhong. "A versatile MOF-based trap for heavy metal ion capture and dispersion." *Nature Communications* 9, no. 1 (2018): 187.

Raghav, Sapna, Jyoti Raghav, Ma'aruf Abdulmumin Muhammad, Madhur Babu Singh, Pallavi Jain, Dinesh Kumar. "Recent advancements in synthesis and applications of covalent organic frameworks." *Applied Materials Today*, 44 (2025): 102694.

Ram, Bhagat, and Ghanshyam S. Chauhan. "New spherical nanocellulose and thiol-based adsorbent for rapid and selective removal of mercuric ions." *Chemical Engineering Journal* 331 (2018): 587–596.

Ritchie, Lyndsey K., Abbie Trewin, Aida Reguera-Galan, Tom Hasell, and Andrew I. Cooper. "Synthesis of COF-5 using microwave irradiation and conventional solvothermal routes." *Microporous and Mesoporous Materials* 132, no. 1–2 (2010): 132–136.

Sajid, Hasnain. "Effect of interlayer slipping on the geometric, thermal and adsorption properties of 2D covalent organic frameworks: a comprehensive review based on computational modelling studies." *Physical Chemistry Chemical Physics* 26, no. 11 (2024): 8577-8603.

Smith, Frances N., Christopher D. Taylor, Wooyong Um, and Albert A. Kruger. "Technetium incorporation into goethite (α-FeOOH): an atomic-scale investigation." *Environmental Science & Technology* 49, no. 22 (2015): 13699–13707.

Stegbauer, Linus, Katharina Schwinghammer, and Bettina V. Lotsch. "A hydrazone-based covalent organic framework for photocatalytic hydrogen production." *Chemical Science* 5, no. 7 (2014): 2789–2793.

Sud, Dhiraj, Garima Mahajan, and Maninder P.P. Kaur. "Agricultural waste material as potential adsorbent for sequestering heavy metal ions from aqueous solutions–a review." *Bioresource Technology* 99, no. 14 (2008): 6017–6027.

Sun, Bing, Jian Liu, Anmin Cao, Weiguo Song, and Dong Wang. "Interfacial synthesis of ordered and stable covalent organic frameworks on amino-functionalized carbon nanotubes with enhanced electrochemical performance." *Chemical Communications* 53, no. 47 (2017): 6303–6306.

Sun, Tingting, Zhi Wang, Yuhui Wang, Qingmei Xu, Kang Wang, and Jianzhuang Jiang. "Porphyrin-Based Covalent Organic Frameworks for CO_2 Photo/Electro-Reduction." *Angewandte Chemie* (2025): e202422814.

Tao, Yuan, Xiao Hong Xiong, Jian Bo Xiong, Li Xiao Yang, Ya Ling Fan, Han Feng, and Feng Luo. "High-performance removal of mercury ions (II) and mercury vapor by SO_3^--anchored covalent organic framework." *Journal of Solid-State Chemistry* 282 (2020): 121126.

Uddin, Mohammad Kashif. "A review on the adsorption of heavy metals by clay minerals, with special focus on the past decade." *Chemical Engineering Journal* 308 (2017): 438–462.

Vardhan, Harsh, Grace Rummer, Angela Deng, and Shengqian Ma. "Large-scale synthesis of covalent organic frameworks: challenges and opportunities." *Membranes* 13, no. 8 (2023): 696.

Velempini, Tarisai, and Kriveshini Pillay. "Sulphur functionalized materials for Hg (II) adsorption: a review." *Journal of Environmental Chemical Engineering* 7, no. 5 (2019): 103350.

Waller, Peter Janes. "Conversion of Linkages in Covalent Organic Frameworks." PhD diss., UC Berkeley, 2019.

Wan, Shun, Jia Guo, Jangbae Kim, Hyotcherl Ihee, and Donglin Jiang. "A belt-shaped, blue luminescent, and semiconducting covalent organic framework." *Angewandte Chemie International Edition* 47, no. 46 (2008): 8826–8830.

Wang, Guang-Bo, Sha Li, Cai-Xin Yan, Fu-Cheng Zhu, Qian-Qian Lin, Ke-Hui Xie, Yan Geng, and Yu-Bin Dong. "Covalent organic frameworks: emerging high-performance platforms for efficient photocatalytic applications." *Journal of Materials Chemistry A* 8, no. 15 (2020a): 6957–6983.

Wang, Han, Zhuotong Zeng, Piao Xu, Lianshan Li, Guangming Zeng, Rong Xiao, Zhiyong Tang et al. "Recent progress in covalent organic framework thin films: fabrications, applications and perspectives." *Chemical Society Reviews* 48, no. 2 (2019): 488–516.

Wang, Jianlong, and Shuting Zhuang. "Covalent organic frameworks (COFs) for environmental applications." *Coordination Chemistry Reviews* 400 (2019): 213046.

Wang, Longlong, Haomiao Xu, Yixiang Qiu, Xiaoshuang Liu, Wenjun Huang, Naiqiang Yan, and Zan Qu. "Utilization of Ag nanoparticles anchored in covalent organic frameworks for mercury removal from acidic wastewater." *Journal of Hazardous Materials* 389 (2020b): 121824.

Wang, Tian-Xiong, Hai-Peng Liang, Dejene Assefa Anito, Xuesong Ding, and Bao-Hang Han. "Emerging applications of porous organic polymers in visible-light photocatalysis." *Journal of Materials Chemistry A* 8, no. 15 (2020c): 7003–7034.

Wang, Xiaolan, Ruiyang Ma, Lin Hao, Qiuhua Wu, Chun Wang, and Zhi Wang. "Mechanochemical synthesis of covalent organic framework for the efficient extraction of benzoylurea insecticides." *Journal of Chromatography A* 1551 (2018): 1–9.

Wu, Yihan, Hongwei Pang, Yue Liu, Xiangxue Wang, Shujun Yu, Dong Fu, Jianrong Chen, and Xiangke Wang. "Environmental remediation of heavy metal ions by novel-nanomaterials: a review." *Environmental Pollution* 246 (2019): 608–620.

Xu, Wentao, Xuejiao Sun, Mianli Huang, Xiaoyang Pan, Xiaoping Huang, and Huaqiang Zhuang. "Novel covalent organic framework/PVDF ultrafiltration membranes with antifouling and lead removal performance." *Journal of Environmental Management* 269 (2020): 110758.

Xu, Yueshuai, Guanshi Ren, Dianqi Zhang, Lishui Sun, and Yingjie Zhao. "Fully conjugated covalent organic frameworks: synthesis, structures and applications." *Chinese Journal of Chemistry* 41, no. 23 (2023): 3447–3472.

Xue, Hui, Qihui Chen, Feilong Jiang, Daqiang Yuan, Guangxun Lv, Linfeng Liang, Luyao Liu, and Maochun Hong. "A regenerative metal–organic framework for reversible uptake of Cd (II): from effective adsorption to in situ detection." *Chemical Science* 7, no. 9 (2016): 5983–5988.

Yagub, Mustafa T., Tushar Kanti Sen, Sharmeen Afroze, and Ha Ming Ang. "Dye and its removal from aqueous solution by adsorption: a review." *Advances in Colloid and Interface Science* 209 (2014): 172–184.

Yang, Cheng-Hao, Jo-Shu Chang, and Duu-Jong Lee. "Covalent organic framework EB-COF: Br as adsorbent for phosphorus (V) or arsenic (V) removal from nearly neutral waters." *Chemosphere* 253 (2020): 126736.

Yang, Seung-Tae, Jun Kim, Hye-Young Cho, Sangho Kim, and Wha-Seung Ahn. "Facile synthesis of covalent organic frameworks COF-1 and COF-5 by sonochemical method." *RSC Advances* 2, no. 27 (2012): 10179–10181.

Yang, Tianzi, Weimin Zhang, Haiyan Liu, and Yadan Guo. "Enhanced removal of U (VI) from aqueous solution by chitosan-modified zeolite." *Journal of Radioanalytical and Nuclear Chemistry* 323, no. 2 (2020): 1003–1012.

Yoo, JongTae, SuHyun Lee, Shinsuke Hirata, ChaeRin Kim, Chang Kee Lee, Tomohiro Shiraki, Naotoshi Nakashima, and Jin Kie Shim. "In situ synthesis of covalent organic frameworks (COFs) on carbon nanotubes and graphenes by sonochemical reaction for CO_2 adsorbents." *Chemistry Letters* 44, no. 4 (2015): 560–562.

Yu, Yanxin, Guoliang Li, Jianghua Liu, and Daqiang Yuan. "A recyclable fluorescent covalent organic framework for exclusive detection and removal of mercury (II)." *Chemical Engineering Journal* 401 (2020): 126139.

Zha, Jianrui, Yaji Huang, Wenqing Xia, Zhipeng Xia, Changqi Liu, Lu Dong, and Lingqin Liu. "Effect of mineral reaction between calcium and aluminosilicate on heavy metal behavior during sludge incineration." *Fuel* 229 (2018): 241–247.

Zhang, Hanyin, Yubo Geng, Jin Huang, Zixiao Wang, Kun Du, and Haoyuan Li. "Charge and mass transport mechanisms in two-dimensional covalent organic frameworks (2D COFs) for electrochemical energy storage devices." *Energy & Environmental Science* 16, no. 3 (2023): 889–951.

Zhang, Ting, Congwei Gao, Wei Huang, Yuling Chen, Yang Wang, and Jianmin Wang. "Covalent organic framework as a novel electrochemical platform for highly sensitive and stable detection of lead." *Talanta* 188 (2018): 578–583.

Zhang, Ting, Guolong Xing, Weiben Chen, and Long Chen. "Porous organic polymers: a promising platform for efficient photocatalysis." *Materials Chemistry Frontiers* 4, no. 2 (2020): 332–353.

Zhao, Guixia, Xiubing Huang, Zhenwu Tang, Qifei Huang, Fenglei Niu, and Xiangke Wang. "Polymer-based nanocomposites for heavy metal ions removal from aqueous solution: a review." *Polymer Chemistry* 9, no. 26 (2018): 3562–3582.

Zhong, Xin, Zhipeng Lu, Wen Liang, and Baowei Hu. "The magnetic covalent organic framework as a platform for high-performance extraction of Cr (VI) and bisphenol a from aqueous solution." *Journal of Hazardous Materials* 393 (2020): 122353.

Zhu, Donghai, Shuangxi Zhou, Ziming Zhou, Rui Li, Jia Ye, Xu Ziyu, Shenjie Lan, Yunhai Zhang, Shiyu Miao, and Wei Wang. "Highly efficient and selective removal of Cr (VI) by covalent organic frameworks: structure, performance and mechanism." *Colloids and Surfaces A: Physicochemical and Engineering Aspects* 600 (2020): 124910.

Zhu, Hongshan, Jinyun Yuan, Xiaoli Tan, Wenhua Zhang, Ming Fang, and Xiangke Wang. "Efficient removal of Pb^{2+} by Tb-MOFs: identifying the adsorption mechanism through experimental and theoretical investigations." *Environmental Science: Nano* 6, no. 1 (2019): 261–272.

11 Carbon Dioxide Conversion and Utilization with Covalent Organic Framework

Vaibhav Sharma, Shobhit Saini, Piyush Gupta, and Pallavi Jain

11.1 INTRODUCTION

In the 20th century, industrialization and urbanization have emitted substantial greenhouse gases into the nature. A main problem of this century is the regulation and decrease in the concentration of greenhouse gas emissions, which are linked to global warming and climate change. Carbon dioxide (CO_2) is the principal contributor of climate change (Alenazi et al., 2025). In 2024, about 41.6 billion tons CO_2 are released into our environment across the world. Emissions of CO_2 from the energy sector and industrial activities are increased by 0.9% in 2022. In 2023, the level of CO_2 due to the energy sector increased by 1.1%. The countries are implementing policies to restrict the production of CO_2 and other greenhouse gases, which helps in the reduction of pollution caused by CO_2. The scientists and researchers are actively engaged in decreasing the excess CO_2 through various approaches, including capture/sequestration/conversion/photoreduction (Peter, 2018). Such solutions have been suggested to solve this issue and will be important unless clean energy can adequately fulfill the substantial demands of societal development. The advancement of high-performance materials capable of capturing CO_2 in a sustainable manner is vital for the success of carbon-capturing and sequestration technology (Li et al., 2023). Metal-organic frameworks (MOFs) render superior properties like CO_2 capturing ability, high selectivity (Sel) for CO_2, remarkable recyclability, facile pore functionalization, consistent pore dimensions, minimal energy requirements for regeneration, and excellent stability than previously documented porous materials in CO_2 capture (Sumida et al., 2012). Many MOFs showed lower CO_2 capturing properties under humid condition as water may affect the chemical stability of MOFs. Particularly, most of the high-performing MOFs had problems with repeated usage in moist conditions (Mahajan and Lahtinen, 2022). Covalent organic frameworks (COFs) could be utilized as adsorbents for the collection and capture of CO_2 owing to their increased surface area and excellent stability (Li et al., 2023). CO_2 binding moieties were added in the COF, which helps in the preferential adsorption of CO_2.

DOI: 10.1201/9781003605386-11

The modification in the pore size of COFs is associated with significant CO_2 adsorption and helps in the preparation of heterogeneous catalysts for the conversion of CO_2 into nontoxic and value-added products. CO_2 could be fixed to epoxides for preparing cyclic carbonates and reducing CO_2 into CO also takes place (Ozdemir et al., 2019). COFs serve as a versatile material, demonstrating significant contributions in the photo/electrocatalytic process of CO_2 reduction reaction. They are characterized by good porosity and greater crystallinity compared to various porous materials, typically synthesized by dynamic covalent chemistry principles. A range of techniques has been utilized in the synthesis of COFs. The solvothermal method is the best and commonly employed technique for the preparation of COFs (Raghav et al., 2005). Due to its organized structure, extensive surface area, substantial pores, superior thermal stability, consistent topologies, and adjustable functions, COFs are anticipated to provide an optimal platform for the adsorption, diffusion, and activation of CO_2 (Sarkar et al., 2022). The applications of COFs in the process of CO_2 conversion and capturing are discussed in depth in this chapter. It also discusses the function of various COFs in the transformation of CO_2 to cyclic carbonates. COFs are utilized in the photoreduction and electrocatalytic conversion of CO_2. In addition, this chapter covers the most significant challenges that are associated with COFs for the purpose of CO_2 capture and conversion.

11.2 COFs FOR CO_2 CAPTURE VIA ADSORPTION

CO_2 is recognized for its heat-trapping capability, which makes a serious impact on global warming. Consequently, the process of carbon capturing from CO_2 is becoming increasingly vital to mitigate and avert the detrimental effects of climate change. COFs are crystalline porous materials that highlight the capacity to adsorb substantial amounts of CO_2. They had well-defined pore architectures enabling CO_2 adsorption. Attractive for sustainable CO_2 application, COFs also provide outstanding stability and recyclability. Their performance in photocatalytic and electrocatalytic CO_2 reduction has been much improved by recent developments in designing COFs with enhanced charge transfer, light-harvesting capabilities, and cooperative catalytic sites, so positioning them as fundamental materials for carbon capture (Dautzenberg et al., 2023). Efficient CO_2 capture is accomplished by transforming a traditional 2D COF into a superior material characterized by substantial capacity and great Sel. The CO_2 capturing properties could be enhanced when nanopore walls were modified with functional groups, resulting in a novel approach to designing COFs for gas retention and separation (Huang et al., 2015). COF-102 at 35 bar had an adsorption efficiency of 72 mg/g for hydrogen at 77 K, 187 mg g^{-1} for methane at 298 K, and 1,180 mg g^{-1} for CO_2 at 298 K (Furukawa and Yaghi, 2009). The microwave-assisted solvothermal method was used for preparing 2D TpPa-COF (MW), resulting in dramatically reduced time and good yield. It exhibited exceptional porosity, characterized by a substantial BET surface area and elevated crystallinity and had outstanding performance in CO_2 storage and CO_2/N_2 separation under low-pressure conditions. It demonstrated an exceptional CO_2 adsorption efficiency of 5.1 mmol g^{-1} at 273 K under 1 bar (Wei et al., 2015). Zn-modified 1,3,5-triformylbenzene (AB-COF) could adsorb 4.68 mmol g^{-1} of CO_2 at 273 K (Stegbauer et al., 2015).

At 1 bar, PdII/TAT-TFP demonstrated CO_2 uptake of 11 wt % at 273 K and 7.5 wt % at 298 K (Kaleeswaran et al., 2017). COF-5 nanosheet was prepared by using the sonochemical process, demonstrating strong CO_2 absorption capabilities due to its distinctive structure. Mixed matrix membranes (MMMs) were synthesized by adding COF-5 in Pebax-1657 matrix. This helped in increasing the permeability for CO_2. MMM prepared from 0.4 wt % COF-5 filler showed approx. 493 Barrer CO_2 permeability. MMM with COF-5 increased CO_2/N_2 selectivity from 31.3 to 49.3 (Duan et al., 2019). 2D hydrazone-linked COF, prepared by using the solvothermal method, exhibited CO_2 capturing properties of 0.9 mmol g^{-1} at 1 bar and 298 K (Bagherian et al., 2021). COF-609 demonstrated 1,360 times increase in CO_2 uptake ability as compared to the unmodified framework; moreover, 29% improvement was also observed in humid conditions (Lyu et al., 2022). 69,840 hypoCOFs were utilized for capturing CO_2. CO_2/H_2 adsorption selectivities were 6.13–742 (6.39–954) and CO_2 working capacities were 0.07–8.68 mol kg^{-1} (0.01–3.92 mol kg^{-1}) (Aksu et al., 2022). The adsorption efficiency of [AeImBr]X%-TAPT-COFs improved from 62.6 to 117.4 mg g^{-1} for CO_2, and CO_2 was stored around the catalytic active sites (Yin et al., 2022). The CO_2 capturing properties of TAPT-BP^{2+}-COF increased by 55.6% due to imine and triazine groups, –OH groups, charged framework, and optimal pore size, hence facilitating adequate CO_2 availability near the catalytic active sites (Yin et al., 2023b). TpPa-1, containing keto-enamine moieties, demonstrated a notable CO_2 absorption of 0.6 mmol g^{-1} and a CO_2/N_2 sorption Sel of 114 (Wang et al., 2022). At 1 bar, Me_3TFB-(NH_2)2BD adsorbed CO_2, 1.12 ± 0.26 at 273 K and 0.72 ± 0.07 mmol g^{-1} at 295 K (Dautzenberg et al., 2023). COF-709 (a type of polyethylenimine-functionalized COF) had 0.48 and 1.24 mmol g^{-1} CO_2 capturing efficiency in dry and 75% humid environment at 25°C, and the concentration of CO_2 was 400 ppm (Li et al., 2024). COF-999 is capable of capturing CO_2 from ambient air. This COF exhibited around 0.96 and 2.05 mmol g^{-1} capturing capacity in dry and at 50% relative humid condition, the concentration of CO_2 was 400 ppm (Zhou et al., 2024). UPC-6Li, UPC-6Na, and UPC-6K had roughly 221.64, 188.28, and 140.50 cm^{-1} CO_2 adsorption capacity, respectively. Since the Li atom had a smaller atomic radius and required less adsorption space, producing a maximum CO_2 capture capacity of 221.64 cm^{-3} (Zhang et al., 2024). CO_2 absorption capacity of imine-bonded COFs was dramatically raised from approximately 55 to around 235 mg g^{-1} (Chen et al., 2025). Table 11.1 highlights the CO_2 capturing capacity of different COFs.

11.3 COFs USED IN CONVERSION OF CO_2

Because of their high surface area, adjustable porosity, and flexible functionalizing, COFs have become attractive substances for CO_2 conversion. While specialized catalytic sites such as metal nanoparticles, organocatalysts, or photoactive moieties enhance their effectiveness in converting CO_2 into CO. Their ordered architectures help in efficient CO_2 adsorption. Through optimization of charge separation, light absorption, and electron transfer processes, COFs enable selective transformation of CO_2 into CO.

2D metalloporphyrin-based COF composites were prepared and used in catalyzing the coupling of CO_2 and epoxides to form cyclic carbonates. Co-catalyst helped

TABLE 11.1
CO_2 Capturing Capacity of Different COFs

COFs	Key Features	References
COF-102	1,180 mg g⁻ CO_2 was captured at 298 K	Furukawa and Yaghi (2009)
TpPa-COF (MW)	CO_2 uptake of 5.1 mmol g^{-1} at 273 K under 1 bar	Wei et al. (2015)
Zn-doped AB-COF	4.68 mmol g^{-1} CO_2 capture at 273 K	Stegbauer et al. (2015)
Pd^{II}/TAT-TFP	About 11 and 7.5 wt % of CO_2 adsorbed at 273 and 298 K at 100 kPa	Kaleeswaran et al. (2017)
MMM (0.4 wt % COF-5)	~493 Barrer permeability for CO_2	Duan et al. (2019)
COF-609	1,360-fold increase in CO_2 uptake	Lyu et al. (2022)
$[AeImBr]_{X\%}$-TAPT-COFs	CO_2 capturing capacity increased from 62.6 to 117.4 mg g^{-1}	Yin et al. (2022)
TAPT-BP^{2+}-COF	CO_2 capturing capacity increased by 55.6% due to its CO_2-philic groups	Yin et al. (2023b)
TpPa-1 modified with keto-enamine moieties	0.6 mmol g^{-1} CO_2 adsorbed	Wang et al. (2022)
COF-709	0.48 and 1.24 mmol g^{-1} CO_2 was captured in dry conditions and 75% relative humidity, respectively	Li et al. (2024)
COF-999	CO_2 capturing of 0.96 and 2.05 mmol g^{-1} occur in dry conditions and 50% relative humid condition, respectively	Zhou et al. (2024)
UPC-6 Li	~221.64 $cm^3 cm^{-3}$	Zhang et al. (2024)

in achieving the selectivity of about 100% and activity (99.2%) for the synthesis of 1,2-butylene carbonate on COF-366-Zn under mild reaction conditions (Xu et al., 2017). COF-JLU7 could adsorb 151 mg/g CO_2 at 273 K and 1 bar pressure. COF-JLU7 worked as a catalyst for the conversion of CO_2 into cyclic carbonate via cycloaddition reaction with epoxides under mild conditions (Zhi et al., 2018). Covalent ionic frameworks showed significant efficacy in catalyzing the cycloaddition of C_3H_6O with CO_2 to produce $C_4H_6O_3$. Zn–CIF_2–C_2H_4 exhibited adequate stability to achieve over 94% propylene carbonate yield after six cycles. Moreover, Zn–CIF_2–C_2H_4 had the capability to catalyze the cycloaddition of various epoxides with CO_2 (Liu et al., 2019). COF modified with (–OCH_3) and (–OH) groups, namely OMe–OH–TPBP–COF, worked as a metal-free heterogeneous catalyst, and it could be utilized for cycloaddition of CO_2 under mild condition (Yang et al., 2020). AMIMBr@H_2P-DHPh COF helped in the cycloaddition reactions between epoxide and CO_2. This COF was prepared by using 1-alkyl-3-methylimidazolium-based ionic liquid and acted as a heterogeneous catalyst in the reaction of CO_2 and epoxides. There was no need for solvent and co-catalyst in this reaction (Zhang et al., 2020). The cycloaddition of CO_2 and epoxides took place by using metalloporphyrin-comprising Co(II)@TA-TF COF. At 313 K and 0.1 MPa, this COF had conversion efficiency of 92% for chloropropene carbonate and 99% for C_3H_6 carbonate, which was greater than the most reported COF and MOF catalytic systems (Li et al., 2021). Porphyrin and imidazoliummionic liquid decorated and quinolone linked COF (named as COF-P1-1) was prepared via three-component one-pot Povarov reaction process one pot Povarov method. The metallization with Co(II) was done to prepare COF-PI-2, which had a good photothermal conversion efficiency. The integration of Co(II)–porphyrin with imidazolium-IL facilitated highly effective CO_2 cycloaddition (Ding et al., 2021). Pyridine-modified COFs with varying thickness were used in making M@COF hybrid materials on the surface of NH_2-UiO-66. M@COF-0.15-Br was used in the conversion of CO_2 into mono- and poly-cyclic carbonates (Liu et al., 2024).

11.4 COFs FOR PHOTOREDUCTION OF CO_2

COFs are a new type of porous crystalline substances, which consist of organic linking units interconnected by stronger covalent bonds. COFs show the semiconductive properties and improved catalytic efficacy in the CO_2 photoreduction process due to their denser stacked layer structure. This perspective aims to elucidate the CO_2 photoreduction initiated by COFs, which are emerging as advanced photocatalytic platforms (Nguyen and Alzamly, 2021). MOFs and COFs work as suitable materials for understanding the important mechanism that enhances CO_2 photoreduction and helps in investigating the catalytic mechanisms. MOFs and COFs have garnered significant attention for CO_2 photocatalysis due to their varied precursors, precisely defined and customizable structures, numerous exposed catalytic active sites, and elevated surface areas. The precise orientation and homogeneous integration of photosensitizers and catalytic active sites in MOFs and COFs enhances light harvesting and charge separation (Yin et al., 2023a). CO_2 can be converted into water and methanol by using azine-based COFs under visible light irradiation, which act as a photocatalyst. Azine-based COFs are metal-free organic semiconductors (Fu et al., 2018). Ni-TpBpy was used to generate CO from CO_2 reduction reaction under visible light. It produced 4,057 μmol g^{-1} of CO in 5 hours and 96% Sel was achieved for H_2 evolution. Seventy-six percent Sel for CO is achieved when the partial pressure of CO_2 decreases to 0.1 atm (Zhong et al., 2019). Photoconversion of CO_2 took place in water. sp_2c-COFdpy-Co showed the highest CO evolution (17.93 mmol g^{-1}) in water with 81.4% Sel (CO_2 photoreduction to CO) (Xiang et al., 2020). Van der Waals heterojunction (vdWH) catalyst was synthesized from 2D carbon nitride (2D CN) and triazine-based COF (COF-TD). 2D CN COF had a more CO evolution rate, which was 9.2 and 3.3 times higher than 2D CN and COF-TD, respectively (Song et al., 2021). Single atom sites were doped in the triazine-based COF, i.e., SAS/Tr-COF, and helped in the reduction of CO_2. Fe SAS/Tr-COF catalyst attained a CO production rate of 980.3 μmol $g^{-1}h^{-1}$ with a Sel of 96.4%, exceeding Tr-COF by more than 26 times under visible light (Ran et al., 2022). A heterojunction responsive to visible light (COF-318/TNF) was constructed by integrating TiO_2 nanofibers (TNFs) with COF-318 for artificial photosynthesis, both CO_2 reduction and H_2O oxidation took place at the same time, and there was no need for sacrificial agents, photosensitizers, or cocatalysts. COF-318/TNF-15 had 70.1 ± 0.7 μmol·g^{-1}·h^{-1} conversion rate and helped in the reduction of CO_2 into CO, and the conversion rate was 12.7 times greater than that of COF-318, with a CO:O_2 ratio of around 2:1 (Gao et al., 2022). Perylene-modified COFs showed significantly improved performance for photocatalytic CO_2 reduction, with an efficiency that was eight times greater than that of unmodified COFs (Lin et al., 2022). NJU-318 and NJU-319 Fe were prepared by combining hexaphenyl-triphenylene (HTP) units with pyrene- or iron-porphyrin-based connectors. To improve visible light absorption and CO_2 reduction by visible light, HTP units in the COFs were modified and converted into pNJU-318 and pNJU-319Fe. pNJU-319Fe could generate about 688 μmol g^{-1} CO, which was 2.5-fold more than unmodified NJU-319Fe (Dong et al., 2023). 1D PyTTA-COF showed an improvement of CO_2 photoreduction efficacy under visible-light exposure when complexed with Co ions, producing an exceptional CO evolution of 1,003 μmol g^{-1} over 8 hours, exceeding the performance of the analogous 2D variant by a factor of 59 (Zou et al., 2023). The OMHS-COF-Co material served as a photocatalyst for reducing CO_2 into CO. It achieved a notable CO generation rate of 15,874 μmol $g^{-1}h^{-1}$, a Sel of 92.4%, and exceptional cycle stability (Zheng et al., 2024). En-COF-TAPB-TDOEB demonstrated a peak in photogenerated electron density at the reduction site owing to an extra electric field, whose polarization direction aligns with that of the fundamental unit of COFs. En-COF-TAPB-TDOEB also demonstrated a rise in CO_2 photoreduction efficiency compared to Am-COF-TAPB-TFB and Pr-COF-TFPB-TAB (Bai et al., 2024).

11.5 COFs AS CO_2 REDUCTION ELECTROCATALYSTS

MOFs and COFs are crystalline materials distinguished by their hierarchical porosity structure. The swift progress in MOFs and COFs has significant possibilities for diversifying electrocatalysts for CO_2 reduction processes (Mushtaq et al., 2024). M-TTCOFs can function as electrocatalysts exhibiting a high faradaic efficiency (FE) for CO (91.3%, −0.7 V) and demonstrate significant cycle stability beyond 40 hours. Furthermore, upon exfoliation, 90% FE was achieved at −0.6 to −0.9 while 99.7% FE at −0.8 V by ~5 nm nanosheets of Co-TTCOF (Zhu et al., 2020). The cobalt(II) phthalocyanine catalysts have phenazine linkages, which help in the formation of a 2D tetragonal structure that endures boiling water, acidic, or basic environments. Pi-conjugated 2D COF with aligned metallophthalocyanine electrocatalytic sites could improve the transfer of pi electrons to CO_2 reduction sites. The 2D lattice catalyzed two-electron reduction of CO_2 to CO in water to achieve 320,000 h^{-1} turnover number and an FE of 96% (Huang et al., 2020). NiPc and CoPc-TFPN COF were the single site catalysts, which had an FE of CO 99.8(±1.24)% and 96.1(±1.25)%, respectively (Lu et al., 2021). CoPc-PI-COF-1 and CoPc-PI-COF-2 were the polyimide-lined phthalocyanine COF in which AA stacking was present in their structure and attained thermal stability at 300°C. These COFs were prepared by the solvothermal process. Their conductivities were 3.7×10^{-3} and 1.6×10^{-3} S m^{-1}, respectively. Both COFs attained a CO_2-to-CO FE of 87%–97% at −0.60 to −0.90 V in 0.5 M $KHCO_3$. CoPc-PI-COF-1 exhibited a rise in current density (−21.2 mA cm^{-2} at −0.90 V) attributed to enhanced conductivity and attained a turnover frequency of 2.2 s^{-1} over 40 hours (Han et al., 2021). CoPc-PI-COF-3 doped with carbon black and acts as an electrocatalytic cathode in the conversion of CO_2 into CO. This COF had an FE of 88%–96% at −0.60 to −1.00 V (vs. RHE) and jCO of −31.7 mA cm^{-2} at −0.90 V (Han et al., 2022). Various honeycomb-structured porous crystalline hetero-electrocatalysts (MCH-X) were produced for the electrocatalytic

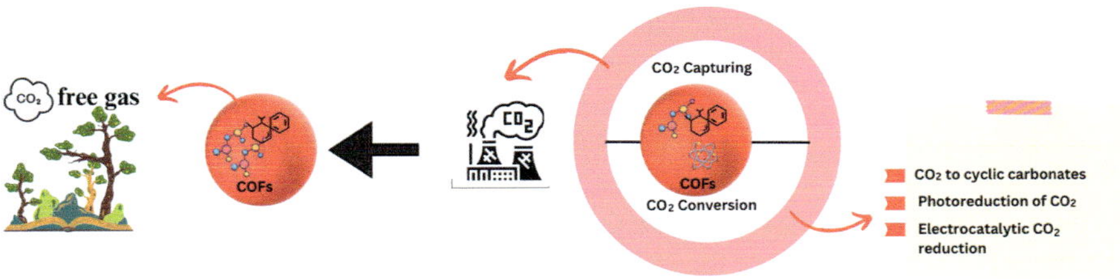

FIGURE 11.1 COFs utilization in the process of capturing and conversion of CO_2.

conversion of CO_2 to CH_4. Their distinctive heterostructures, integrating a porous MOF template with an ultrathin COF covering, improve CO_2 adsorption and activation. MCH-3 catalyst demonstrated −398.1 mA cm^{-2} current density and 76.7% FE for CH_4 (Yang et al., 2022). COFs serve as promising electrocatalysts for carbon dioxide reduction (CO_2RR) owing to their adjustable topologies and porosity. Phthalocyanine–porphyrin COFs (CoPc-2H_2Por and CoPc-H_2Por) were produced by imidization utilizing CoTAPc and porphyrin-based linkers. Both COFs attained over 90% Faraday efficiency with significant stability, but CoPc-2H_2Por exhibited reduced overpotential, elevated current density, and enhanced CO Sel. The improved CO_2RR performance resulted from an increased pore size and conjugated structure, which boosted electron transfer, CO_2 adsorption, and accelerated reaction kinetics (Yuan et al., 2022). The photocoupled CO_2 reduction occurs to get CO by using Ru(bpy)$_3$Cl$_2$, which serves as a potent photosensitive donor. Co-Bpy-COF-Ru1/2 had a CO FE of 96.7% at −0.7 V and a CO partial current density of 16.27 mA cm^{-2} at −1.1 V under light assistance (Wu et al., 2023). Heme and chlorophyll increased the CO_2 reduction processes. COFs doped with piperazine molecules worked as flexible supports for single-atom catalysts. Therefore, COFs showed greater conductivity and structural stability under optimum conditions, and the inclusion of $-CH_2NH_2$ greatly enhanced the effectiveness of CO_2 reduction processes (Xie et al., 2024). Figure 11.1 shows the CO_2 capturing and conversion of CO_2 by COFs.

11.6 CHALLENGES AND FUTURE DIRECTIONS

The Sel for CO_2 remains insufficient, and the capture capacity is comparatively low. The main problem for efficient CO_2 capturing is the high Sel needed due to the low level of CO_2. The transformation of CO_2 into beneficial organic compounds remains the primary issue due to the high energy required for the cleavage of C–O bonds. The efficacy of energy utilization may be improved by adding COFs with metals or metal oxides, hence augmenting the catalytic properties of COFs. The unmodified COFs have relatively low photocatalytic properties. The band gaps, porous morphology, and modified surface sites may increase the rate of photocatalytic conversion of CO_2. The reliability and durability of nanoparticles must be taken into account. The toxic nature of the COFs must be determined for their utilization, as they are discharged into nature after consumption. COFs with varying characteristics may currently be produced in the laboratory. Nevertheless, the cost of preparing COFs remains high (about \$3,500 kg^{-1} COFs), which is expensive for COF applications. Cost-effective synthesis methods for the commercial manufacturing of COFs must be developed for their practical use.

11.7 CONCLUSION

Because of their highly flexible structure and functional characteristics, COFs have become interesting materials for CO_2 capture. Many COFs show good adsorption capacities; some show amazing increases via structural modification, metal doping, and the addition of functional groups. MMM with COFs enhanced gas permeability, hence extending their application potential. Ionic and amine-functionalized COFs further improved CO_2 affinity. Furthermore, humidity-based COFs revealed that the environmental factors play an important role in adsorption efficiency. These developments highlight how flexible COFs are in maximizing CO_2 capture, which renders them a better choice for environmentally friendly carbon capture technology. COFs are the best catalysts due to their great Sel, stability, and efficiency. COFs help in the process of cycloaddition, photoreduction, and electrocatalysis of CO_2. Their flexible structure, high porosity, and integrated active sites improve CO_2 adsorption and activation, thereby enabling effective conversion into useful products including cyclic carbonates, CO, and CH_4. By means of optimal light absorption, charge separation, and electron transfer, photocatalytic CO_2 reduction with COFs gains great Sel and activity free from sacrificial agents. Analogous to high FE, long-term stability, and improved current densities are shown by electrocatalytic CO_2 reduction with COFs, especially those including metal centers like Co, Ni, and Ru. By means of electron transfer and H_2 evolution suppression, novel heterostructures like MOF@COF hybrids enhance CO_2 reduction reaction performance even more. These results highlight the possibilities of COFs in sustainable CO_2 utilization, therefore making them a preferable material for scalable catalytic utilization in carbon capture and conversion systems.

REFERENCES

Aksu, G.O., I. Erucar, Z.P. Haslak, and S. Keskin. 2022. Accelerating discovery of COFs for CO_2 capture and H_2 purification using structurally guided computational screening. *Chemical Engineering Journal*. 427:131574. https://doi.org/10.1016/J.CEJ.2021.131574.

Alenazi, M.H., A. Helal, M.Y. Khan, A. Khalil, A. Khan, M. Usman, and M.H. Zahir. 2025. Covalent organic frameworks (COFs) for CO_2 utilizations. *Carbon Capture Science & Technology*. 14:100365. https://doi.org/10.1016/J.CCST.2025.100365.

Bagherian, N., A.R. Karimi, and A. Amini. 2021. Chemically stable porous crystalline macromolecule hydrazone-linked covalent organic framework for CO_2 capture. *Colloids and Surfaces A: Physicochemical and Engineering Aspects*. 613:126078. https://doi.org/10.1016/J.COLSURFA.2020.126078.

Bai, J., M. Zhang, F. Si, Li, Yan, G. Liang, T. Hou, and Y. Li. 2024. Directional transport of photogenerated electrons to reduction sites in covalent organic frameworks by microenvironment modulation for CO_2 photoreduction. *Advanced Functional Materials*. 35(15):2420218. https://doi.org/10.1002/ADFM.202420218.

Chen, H., J. Qin, X. Ruan, Q. Zhang, H. Zhu, and S. Zhu. 2025. Linkage engineering of covalent-organic frameworks for CO_2 capture. *Separation and Purification Technology*. 354:129378. https://doi.org/10.1016/J.SEPPUR.2024.129378.

Dautzenberg, E., G. Li, and L.C.P.M. de Smet. 2023. Aromatic amine-functionalized covalent organic frameworks (COFs) for CO_2/N_2 Separation. *ACS Applied Materials and Interfaces*. 15(4): 5118–5127. https://doi.org/10.1021/ACSAMI.2C17672.

Ding, L.G., B.J. Yao, W.X. Wu, Z.G. Yu, X.Y. Wang, J.L. Kan, and Y.D. Bin. 2021. Metalloporphyrin and ionic liquid-functionalized covalent organic frameworks for catalytic CO_2 cycloaddition via visible-light-induced photothermal conversion. *Inorganic Chemistry*. 60(16): 12591–12601. https://doi.org/10.1021/ACS.INORGCHEM.1C01975.

Dong, P., X. Xu, R. Luo, S. Yuan, J. Zhou, and J. Lei. 2023. Postsynthetic annulation of three-dimensional covalent organic frameworks for boosting CO_2 photoreduction. *Journal of the American Chemical Society*. 145(28):15473–15481. https://doi.org/10.1021/JACS.3C03897.

Duan, K., J. Wang, Y. Zhang, and J. Liu. 2019. Covalent organic frameworks (COFs) functionalized mixed matrix membrane for effective CO_2/N_2 separation. *Journal of Membrane Science*. 572: 588–595. https://doi.org/10.1016/J.MEMSCI.2018.11.054.

Fu, Y., X. Zhu, L. Huang, X. Zhang, F. Zhang, and W. Zhu. 2018. Azine-based covalent organic frameworks as metal-free visible light photocatalysts for CO_2 reduction with H_2O. *Applied Catalysis B: Environmental*. 239:46–51. https://doi.org/10.1016/J.APCATB.2018.08.004.

Furukawa, H. and O.M. Yaghi. 2009. Storage of hydrogen, methane, and carbon dioxide in highly porous covalent organic frameworks for clean energy applications. *Journal of the American Chemical Society*. 131(25):8875–8883. https://doi.org/10.1021/JA9015765.

Gao, Y., Z. Tan, R. Yang, G. Huang, and J. Bi. 2022. Integrating polyarylether-COFs with TiO_2 nanofibers for enhanced visible-light-driven CO_2 reduction in artificial photosynthesis. *Applied Surface Science*. 605:154605. https://doi.org/10.1016/J.APSUSC.2022.154605.

Han, B., X. Ding, B. Yu, H. Wu, W. Zhou, W. Liu, C. Wei, B. Chen, D. Qi, H. Wang, K. Wang, Y. Chen, B. Chen, and J. Jiang. 2021. Two-dimensional covalent organic frameworks with Cobalt(II)-phthalocyanine sites for efficient electrocatalytic carbon dioxide reduction. *Journal of the American Chemical Society*, 143(18): 7104–7113. https://doi.org/10.1021/JACS.1C02145.

Han, B., Y. Jin, B. Chen, W. Zhou, B. Yu, C. Wei, H. Wang, K. Wang, Y. Chen, B. Chen, and J. Jiang. 2022. Maximizing electroactive sites in a three-dimensional covalent organic framework for significantly improved carbon dioxide reduction electrocatalysis. *Angewandte Chemie International Edition*. 61(1):e202114244. https://doi.org/10.1002/ANIE.202114244.

Huang, N., K.H. Lee, Y. Yue, X. Xu, S. Irle, Q. Jiang, and D. Jiang. 2020. A stable and conductive metallophthalocyanine framework for electrocatalytic carbon dioxide reduction in water. *Angewandte Chemie*. 132(38): 16730–16736. https://doi.org/10.1002/ANGE.202005274.

Huang, N., X. Chen, R. Krishna, and D. Jiang. 2015. Two-dimensional covalent organic frameworks for carbon dioxide capture through channel-wall functionalization. *Angewandte Chemie International Edition*, 54(10): 2986–2990. https://doi.org/10.1002/ANIE.201411262.

Kaleeswaran, D., R. Antony, A. Sharma, A. Malani, and R. Murugavel. 2017 Catalysis and CO_2 capture by palladium-incorporated covalent organic frameworks. *ChemPlusChem*. 82(10):1253–1265. https://doi.org/10.1002/CPLU.201700342.

Li, H., A. Dilipkumar, S. Abubakar, and D. Zhao. 2023. Covalent organic frameworks for CO_2 capture: from laboratory curiosity to industry implementation. *Chemical Society Reviews*. 52(18):6294–6329. https://doi.org/10.1039/D2CS00465H.

Li, H., Z. Zhou, T. Ma, K. Wang, H. Zhang, A.H. Alawadhi, and O.M. Yaghi. 2024. Bonding of polyethylenimine in covalent organic frameworks for CO_2 capture from air. *Journal of the American Chemical Society*. 146(51):35486–35492. https://doi.org/10.1021/JACS.4C14971.

Li, Y., J. Zhang, K. Zuo, Z. Li, Y. Wang, H. Hu, C. Zeng, H. Xu, B. Wang, and Y. Gao. 2021. Covalent organic frameworks for simultaneous CO_2 capture and selective catalytic transformation. *Catalysts*. 11(9): 1133. https://doi.org/10.3390/CATAL11091133.

Lin, H., Y. Liu, Z. Wang, L. Ling, H. Huang, Q. Li, L. Cheng, Y. Li, J. Zhou, K. Wu, J. Zhang, and T. Zhou. 2022. Enhanced CO_2 photoreduction through spontaneous charge separation in end-capping assembly of heterostructured covalent-organic frameworks. *Angewandte Chemie*. 134(50):e202214142. https://doi.org/10.1002/ANGE.202214142.

Liu, J., G. Zhao, O. Cheung, L. Jia, Z. Sun, and S. Zhang. 2019. Highly porous metalloporphyrin covalent ionic frameworks with well-defined cooperative functional groups as excellent catalysts for CO_2 cycloaddition. *Chemistry – A European Journal*. 25(38):9052–9059. https://doi.org/10.1002/CHEM.201900992.

Liu, P., K. Cai, D.J. Tao, and T. Zhao. 2024. The mega-merger strategy: M@COF core-shell hybrid materials for facilitating CO_2 capture and conversion to monocyclic and polycyclic carbonates. *Applied Catalysis B: Environmental*. 341:123317. https://doi.org/10.1016/J.APCATB.2023.123317.

Lu, M., M. Zhang, C. Liu, G.J. Liu, L.J. Shang, M. Wang, J.N. Chang, S.L. Li, and Y.Q. Lan. 2021. Stable dioxin-linked metallophthalocyanine covalent organic frameworks (COFs) as photo-coupled electrocatalysts for CO_2 reduction. *Angewandte Chemie*. 133(9):4914–4921. https://doi.org/10.1002/ANGE.202011722.

Lyu, H., H. Li, N. Hanikel, K. Wang, and O.M. Yaghi. 2022. Covalent organic frameworks for carbon dioxide capture from air. *Journal of the American Chemical Society*. 144(28):12989–12995. https://doi.org/10.1021/JACS.2C05382.

Mahajan, S. and M. Lahtinen. 2022. Recent progress in metal-organic frameworks (MOFs) for CO_2 capture at different pressures. *Journal of Environmental Chemical Engineering*. 10(6):108930. https://doi.org/10.1016/J.JECE.2022.108930.

Mushtaq, N., A. Ahmad, X. Wang, U. Khan, and J. Gao. 2024. MOFs/COFs hybrids as next-generation materials for electrocatalytic CO_2 reduction reaction. *Chemical Engineering Journal*. 486:150098. https://doi.org/10.1016/J.CEJ.2024.150098.

Nguyen, H.L. and A. Alzamly. 2021. Covalent organic frameworks as emerging platforms for CO_2 photoreduction. *ACS Catalysis*, 11(15): 9809–9824. https://doi.org/10.1021/ACSCATAL.1C02459.

Ozdemir, J., I. Mosleh, M. Abolhassani, L.F. Greenlee, R.R Beitle, and M.H. Beyzavi. 2019. Covalent organic frameworks for the capture, fixation, or reduction of CO_2. *Frontiers in Energy Research*. 7:460449. https://doi.org/10.3389/FENRG.2019.00077.

Peter, S.C. 2018. Reduction of CO_2 to chemicals and fuels. A solution to global warming and energy crisis. *ACS Energy Letters*. 3(7):1557–1561. https://doi.org/10.1021/ACSENERGYLETT.8B00878.

Raghav, S., J. Raghav, M.A. Muhammad, M. B. Singh, P.Jain, D. Kumar. 2025. Recent advancements in synthesis and applications of covalent organic frameworks. *Applied Materials Today* 44(June): 102694. doi:10.1016/j.apmt.2025.102694

Ran, L., Z. Li, B. Ran, J. Cao, Y. Zhao, T. Shao, Y. Song, M.K.H. Leung, L.J. Sun, and J. Hou. 2022. Engineering single-atom active sites on covalent organic frameworks for boosting CO_2 photoreduction. *Journal of the American Chemical Society*. 144(37):17097–17109. https://doi.org/10.1021/JACS.2C06920.

Sarkar, P., I.H. Chowdhury, S. Das, and S.M. Islam. 2022. Recent trends in covalent organic frameworks (COFs) for carbon dioxide reduction. *Materials Advances*. 3(22):8063–8080. https://doi.org/10.1039/D2MA00600F.

Song, X., Y. Wu, X. Zhang, X. Li, Z. Zhu, C. Ma, Y. Yan, P. Huo, and G. Yang. 2021. Boosting charge carriers separation and migration efficiency via fabricating all organic van der Waals heterojunction for efficient photoreduction of CO_2. *Chemical Engineering Journal*. 408:127292. https://doi.org/10.1016/J.CEJ.2020.127292.

Stegbauer, L., M.W. Hahn, A. Jentys, G. Savasci, C. Ochsenfeld, J.A. Lercher, and B.V. Lotsch. 2015. Tunable water and CO_2 sorption properties in isostructural azine-based covalent organic frameworks through polarity engineering. *Chemistry of Materials*. 27(23):7874–7881. https://doi.org/10.1021/ACS.CHEMMATER.5B02151.

Sumida, K., D.L. Rogow, J.A. Mason, T.M. McDonald, E.D. Bloch, Z.R. Herm, T.H. Bae, and J.R. Long. 2012. Carbon dioxide capture in metal-organic frameworks. *Chemical Reviews*. 112(2):724–781. https://doi.org/10.1021/CR2003272.

Wang, Y., C. Kang, Z. Zhang, A.K. Usadi, D.C. Calabro, L.S. Baugh, Y.D. Yuan, and D. Zhao. 2022. Evaluation of schiff-base covalent organic frameworks for CO_2 capture: Structure-performance relationships, stability, and performance under wet conditions. *ACS Sustainable Chemistry and Engineering*. 10(1): 332–341. https://doi.org/10.1021/ACSSUSCHEMENG.1C06318.

Wei, H., S. Chai, N. Hu, Z. Yang, L. Wei, and L. Wang. 2015. The microwave-assisted solvothermal synthesis of a crystalline two-dimensional covalent organic framework with high CO_2 capacity. *Chemical Communications*. 51(61):12178–12181. https://doi.org/10.1039/C5CC04680G.

Wu, Q.J., D.H. Si, S. Ye, Y.L. Dong, R. Cao, and Y.B. Huang. 2023. Photocoupled electroreduction of CO_2 over photosensitizer-decorated covalent organic frameworks. *Journal of the American Chemical Society*. 145(36):19856–19865. https://doi.org/10.1021/JACS.3C06113.

Xiang, Y., W. Dong, P. Wang, S. Wang, X. Ding, F. Ichihara, Z. Wang, Y. Wada, S. Jin, Y. Weng, H. Chen, and J. Ye. 2020. Constructing electron delocalization channels in covalent organic frameworks powering CO_2 photoreduction in water. *Applied Catalysis B: Environmental*. 274:119096. https://doi.org/10.1016/J.APCATB.2020.119096.

Xie, T., S. Chen, Y. Yue, T. Sheng, N. Huang, and Y. Xiong. 2024. Biomimetic phthalocyanine-based covalent organic frameworks with tunable pendant groups for electrocatalytic CO_2 reduction. *Angewandte Chemie International Edition*. 63(43):e202411188. https://doi.org/10.1002/ANIE.202411188.

Xu, K., Y. Dai, B. Ye, and H. Wang. 2017. Two-dimensional covalent organic framework materials for chemical fixation of carbon dioxide: excellent repeatability and high selectivity. *Dalton Transactions*. 46(33):10780–10785. https://doi.org/10.1039/C7DT02527K.

Yang, F., Y. Li, T. Zhang, Z. Zhao, G. Xing, and L. Chen. 2020. Docking site modulation of isostructural covalent organic frameworks for CO_2 fixation. *Chemistry – A European Journal*. 26(20):4510–4514. https://doi.org/10.1002/CHEM.202000552.

Yang, Y.L., Y.R. Wang, L.Z. Dong, Q. Li, L. Zhang, J. Zhou, S.N. Sun, H.M. Ding, Y. Chen, S.L. Li, and Y.Q. Lan. 2022. A honeycomb-like porous crystalline hetero-electrocatalyst for efficient electrocatalytic CO_2 reduction. *Advanced Materials*. 34(44):2206706. https://doi.org/10.1002/ADMA.202206706.

Yin, H.Q., Z.M. Zhang, and T.B. Lu. 2023a. Ordered integration and heterogenization of catalysts and photosensitizers in metal-/covalent-organic frameworks for boosting CO_2 photoreduction. *Accounts of Chemical Research*. 56(19): 2676–2687. https://doi.org/10.1021/ACS.ACCOUNTS.3C00380.

Yin, M., L. Wang, and S. Tang. 2022. Amino-functionalized ionic-liquid-grafted covalent organic frameworks for high-efficiency CO_2 capture and conversion. *ACS Applied Materials and Interfaces*. 14(50):55674–55685. https://doi.org/10.1021/ACSAMI.2C18226.

Yin, M., L.P. Wang, and S. Tang. 2023b. Stable dicationic covalent organic frameworks manifesting notable structure-enhanced CO_2 capture and conversion. *ACS Catalysis*. 13(19):13021–13033. https://doi.org/10.1021/ACSCATAL.3C02796.

Yuan, J., S. Chen, Y. Zhang, R. Li, J. Zhang, and T. Peng. 2022. Structural regulation of coupled phthalocyanine–porphyrin covalent organic frameworks to highly active and selective electrocatalytic CO_2 reduction. *Advanced Materials*. 34(30):2203139. https://doi.org/10.1002/ADMA.202203139.

Zhang, H., S. Liu, L. Wang, H. Fang, X. Yue, Z. Wang, S. Wei, S. Liu, and X. Lu. 2024. Ultrahigh performance CO_2 capture and separation in alkali metal anchored 2D-COF. *Separation and Purification Technology*. 333:125937. https://doi.org/10.1016/J.SEPPUR.2023.125937.

Zhang, Y., H. Hu, J. Ju, Q. Yan, V. Arumugam, X. Jing, H. Cai, and Y. Gao. 2020. Ionization of a covalent organic framework for catalyzing the cycloaddition reaction between epoxides and carbon dioxide. *Chinese Journal of Catalysis*. 41(3):485–493. https://doi.org/10.1016/S1872-2067(19)63487-X.

Zheng, T., X. Ding, T. Sun, X. Yang, X. Wang, X. Zhou, P. Zhang, B. Yu, Y. Wang, Q. Xu, L. Xu, D. Wang, and J. Jiang. 2024. Nanostructurally engineering covalent organic frameworks for boosting CO_2 photoreduction. *Small*. 20(15):2307743. https://doi.org/10.1002/SMLL.202307743.

Zhi, Y., P. Shao, X. Feng, H. Xia, Y. Zhang, Z. Shi, Y. Mu, and X. Liu. 2018 Covalent organic frameworks: efficient, metal-free, heterogeneous organocatalysts for chemical fixation of CO_2 under mild conditions. *Journal of Materials Chemistry A*. 6(2):374–382. https://doi.org/10.1039/C7TA08629F.

Zhong, W., R. Sa, L. Li, Y. He, L. Li, J. Bi, Z. Zhuang, Y. Yu, and Z. Zou. 2019. A Covalent organic framework bearing single Ni sites as a synergistic photocatalyst for selective photoreduction of CO_2 to CO. *Journal of the American Chemical Society*. 141(18):7615–7621. https://doi.org/10.1021/JACS.9B02997.

Zhou, Z., T. Ma, H. Zhang, S. Chheda, H. Li, K. Wang, S. Ehrling, R. Giovine, C. Li, A.H. Alawadhi, M.M. Abduljawad, M.O. Alawad, L. Gagliardi, J. Sauer, and O.M. Yaghi. 2024. Carbon dioxide capture from open air using covalent organic frameworks. *Nature*. 635:96–101. https://doi.org/10.1038/s41586-024-08080-x.

Zhu, H.J., M. Lu, Y.R. Wang, S.J. Yao, M. Zhang, Y.H. Kan, J. Liu, Y. Chen, S.L. Li, and Y.Q. Lan. 2020. Efficient electron transmission in covalent organic framework nanosheets for highly active electrocatalytic carbon dioxide reduction. *Nature Communications*. 11(1):1–10. https://doi.org/10.1038/s41467-019-14237-4.

Zou, L., Z.A. Chen, D.H. Si, S.L. Yang, W.Q. Gao, K. Wang, Y.B. Huang, and R. Cao. 2023. Boosting CO_2 photoreduction via regulating charge transfer ability in a one-dimensional covalent organic framework. *Angewandte Chemie International Edition*. 62(46):e202309820. https://doi.org/10.1002/ANIE.202309820.

12 Covalent Organic Frameworks for Water Purification Methods

Amrutha V. Nambiar and Rekha Sharma

12.1 INTRODUCTION

In the current context of the escalating water crisis around the globe, the progress of advanced water treatment technologies is of great relevance. This chapter introduces covalent organic frameworks (COFs) as a growing class of polymers that are crystalline and porous materials that are ideal for addressing critical water remediation challenges faced by people all over the world. COFs are well-organized, permeable structures with excellent properties, such as high surface area, adjustable pore size, and dependable thermal and chemical steadiness.

COFs alone may not show features that will solve many challenges like water treatment, energy applications, and batteries; various design strategies have to be put to use to utilize the complete potential of the COFs.

The potential of two-dimensional (2D) COFs is highlighted to have tremendous potential to act as high-performance media in many recent investigations conducted all around the world. Their hydrophilicity, exceptional mechanical and chemical toughness, and inherent flexibility make them promising materials for desalination, evaporation, and solvent/water separations. Figure 12.1 explains the different applications of COFs.

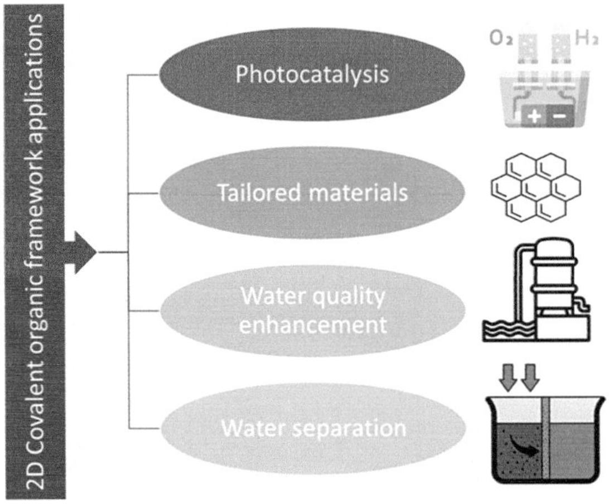

FIGURE 12.1 Applications of 2D COFs (Zhang et al., 2022.)

DOI: 10.1201/9781003605386-12

Photoactive COFs display significant potential for the photocatalytic degradation of organic pollutants. They can effectively produce reactive oxygen species, such as hydroxyl radicals and superoxide anions, upon exposure to light irradiation. These highly reactive species initiate a cascade of redox reactions leading to the oxidative breakdown of organic pollutants into less harmful or even completely neutralized products. Hence, it helps in green water treatment technologies capable of addressing the growing challenges of organic pollution. Recent evidence suggests that certain COFs possess inherent antimicrobial properties (Hou et al., 2024). Such features, which can be an outcome of properties such as surface charge, pore size distribution, and the presence of specific functional groups in the COF structure, appear as promising candidates for antimicrobial applications.

Using different designs and synthesis techniques, COF-based membranes or COMs can be synthesized that show multitasking actions like heavy metal removal, dye removal, and fouling resistance (Liu et al, 2023). Improved permeation flux is also a key advantage.

The topics covered in this chapter involve the design, synthesis methods, assembly techniques, and future perspective of COFs and associated materials.

12.2 DESIGN PRINCIPLES OF COFs: A GENERAL EVALUATION

The design and synthesis of COFs and associated materials are based on reticular chemistry. In reticular synthesis, it is a study of crystalline frameworks and their synthesis that takes the major limelight. The major focus during synthesis is on structural stability and porosity through the effective formation of covalent bonds (Ding and Wang, 2013). The covalent and coordination bonds are the major chemical bonds found in crystalline frameworks. Reactions are also based on the surrounding conditions and operational aspects.

Another branch of chemistry involved in the design and synthesis studies is linkage chemistry. It offers unique features to the COFs and also broadens the possibilities of features that can create highly specific COF designs. Some linkages, like the azo linkage, influence the photocatalytic properties of the COFs, thereby enhancing the water treatment applications. In Table 12.1, certain linkages and their examples are stated.

Through employing reticular and linkage chemistry, the maintenance of functionality in one-dimensional, 2D, and two-dimensional (3D) COF structures helps in

TABLE 12.1
Different Types of Linkages and Examples

Linkage	Chemical Structure	COF Example	Reference
Boroxine	B_3O_3	COF-1, COF-5	Shi et al. (2020)
Imine	–C=N–	COF-300, TpOMe-DAQ	Halder et al. (2018)
β-ketoenamine	–C(=O)–CH=C(OH)–	DAAQ-TFP	Kandambeth et al. (2012)
Triazine	C_3N_3	TpDTz COF	Machado et al. (2021)

achieving the desired porosity, which is an important milestone in water treatment technologies (Mohammed and Shetty, 2021). Some studies have reported the use of thiol and thioether chelating arms for the synthesis of COFs for the removal of mercury from water streams (Xia et al., 2021). These applications of tailored structures are not only for water treatment applications but also for other biological activities like drug delivery (Khan et al., 2022; Rejali et al., 2023).

12.2.1 Design Principles of COFs for Water Treatment

This section covers in detail the major considerations made during the design and synthesis of COF for water treatment applications, which are the best materials about which research is being conducted. The control of the aperture size using the hydrophilic functional groups showed very high salt rejection rates in desalination plants (Zhang et al., 2017). Similarly, the high crystalline nature of COFs also acts as excellent membrane materials for water treatment technologies.

The usage of molecular dynamics and tempered meta dynamics on the COFs improved the absorption of pharmaceutical pollutant molecules from the wastewater stream. It was also observed that the pore-based structure of COFs has improved the metal pollutant absorption. COFs have also been found to have predictable atomic distribution, hence making the application of molecular dynamics for designing a specific material in a specific application (Akhzari et al., 2024).

Additionally, steric hindrance, which is attained by the introduction of bulky groups near reactive sites, can reduce hydrolytic degradation, especially under alkaline conditions.

Weaker interlayer forces, including C–H···N hydrogen bonding, can satirically and hydrophobically induce effects, thereby retaining the framework's structural integrity (Liu et al., 2023). Also, the supramolecular coordination of metal ions into the open channel of the COF can suppress bond rotation and promote overall stability.

Hybrid strategies, in which COFs are incorporated into higher-level materials, provide a practical means to achieve overall stability. The multilayer postsynthetic modification (PSM) has been studied by researchers, which proved to give the resulting COFs high stability (Liu et al., 2023). It is thus imperative to achieve a balance between reversibility and stability in striving for the successful development of long-lived COF-based materials (Figure 12.2).

COFs designed for water purification are evaluated under three key aspects: aqueous stability, controlled aperture size, and surface functionalization design. The synthesis of hydrophilic polyvinylidene fluoride membranes improved the water flux and antifouling properties compared to sodium alginate, a natural organic membrane. However, the synthesis of the water-soluble polymer may cause the removal of the top layer of the membrane over time due to elution (Yuan et al., 2020). In a study of COF-based removal of heavy metals and organic pollutants, adsorption kinetic studies suggested second-order reactions, which indicate the surface reactions, hence confirming the surface functionalization phenomenon. Also, one of the major advantages of COF is that pore-wall modification can be introduced without compromising the structural integrity of the COF (Machado et al., 2022).

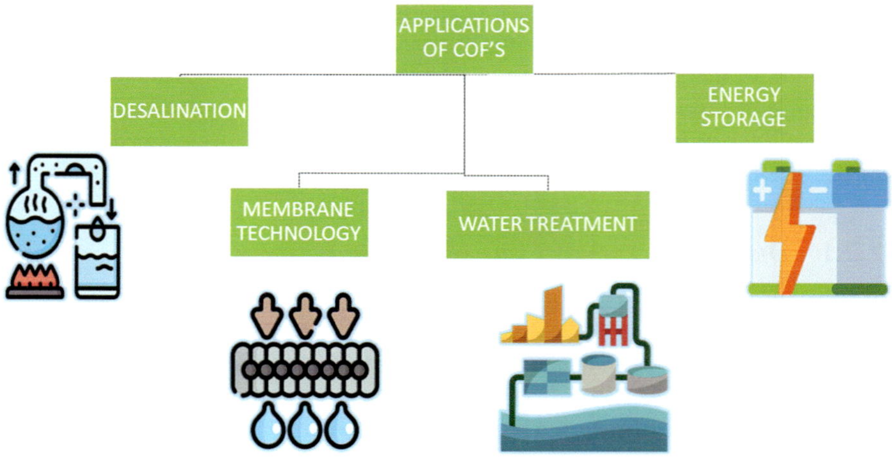

FIGURE 12.2 Schematic presentation of applications of COFs.

12.2.2 Stability in Aqueous Environments

The pH conditions in the aqueous environment varied from industry to industry, which also affected the COFs by damaging the weaker bonds in the molecular framework. Stability was seen to be enhanced in COFs that had embedded alkylation. Still, alkylation showed a gradual hydrolysis. Simple hydrolysis itself posed a great challenge to COFs in water treatment applications. The solubility of almost all things in water also affects the performance of COFs in due course of time. Improving the bond topologies and focusing on the intricate designs of COF during the synthesis can almost provide a solution to this problem, though there are more studies to be conducted.

12.2.3 Aperture Size Control

The control over the pore size can significantly increase the chances of improved efficiency of COFs. We must pre-design the molecular structure with mesopores, which allows the passage of organic, inorganic, or biological units smoothly. This also increases the range of applications where COFs can be used successfully. Desalination studies have impacted in a better way through the control over the aperture size of the COFs. Smaller pore sizes and precisely spaced layers also provided a high surface area for the improved desalination process. High ion rejection and water fluxes were observed in the saline water purification domain.

12.2.4 Surface Functionalization

The careful introduction of functional groups can control the acidic or basic moieties incorporated in the COFs (Liu et al., 2021). In 2D COFs, the structures are based on the single-plane covalent bonds and a linkage with noncovalent bonds to form COF structures. Whereas in 3D COF, there are channel and cage-like structures, and active reaction sites are found in bulk, to accommodate larger molecules of pollutants. Also, some studies imply that the application of polyaromatic hydrocarbon bonds can provide bigger dipole moments and hence offer rigidity to the structure.

12.3 SYNTHESIS AND PROCESSING OF COFs FOR WATER PURIFICATION

To address the disadvantages of COFs, the synthesis can be modified to accommodate the rising needs of the water treatment industry. Right from the selection of the monomer to the formation of a 2D or 3D cage can alter the course of pollutant removal from the water stream (Altintas et al., 2022; Zhang et al., 2020). The following sections describe the formation of COFs using methods based on conventional methods as well as modified ones.

12.3.1 Solvothermal Synthesis

Solvothermal synthesis is a technique where a mixture of a monomer and a solvent is covered in a temperature- and pressure-regulated vessel or a hydrothermal flask for the synthesis of COFs. Most of the COFs synthesized are in powder form or as films. A solvothermal synthesis involves a typical ligand exchange or a Schiff's base reaction at 80°C–120°C depending on the chemical reactivity of the building blocks of the COFs. The reaction time ranges from hours to days for an optimal synthesis. The selection of a suitable solvent can drastically affect the reaction products like crystal growth and nucleation. However, the major disadvantages are slow reactivity, high temperature, and pressure requirements (Akinnawo, 2024).

12.3.2 Microwave-Assisted Synthesis

Microwave irradiation is used for the synthesis to form improved crystal growth. Microwave-assisted synthesis also helps reduce the reaction time, which is a critical economic advantage in the water treatment industry (Rodríguez-Carríllo et al., 2023). Researchers have reported 200 times faster reaction rates as compared to the conventional technique of solvothermal. The microwave-assisted synthesis also offers control over the reaction and regulates the stability.

12.3.3 Interfacial Polymerization

In this type of polymerization, two thermodynamically different phases are mixed, which forms an interface that acts as a reaction site. The monomers are dissolved in suitable liquid phases and materials that are synthesized. In another method, monomers are dissolved in separate liquid phases and create homogeneous films through polymerization. This type of interface can alter the morphology of COFs by creating very thin materials or films. This technique is mostly applied to create metal-organic frameworks (MOFs) and other hybrid materials (Wang et al., 2018). Thin materials can participate actively in the nanofiltration of water and hence hold great importance in water purification technology.

12.3.4 Postsynthetic Modification of Outcome-Based Membranes

In PSM, new reactive groups are introduced on an existing molecular skeleton of COFs and transform the chemical activity of the material. This technique helps in improving the functionality of the COFs. The active and strong covalent bonds present in COFs help hold metals in the binding sites formed through the modification. PSM also assists in incorporating the reversible reaction in COFs, which helps with improved life expectancy of COFs in water treatment technology. PSM is also used widely for catalysis since it showed an improved catalytic rate in reactions (Khan et al., 2022).

In water treatment technology, specifically in desalination, adjusting the pore size from nanoscale to submeter nanoscale can correlate well with the kinetics of the reaction and hence shows improved salt rejection rates. However, creating a flawless material using multiple reactions is still a challenge in his method (Zheng et al., 2022). In this approach, pure COF nanosheets are first prepared by interfacial polymerization and followed by functional group switching (e.g., cyano to amidoxime or carboxyl groups) for enhanced performance in some applications like carbon capture and so on.

Development of a hybrid structure, for example, COF–MOF hybrids, is an emerging strategy for bringing together the best properties of both materials of a cause. Some researched methods involve the MOF-first approach (sowing COFs onto presynthesized MOFs), the COF-first approach (employing COFs as seeds for MOF growth), and postsynthetic linkage (mating presynthesized COFs and MOFs). The materials proved to be very useful in various water treatment processes, ranging from toxin detection, adsorption-based removal, and photodegradation through catalytic activity (Ahmadijokani et al., 2024).

The development of COF synthesizing technology also requires to consider the sustainability and reusability factor. For instance, water as a reaction solvent has been explored for the formation of safer, cleaner, and less costly synthesis procedures that are economical for business production (Cui and Fu, 2022). These new trends in COF processing techniques are paving the path for the design and construction of highly efficient and engineered materials for water purification and other critical uses that are water-based and perform well in aqueous media providing sustainable solutions.

12.3.5 Membrane Fabrication

COFs are currently very popular materials in the water treatment process, specifically in membrane forms. Hence, their fabrication techniques are very relevant. Certain configurations, namely three of them, have been developed: supported membranes, free-standing membranes, and mixed matrix membranes (MMMs). In supported membranes, the growth or deposition of COFs on porous supports improves mechanical stability and enables integration into standard filtration systems of water treatment systems. This technique is used to develop thin, highly selective layers without compromising molecular and structural stability. Free-standing membranes, on the other hand, are free-standing COF films prepared for particular applications where substrate interference is undesirable and of the least importance (Karmakar and Li, 2020; Yu et al., 2023).

MMMs are a hybrid approach, and COF particles are implanted in polymer matrices to yield composite membranes with boosted properties. The membranes have exceptional properties, such as flexibility and postsynthetic modifiability. These synergistic hybrid membranes control the porosity and selectivity of COFs and the mechanical stability of polymers, which are very critical for water-based processes. Specifically, MMMs based on COFs have revealed exceptional enhancement in water permeation and rejection of model foulants in ultrafiltration operations. For example, it has been demonstrated by some researchers that the incorporation of COF nanoparticles into polyether sulfone membranes can lead to improved hydrophilicity, fouling resistance, and separation efficiency (Arqueros et al., 2021; Mohammed and Shetty, 2021). These progressive technologies in the synthesis of COFs have given endless possibilities of combinations for tailoring to the needs of the purification process, irrespective of the pollutant or the medium of pollution.

12.3.6 Macroscopic Architectures

Various macroscopic architecture assemblies of COFs have significantly improved their potential applications in water treatment technologies. The emergence of three major forms has been very promising. They are foams, thick films, and thin films. Studies conducted by some researchers explain that COF-based foams possess better surface area-to-volume ratios and enhanced mass transfer properties, interpreting them as highly efficient for adsorption-based purification processes (Xia et al., 2021). These are major 3D structures that facilitate the rapid sorption of pollutants from water solutions, which drastically improve the efficiency of water treatment processes and control the generation of by-products. Another important macroscopic morphology of thick COF films is that they offer improved mechanical stability required in pressure-driven membrane applications.

The hybridization of MOFs and aldehyde-based material resulted in the synthesis of an orderly spherical shape and nanosized particles, which showed an efficient removal of cationic dyes from the water stream (Firoozi et al., 2020).

Heterogeneous catalysts offer COFs the ability to act as excellent acid-solid catalysts by incorporating acids into the COF pores. COFs can also be transformed into a basic-solid catalyst, which can be synthesized through the PSM approach (Alsudairy et al., 2023).

12.4 SURFACE FUNCTIONALIZATION THROUGH POSTSYNTHETIC MODIFICATION

For significant structural modification, reversible bond formation is a step. The inclusion of metals, covalent bonds, and building block exchange is crucial for improving the surface functionality. The application of de novo synthesis helps in the PSM of COFs by creating new materials with enhanced properties (Segura et al., 2019).

PSM can also help in the embedding of metal nanoparticles in the surface cavities of COFs for various applications (Chen et al., 2023; Segura et al., 2019). PSM also supports different applications like energy storage and supercapacitors by providing active redox reaction sites for energy formation (Kumar et al., 2021). The careful selection of the surface modifications can improve a lot of properties of the COFs and enhance their applications in various fields.

12.5 HYBRID STRUCTURES

The synthesis of hybrid structures can help researchers to overcome the challenges that the COFs cannot solve alone. The COF–MOF hybrids proved to influence in setting unique standards in performance characteristics (Han et al., 2023).

Majorly, there are three primary methods in the synthesis of COF–MOF hybrids. The first method involves the growth of COF particles on pre-synthesized MOF substrates. This helps researchers control the core MOF layer and the functional COF layer in water treatment applications. In the second method, MOF layer is developed over a core layer of COF, which specifically controls the pore size and chemical functionalities. As for the third method, a postsynthetic linkage is introduced on the synthesized COF–MOF hybrid, which can highly influence the photocatalytic performance of water treatment systems.

These hybrid materials are found to perform promisingly in numerous applications of water treatment. In sensing water pollutants, COF–MOF hybrid materials can utilize the high surface area and tunable pore structure of both components to produce highly sensitive and selective sensors toward water pollutants. The interplay of MOF and COF architectures is capable of enhancing adsorption selectivity and capacity toward specific pollutants and making the hybrids highly efficient in the adsorptive removal of water pollutants.

Catalytic photodegradation is one of the fields where COF–MOF hybrids have shown specific promise. The synergistic effect of combining MOFs' metal centers with COFs' organic linkers can result in enhanced photocatalytic activity for the degradation of organic water pollutants. Such improved performance is typically acclaimed based on enhanced charge separation and transfer within the hybrid structure, which results in more efficient photocatalytic reactions.

The development of some hybrid materials also faces some setbacks like limited stability in water, which is a critical challenge to the water treatment industry. Similarly, when we compare the functional efficiency of COF with MOF, it lacks the functional diversity that is easily offered by MOFs. Thus, the thorough mixing of the materials has to be conducted for efficient materials synthesis with enhanced functional properties (Rodríguez-Camargo et al., 2024).

Studies are being conducted on various combinations of COFs and MOFs to achieve optimal performance for water treatment applications. As evidence, scientists have fabricated COF–MOF materials to be more selective in the decontamination of heavy metals from water together with increased adsorption capacity. Some others have studied the improved catalysis of water by incorporating a hierarchical pore size study of structures within the hybrid for efficient mass transfer (Aslam et al., 2024).

Through structural synthesis and design, COF–MOF hybrids are becoming more sophisticated. With researchers attempting ways of regulating a 3D arrangement into the subunits of hybrid materials, they can be precisely designed for addressing specific water treatment challenges (Vardhan et al., 2023). Studies are also being made to develop simpler methods for the synthesis of these hybrid materials to improve the economic and commercial potential. As many studies are being conducted in this area, COF-based materials will have high potential in solving global water challenges soon.

12.6 APPLICATION OF COFs IN WATER PURIFICATION

In water treatment methods like desalination, COFs have shown great potential in enhancing the functional properties of conventional technologies. Exploring the influence of various COF characteristics on membrane-based desalination units, researchers aim to develop more effective salt rejection materials (Dubey et al., 2024). The inclusion of COFs in desalination techniques like the electrodes, membranes, and solar absorbers has a promising potential to introduce significant gains in performance efficiency.

COFs display effective adsorption in the removal of organic contaminants from wastewater. The principle of size exclusion is employed through the use of homogenized pore size and structurally designed porosity. Moreover, the surface properties are also adjusted. The functionalized COF structures fit a particular organic contaminant for greater selectivity and capacity toward adsorption.

COFs show a superior performance in heavy metal removal and toxic ion adsorption. The activities of certain functional groups in the COF allow selective ion and other inorganic adsorption. The synthesis of MOF–COF hybrids merges the stability of COF with the adsorption efficiency of toxic ions from polluted water.

The photocatalytic oxidation of organic pollutants is an alternative area where COFs have shown good results that can be studied further. By incorporating photocatalytic units into COF scaffolds, researchers have developed materials that can oxidize stable organic pollutants when illuminated, offering another route for water purification besides simple filtration or adsorption.

The application of COFs in membrane technologies has led to tremendous advances in water treatment. COMs, whether as supported architecture, freestanding films, or mixed matrix membranes, have shown improved water flux and model foulant rejection in various filtration operations, including nanofiltration and ultrafiltration.

Recent developments in COF processing techniques, such as the creation of COF foams and thick films, have unlocked their potential for more upscale water treatment processes. Their macroscopic counterparts possess superior mass transfer properties and mechanical stability, which are necessary for large-scale water treatment systems.

With progressive research in this field, the focus now is on developing more sustainable and scalable COF-based materials for water treatment. Efforts are being made to improve the aqueous stability of COFs, which is essential for their sustained performance in water purification processes. Furthermore, progress in hybrid material development, e.g., composites of COFs and MOFs, is opening up new avenues for the treatment of intricate water treatment challenges through synergistic utilization of properties of different classes of materials (Table 12.2).

TABLE 12.2
Key Findings of Research Conducted on Water Remediation

S. No	COFs Material	Fabrication Method	Removed Pollutants (Heavy Metal and Radionuclides)	Adsorption Capacity (mg g^{-1})	Reference
1	COF-COOH	Solid-phase extraction	Uranium (VI)	211.0	Li et al. (2015)
2	COF's	Room temperature	Nitro aromatic compounds	239.9	Gao et al. (2019)
3	COF-S-SH	Postsynthesis modification	Hg^{2+}	586.3	Wang et al. (2023b)
4	COF-OH-SH	Postsynthesis modification	Hg^{2+}	535.5	Wang et al. (2023b)
5	TG-PD-COF	Sonochemical	Perfluorooctanoic acid (PFOA)	2,600.0	Jrad et al. (2024)
6	Zn-MOF@COF	Solvothermal	Cu^{2+}	500.0	Kumar et al. (2024)
7	Zn-MOF@COF	Solvothermal	Methylene blue	26.3	Kumar et al. (2024)
8	PCN-224-MAA(O)	Postsynthesis modification	Hg^{2+}	843.6	Lin et al. (2019)
9	PCN-224-MAA(P)	Postsynthesis modification	Hg^{2+}	138.5	Lin et al. (2019)
10	COF-TP (LINEAR DIAMINE)	Polymerization (room temperature)	Pb^{2+}	140.0	Li et al. (2019)
11	COF-TF (AROMATIC DIAMINE)	Polymerization (room temperature)	Pb^{2+}	185.7	Li et al. (2019)
12	COF-BTA-DHBZ	Solvothermal	Cr (VI)	384.0	Cui et al. (2019)
13	SCU-COF-2	Solvothermal	I_2	980.0	He et al. (2021)
14	SCU-COF-2	Solvothermal	CH_3I	1,450.0	He et al. (2021)
15	QTD-COF	Polymerization	I_2	629.0	Guo et al. (2020)
16	TAPB-DMTTPA-COF	Solvothermal	Hg^{2+}	734.0	Huang et al. (2017)
17	Ui066-PYDC	Hydrothermal	I_2	1,250.0	Wang et al. (2017)
18	γ-Fe_2O_3@CTF-1	Ionothermal	As (III)	198.0	Gendy et al. (2021)
19	TpPa-NH_2@EDTA	Postsynthesis modification	Ag^+, Pd^{2+}, Fe^{3+}, Cr^{3+}, Cu^{2+}, Ni^{2+}	50.0	Jiang et al. (2019)
20	$Ni_{0.6}Fe_{2.4}O_4$-HT-COF	Room temperature	Pb^{2+}	411.8	Wang et al. (2023a)
21	PAN-BT-DG	Condensation and electrospinning	Cr (VI)	173.0	Herath et al. (2022)

(*Continued*)

TABLE 12.2 (*Continued*)
Key Findings of Research Conducted on Water Remediation

S. No	COFs Material	Fabrication Method	Removed Pollutants (Heavy Metal and Radionuclides)	Adsorption Capacity (mg g^{-1})	Reference
22	COF's-V linked COF@COOH	Room temperature and TC grafting	Cu^{2+}	19.3	Zhang et al. (2025)
23	COF's-V linked COF@COOH	Room temperature and TC grafting	Zn^{2+}	13.0	Zhang et al. (2025)
24	SNN-MIL-125(Ti)@ Fe_3O_4	Postsynthesis modification	Hg^{2+}	669.0	Li et al. (2023)
25	COF-SH	Postsynthesis modification	Hg^{2+}	1,283.0	Ma et al. (2021)

COFs are very efficient and versatile in the removal of different concentrations of pollutants from aqueous solutions obtained from sources like poultry farms and tannery distilleries. COF-COOH is found to be promising for uranium (VI) removal with an adsorption capacity of 211.0 mg g^{-1}, while room-temperature-synthesized COFs remove nitroaromatic compounds efficiently (Guo et al., 2020). Removal of mercury (Hg^{2+}) is accomplished by COFs like COF-S-SH, COF-OH-SH (Wang et al., 2017), PCN-224-MAA(O) (Lin et al., 2019), TAPB-DMTTPA-COF (Huang et al., 2017), and COF-SH (Ma et al., 2021), with capacities of 535.5–1283.0 mg g^{-1}. Lead (Pb^{2+}) was removed by COF-TP and COF (Li et al., 2019). Similarly, lead was also removed by $Ni_{0.6}Fe_{2.4}O_4$-HT-COF (Wang et al., 2023a), with an efficiency of 411.8 mg g^{-1}. In another study, removal of chromium (Cr (VI)) was accomplished by COF-BTA-DHBZ (Cui et al., 2019) and PAN-BT-DG (Herath et al., 2022). Iodine and methyl iodide uptake was also accomplished by SCU-COF-2 (He et al., 2021), QTD-COF (Guo et al., 2020), and Ui066-PYDC (Wang et al., 2017) with absorption intakes up to 1,450.0 mg g^{-1}. Some other heavy metal pollutants like perfluorooctanoic acid, Cu^{2+}, methylene blue, and As (III), were removed by surface-functionalized COFs (Gendy et al., 2021; Jrad et al., 2024; Kumar et al., 2024). Multiple metals were removed (Jiang et al., 2019) by TpPa-NH_2@EDTA and COF's-V linked COF@COOH (Zhang et al., 2025). Studies also concentrated on the dominance of the adsorption capacity of the COFs. The emphasis on various applications of COFs and associated materials serves as a relevant requirement for the studies of water remediation techniques.

12.7 CHALLENGES AND FUTURE PERSPECTIVES

COFs are undoubtedly the most versatile materials for water treatment technologies and other applications. The improvement of their features and the creation of an economical and stable process for the industry to embark on a sustainable journey together with solving the water purification challenges are also crucial. While laboratory-scale synthesis has seen significant development, scaling the processes to industrial production remains challenging. In addition, improvement of the stability of COFs, particularly under aqueous environments, is critical for their future applications in water treatment and other liquid-phase processes.

The synthesis and structural construction of 3D COFs are governed by unique challenges like control over framework topology and prevention of interpenetration, which can limit porosity and accessibility to active sites. To overcome these and enhance structural diversity in 3D COFs, researchers are exploring novel building blocks and connections. Tuning the crystallinity and long-range order of COFs is essential for their high-performance application in a wide range of applications.

Looking ahead, several exciting avenues are emerging in COF research. The synergy of high-throughput computational screening and machine learning will accelerate the discovery of new COF structures and predict their properties, allowing for more rational design methodologies. Robotic and automated synthesis and characterization can streamline the development process and allow for rapid optimization of COFs for specific applications. The application of artificial intelligence in the analysis of large set of data will enable researchers to design the COFs with greater efficiency (Rabiee, 2023).

The synthesis of COF-based composites, such as hybrids of COF and MOFs, is an emerging frontier that can potentially combine the strengths of various classes of materials to address upcoming challenges in catalysis, energy storage, and environmental pollution remediation. Researchers also seek postsynthetic strategies to tailor the properties of COFs and introduce new functions to push their potential application domains (Raghav et al., 2025).

For catalysis, the preparation of COFs with large densities of catalytic active sites and hierarchical pore structure is highly required to mimic the complexity and effectiveness of biological catalysts. In energy applications, improving the conductivity and charge transportation ability of COFs is a primary research topic, particularly for electrochemical energy storage and conversion devices.

Environmental applications of COFs, especially for water purification, are motivating the development of more stable and selective materials for the removal and degradation of pollutants. The preparation of COMs with excellent permeability and selectivity is a promising means to address the global challenge in water purification.

With the development of the field, sustainability concerns in COF synthesis and processing will become increasingly important. This includes more sustainable synthesis methods, exploring bio-based precursors, and considering the entire lifecycle of COF materials.

While COFs are faced with several challenges, the future for these materials is bright. With the advances in computational modeling along with the growth of structure–property tailoring, COFs can expand their applications in various fields other than water treatment.

12.8 CONCLUSION

COFs will lead the next generation of water treatment solutions of obvious reasons like providing unique properties that can address specific water pollution challenges effectively. This also minimizes the steps in water treatment, thereby providing a sustainable and economic solution for water treatment challenges of many countries. The unique characteristics of COFs and subsidiary materials based on COFs make them superior to other counterparts in the water treatment studies. COFs also demonstrate great potential in the process of reverse osmosis and desalination, thereby solving the major water scarcity issues in arid regions of the Middle East and North Africa.

The reticular synthesis based on secondary building units and their assembly in order to synthesize new COFs are being studied by researchers. Some of them have reported studies on the synthesis of crystalline solids. The progress in the development of COF-based materials is increasing day by day.

Focusing on the sustainability aspect and creating solutions feasible for application in industries struggling with attaining permanent and economic solutions is the need of the hour. The major challenges, like fouling, recyclability, and gas evolution in the wastewater industry, can also be addressed by effective synthesis of the COFs and associated materials. The limitations of 2D COFs should also be studied for the COFs in other fields, like energy storage and gas separation, which can also be conducted by effectively using these materials to their full potential.

REFERENCES

Ahmadijokani, Farhad, Ahmadreza Ghaffarkhah, Hossein Molavi, Subhajit Dutta, Yi Lu, Stefan Wuttke, Milad Kamkar, Orlando J. Rojas, and Mohammad Arjmand. "COF and MOF hybrids: Advanced materials for wastewater treatment." *Advanced Functional Materials* 34, no. 43 (2024): 2305527. https://doi.org/10.1002/adfm.202305527

Akhzari, Sajad, Heidar Raissi, and Afsaneh Ghahari. "Architectural design of 2D covalent organic frameworks (COFs) for pharmaceutical pollutant removal." *npj Clean Water* 7, no. 1 (2024): 31. https://doi.org/10.1038/s41545-024-00315-8

Akinnawo, Solomon Oluwaseun. "Covalent organic frameworks: Design, synthesis, characterization, and applications." *ChemPhysMater* 3, no. 1 (2024): 36–63. https://doi.org/10.1016/j.chphma.2023.08.003

Alsudairy, Ziad, Normanda Brown, Allea Campbell, Abrianna Ambus, Bianca Brown, Kayla Smith-Petty, and Xinle Li. "Covalent organic frameworks in heterogeneous catalysis: Recent advances and future perspective." *Materials Chemistry Frontiers* 7, no. 16 (2023): 3298–3331. https://doi.org/10.1039/D3QM00188A

Altintas, Cigdem, Ilknur Erucar, and Seda Keskin. "MOF/COF hybrids as next-generation materials for energy and biomedical applications." *CrystEngComm* 24, no. 42 (2022): 7360–7371. https://doi.org/10.1039/D2CE01296K

Arqueros, Cristina, Félix Zamora, and Carmen Montoro. "A perspective on the application of covalent organic frameworks for detection and water treatment." *Nanomaterials* 11, no. 7 (2021): 1651. https://doi.org/10.3390/nano11071651

Aslam, Muhammad Kashif, Iftikhar Hussain, Ali H. Al-Marzouqi, and Maowen Xu. "Advances in covalent organic frameworks for photocatalytic CO_2 reduction: Strategies and future perspectives." *Nano Research Energy* 4, no. 1 (2024): e9120149. https://doi.org/10.26599/NRE.2024.9120149

Chen, Jia, Yuting Wang, Yongliang Yu, Jianhua Wang, Juewen Liu, Hirotaka Ihara, and Hongdeng Qiu. "Composite materials based on covalent organic frameworks for multiple advanced applications." *Exploration* 3, no. 3 (2023): 20220144. https://doi.org/10.1002/EXP.20220144

Cui, Bingbing and Guodong Fu. "Process of metal–organic framework (MOF)/covalent–organic framework (COF) hybrids-based derivatives and their applications on energy transfer and storage." *Nanoscale* 14, no. 5 (2022): 1679–1699. https://doi.org/10.1039/D1NR07614K

Cui, Fu-Zhi, Rong-Ran Liang, Qiao-Yan Qi, Guo-Fang Jiang, and Xin Zhao. "Efficient removal of Cr (VI) from aqueous solutions by a dual-pore covalent organic framework." *Advanced Sustainable Systems* 3, no. 4 (2019): 1800150.

Ding, San-Yuan and Wei Wang. "Covalent organic frameworks (COFs): From design to applications." *Chemical Society Reviews* 42, no. 2 (2013): 548–568. https://doi.org/10.1039/C2CS35072F

Dubey, Prashant, Vishal Shrivastav, Tribani Boruah, Giorgio Zoppellaro, Radek Zbořil, Aristides Bakandritsos, and Shashank Sundriyal. "Unveiling the potential of covalent organic frameworks for energy storage: Developments, challenges, and future prospects." *Advanced Energy Materials* 14, no. 24 (2024): 2400521. https://doi.org/10.1002/aenm.202400521

Firoozi, Maryam, Zahra Rafiee, and Kheibar Dashtian. "New MOF/COF hybrid as a robust adsorbent for simultaneous removal of auramine O and rhodamine B dyes." *ACS Omega* 5, no. 16 (2020): 9420–9428. https://doi.org/10.1021/acsomega.0c00539

Gao, Manjie, Qifeng Fu, Min Wang, Kailian Zhang, Jing Zeng, Lujun Wang, Zhining Xia, and Die Gao. "Facile synthesis of porous covalent organic frameworks for the effective extraction of nitroaromatic compounds from water samples." *Analytica Chimica Acta* 1084 (2019): 21–32.

Gendy, Eman A., Jerosha Ifthikar, Jawad Ali, Daniel T. Oyekunle, Zouhair Elkhlifia, Irshad Ibran Shahib, Ahmed I. Khodair, and Zhuqi Chen. "Removal of heavy metals by covalent organic frameworks (COFs): A review on its mechanism and adsorption properties." *Journal of Environmental Chemical Engineering* 9, no. 4 (2021): 105687.

Guo, Xinghua, Yang Li, Meicheng Zhang, Kecheng Cao, Yin Tian, Yue Qi, Shoujian Li, Kun Li, Xiaoqi Yu, and Lijian Ma. "Colyliform crystalline 2D covalent organic frameworks (COFs) with quasi-3D topologies for rapid I2 adsorption." *Angewandte Chemie International Edition* 59, no. 50 (2020): 22697–22705.

Halder, Arjun, Meena Ghosh, Abdul Khayum M, Saibal Bera, Matthew Addicoat, Himadri Sekhar Sasmal, Suvendu Karak, Sreekumar Kurungot, and Rahul Banerjee. "Interlayer hydrogen-bonded covalent organic frameworks as high-performance supercapacitors." *Journal of the American Chemical Society* 140, no. 35 (2018): 10941–10945. https://doi.org/10.1021/jacs.8b06460.

Han, Xing, Wenqiang Zhang, Zhijie Chen, Yan Liu, and Yong Cui. "The future of metal-organic frameworks and covalent organic frameworks: Rational synthesis and customized applications." *Materials Horizons* 10, no. 12 (2023): 5337–5342. https://doi.org/10.1039/D3MH01396K

He, Linwei, Long Chen, Xinglong Dong, Shitong Zhang, Mingxing Zhang, Xing Dai, Xiajie Liu et al. "A nitrogen-rich covalent organic framework for simultaneous dynamic capture of iodine and methyl iodide." *Chem* 7, no. 3 (2021): 699–714.

Herath, Amali, Maryam Salehi, and Santa Jansone-Popova. "Production of polyacrylonitrile/ionic covalent organic framework hybrid nanofibers for effective removal of chromium (VI) from water." *Journal of Hazardous Materials* 427 (2022): 128167.

Hou, Yanghui, Peng Zhou, Fuyang Liu, Ke Tong, Yanyu Lu, Zhengmao Li, Jialiang Liang, and Meiping Tong. "Rigid covalent organic frameworks with thiazole linkage to boost oxygen activation for photocatalytic water purification." *Nature Communications* 15, no. 1 (2024): 7350. https://doi.org/10.1038/s41467-024-51878-6

Huang, Ning, Lipeng Zhai, Hong Xu, and Donglin Jiang. "Stable covalent organic frameworks for exceptional mercury removal from aqueous solutions." *Journal of the American Chemical Society* 139, no. 6 (2017): 2428–2434.

Jiang, Yunzhe, Chuanyao Liu, and Aisheng Huang. "EDTA-functionalized covalent organic framework for the removal of heavy-metal ions." *ACS Applied Materials & Interfaces* 11, no. 35 (2019): 32186–32191.

Jrad, Asmaa, Gobinda Das, Nour Alkhatib, Thirumurugan Prakasam, Farah Benyettou, Sabu Varghese, Felipe Gándara, Mark Olson, Serdal Kirmizialtin, and Ali Trabolsi. "Cationic covalent organic framework for the fluorescent sensing and cooperative adsorption of perfluorooctanoic acid." *Nature Communications* 15, no. 1 (2024): 1–17.

Kandambeth, Sharath, Arijit Mallick, Binit Lukose, Manoj V. Mane, Thomas Heine, and Rahul Banerjee. "Construction of crystalline 2D covalent organic frameworks with remarkable chemical (acid/base) stability via a combined reversible and irreversible route." *Journal of the American Chemical Society* 134, no. 48 (2012): 19524–19527. https://doi.org/10.1021/ja308278w

Karmakar, Avishek and Jing Li. "The best of both worlds: An MOP/COF-based hybrid material for highly selective and very fast sequestration of toxic oxoanions from water." *ACS Central Science* 6, no. 9 (2020): 1476–1478. https://doi.org/10.1021/acscentsci.0c01029

Khan, Niaz Ali, Runnan Zhang, Xiaoyao Wang, Li Cao, Chandra S. Azad, Chunyang Fan, Jinqiu Yuan, Mengying Long, Hong Wu, Mark A. Olson, and Zhongyi Jiang. "Assembling covalent organic framework membranes via phase switching for ultrafast molecular transport." *Nature Communications* 13, no. 1 (2022): 3169. https://doi.org/10.1038/s41467-022-30647-3

Kumar, Ajay, Kushal Arya, Sanjay Mehra, Arvind Kumar, Surinder Kumar Mehta, and Ramesh Kataria. "Luminescent Zn-MOF@ COF hybrid for selective decontamination of Cu (II) ions and methylene blue dye in aqueous media." *Separation and Purification Technology* 340 (2024): 126756.

Kumar, Ravinder, Shagufi Naz Ansari, Rakesh Deka, Praveen Kumar, Mohit Saraf, and Shaikh M. Mobin. "Progress and perspectives on covalent-organic frameworks (COFs) and composites for various energy applications." *Chemistry – A European Journal* 27, no. 55 (2021): 13669–13698. https://doi.org/10.1002/chem.202101587

Li, Guiliang, Jianrong Ye, Qile Fang, and Fu Liu. "Amide-based covalent organic frameworks materials for efficient and recyclable removal of heavy metal lead (II)." *Chemical Engineering Journal* 370 (2019): 822–830.

Li, Jing, Guo Lin, Fangguan Tan, Likang Fu, Biao Zeng, Shixing Wang, Tu Hu, and Libo Zhang. "Selective adsorption of mercury ion from water by a novel functionalized magnetic Ti based metal-organic framework composite." *Journal of Colloid and Interface Science* 651 (2023): 659–668.

Li, Juan, Xiaodan Yang, Chiyao Bai, Yin Tian, Bo Li, Shuang Zhang, Xiaoyu Yang et al. "A novel benzimidazole-functionalized 2-D COF material: Synthesis and application as a selective solid-phase extractant for separation of uranium." *Journal of Colloid and Interface Science* 437 (2015): 211–218.

Lin, Daiwu, Xiao Liu, Renliang Huang, Wei Qi, Rongxin Su, and Zhimin He. "One-pot synthesis of mercapto functionalized Zr-MOFs for the enhanced removal of Hg^{2+} ions from water." *Chemical Communications* 55, no. 47 (2019): 6775–6778.

Liu, Minghao, Shuai Yang, Xiubei Yang, Cheng-Xing Cui, Guojuan Liu, Xuewen Li, Jun He, George Zheng Chen, Qing Xu, and Gaofeng Zeng. "Post-synthetic modification of covalent organic frameworks for CO_2 electroreduction." *Nature Communications* 14, no. 1 (2023): 3800. https://doi.org/10.1038/s41467-023-39544-9

Liu, Xiaolu, Hongwei Pang, Xuewei Liu, Qian Li, Ning Zhang, Liang Mao, Muqing Qiu, Baowei Hu, Hui Yang, and Xiangke Wang. "Orderly porous covalent organic frameworks-based materials: Superior adsorbents for pollutants removal from aqueous solutions." *The Innovation* 2, no. 1 (2021): 100076. https://doi.org/10.1016/j.xinn.2021.100076

Liu, Xiaolu, Yang Li, Zhongshan Chen, Hui Yang, Suhua Wang, Zhenwu Tang, and Xiangke Wang. "Recent progress of COFs membranes: Design, synthesis and application in water treatment." *Eco-Environment & Health* 2, no. 3 (2023): 117–130. https://doi.org/10.1016/j.eehl.2023.07.001

Ma, Zhiyao, Fuyang Liu, Nengsheng Liu, Wen Liu, and Meiping Tong. "Facile synthesis of sulfhydryl modified covalent organic frameworks for high efficient Hg (II) removal from water." *Journal of Hazardous Materials* 405 (2021): 124190.

Machado, Tiago F., Filipa A. Santos, Rui F.P. Pereira, Verónica de Zea Bermudez, Artur J.M. Valente, M. Elisa Silva Serra, and Dina Murtinho. "β-Ketoenamine covalent organic frameworks – Effects of functionalization on pollutant adsorption." *Polymers* 14, no. 15 (2022): 3096. https://doi.org/10.3390/polym14153096

Machado, Tiago F., M. Elisa Silva Serra, Dina Murtinho, Artur J.M. Valente, and Mu Naushad. "Covalent organic frameworks: Synthesis, properties and applications – An overview." *Polymers* 13, no. 6 (2021): 970. https://doi.org/10.3390/polym13060970

Mohammed, Abdul Khayum and Dinesh Shetty. "Macroscopic covalent organic framework architectures for water remediation." *Environmental Science: Water Research & Technology* 7, no. 11 (2021): 1895–1927. https://doi.org/10.1039/D1EW00408E

Rabiee, Navid. "Artificial intelligence assisted with designing metal-organic frameworks (MOFs)." *Clinical and Translational Discovery* 3, no. 3 (2023): e207.

Raghav, Sapna, Jyoti Raghav, Ma'aruf Abdulmumin Muhammad, Madhur Babu Singh, Pallavi Jain, Dinesh Kumar. "Recent advancements in synthesis and applications of covalent organic frameworks." *Applied Materials Today,* 44 (2025): 102694. https://doi.org/10.1016/j.apmt.2025.102694

Rejali, Narges Abdolhossein, Mohammad Dinari, and Yong Wang. "Post-synthetic modifications to covalent organic frameworks (COFs) for diverse applications." *Chemical Communications* 59 (2023): 11631–11647. https://doi.org/10.1039/D3CC03091A

Rodríguez-Camargo, Andrés, Kenichi Endo, and Bettina V. Lotsch. "Celebrating ten years of covalent organic frameworks for solar energy conversion: Past, present and future." *Angewandte Chemie* 136, no. 49 (2024): e202413096. https://doi.org/10.1002/anie.202413096

Rodríguez-Carríllo, Cristina, Miriam Benítez, Jamal El Haskouri, Pedro Amorós, and Jose V. Ros-Lis. "Novel microwave-assisted synthesis of COFs: 2020–2022." *Molecules* 28, no. 7 (2023): 3112. https://doi.org/10.3390/molecules28073112

Segura, José L., Sergio Royuela, and M. Mar Ramos. "Post-synthetic modification of covalent organic frameworks." *Chemical Society Reviews* 48, no. 14 (2019): 3903–3945. https://doi.org/10.1039/C8CS00978C

Shi, Xiansong, Dongwei Ma, Fang Xu, Zhe Zhang, and Yong Wang. "Table-salt enabled interface-confined synthesis of covalent organic framework (COF) nanosheets." *Chemical Science* 11, no. 4 (2020): 989–996. https://doi.org/10.1039/C9SC05082E

Vardhan, Harsh, Grace Rummer, Angela Deng, and Shengqian Ma. "Large-scale synthesis of covalent organic frameworks: Challenges and opportunities." *Membranes* 13, no. 8 (2023): 696. https://doi.org/10.3390/membranes13080696

Wang, Rui, Xiansong Shi, Ankang Xiao, Wei Zhou, and Yong Wang. "Interfacial polymerization of covalent organic frameworks (COFs) on polymeric substrates for molecular separations." *Journal of Membrane Science* 566 (2018): 197–204. https://doi.org/10.1016/j.memsci.2018.08.044

Wang, Shuai, Hao Wang, Shixing Wang, Likang Fu, and Libo Zhang. "Novel magnetic covalent organic framework for the selective and effective removal of hazardous metal Pb (II) from solution: Synthesis and adsorption characteristics." *Separation and Purification Technology* 307 (2023a): 122783.

Wang, Wei, Minjuan Gong, Donghai Zhu, Mohammadtaghi Vakili, Zahra Gholami, Huanhuan Jiang, Shuangxi Zhou, and Han Qu. "Post-synthetic thiol modification of covalent organic frameworks for mercury (II) removal from water." *Environmental Science and Ecotechnology* 14 (2023b): 100236.

Wang, Zhe, Ying Huang, Jian Yang, Yongsheng Li, Qixin Zhuang, and Jinlou Gu. "The water-based synthesis of chemically stable Zr-based MOFs using pyridine-containing ligands and their exceptionally high adsorption capacity for iodine." *Dalton Transactions* 46, no. 23 (2017): 7412–7420.

Xia, Zijing, Yusen Zhao, and Seth B. Darling. "Covalent organic frameworks for water treatment." *Advanced Materials Interfaces* 8, no. 1 (2021): 2001507. https://doi.org/10.1002/admi.202001507

Yu, Chen, Jue Zhang, Xiaogang Luo, and Juan Zhang. "Metal organic framework/covalent organic framework composite for solid-phase microextraction of polycyclic aromatic hydrocarbons in milk samples." *Microchemical Journal* 187 (2023): 108388. https://doi.org/10.1016/j.microc.2023.108388

Yuan, Xiao Song, Wei Liu, Wei Ya Zhu, and Xun Xian Zhu. "Enhancement in flux and antifouling properties of polyvinylidene fluoride/polycarbonate blend membranes for water environmental improvement." *ACS Omega* 5, no. 46 (2020): 30201–30209. https://doi.org/10.1021/acsomega.0c04656

Zhang, Fanyu, Jianling Zhang, Bingxing Zhang, Lirong Zheng, Xiuyan Cheng, Qiang Wan, Buxing Han, and Jing Zhang. "CO_2 controls the oriented growth of metal-organic framework with highly accessible active sites." *Nature Communications* 11, no. 1 (2020): 1431. https://doi.org/10.1038/s41467-020-15200-4

Zhang, Kang, Zhongjin He, Krishna M. Gupta, and Jianwen Jiang. "Computational design of 2D functional covalent–organic framework membranes for water desalination." *Environmental Science: Water Research & Technology* 3, no. 4 (2017): 735–743. https://doi.org/10.1039/C7EW00074J

Zhang, Ting, Guang Zhang, and Long Chen. "2D conjugated covalent organic frameworks: Defined synthesis and tailor-made functions." *Accounts of Chemical Research* 55, no. 6 (2022): 795–808. https://doi.org/10.1021/acs.accounts.1c00693

Zhang, Yupei, Shuai Zhang, Yan Lin, Shaohua Wu, Xiang Li, and Chunping Yang. "Simultaneous removal of heavy metals and antibiotics from anaerobically digested swine wastewater via functionalized covalent organic frameworks." *Environmental Research* 272 (2025): 121152.

Zheng, Yu, Jianliang Shen, Jinqiu Yuan, Niaz Ali Khan, Xinda You, Chao Yang, Shiyu Zhang, Ayman El-Gendi, Hong Wu, Runnan Zhang, and Zhongyi Jiang. "2D nanosheets seeding layer modulated covalent organic framework membranes for efficient desalination." *Desalination* 532 (2022): 115753. https://doi.org/10.1016/j.desal.2022.115753

13 Applications of Covalent Organic Frameworks in Gas Adsorption and Storage

Bhawana Jangir, Shubhangi Singhal, Anjali Yadav, Aruna Sharma, and Vishnu Kumar Khandelwal

13.1 INTRODUCTION

The growing overutilization of fossil fuels has contributed to the notable increase in concentrations of greenhouse gases in the atmosphere, like carbon dioxide and methane, which have contributed to global warming. These serious environmental issues around the world have drawn a lot of public and scientific attention. Therefore, the development of materials with a high degree of efficiency for carbon dioxide capture and source for sustainable energy storage, like hydrogen, has been an area of special interest in the quest to satisfy energy and environmental needs. Even though certain technologies, like amine scrubbing, are already accessible for carbon dioxide capture on a large scale, they are plagued by serious corrosion problems and require significant energy to regenerate amine-based aqueous solutions (Rochelle, 2024). Microporous materials, such as zeolites, metal-organic frameworks (MOFs), and activated carbon, have shown potential applications in storage and separation of carbon dioxide via physical adsorption using weak van der Waals forces. However, the majority of currently available adsorbents suffer from instability in the presence of water or the challenge in modifying their functionalities and architecture. Therefore, it is highly desirable to develop a new category of adsorbents that meet the requirements of better stability, enhanced specific surface area, and tuneable structural framework.

Covalent organic frameworks (COFs) are a type of organic polymer possessing porosities that enable precise assembly of atoms as organic building components into long-range ordered layers of two-dimensional or three-dimensional networks (Raghav et al., 2005). Compared to other porous materials with crystallinity, COFs are built from lightweight elements joined by robust covalent bonds, which impart hardness with low density into them. Till now, COFs containing boronate ester (Chavez et al., 2016), borosilicate (Hunt et al., 2008), borazine (Li et al., 2024), triazine (Wei et al., 2019), squaraine (Wang et al., 2023), imine (Vitaku and Dichtel, 2017), hydrazone (Zhuang et al., 2024), imide (Qian et al., 2024), and azine (Geng et al., 2022) linkages

 DOI: 10.1201/9781003605386-13

have been well reported. In comparison to COFs with B–O bond linkages, COFs designed via C–N bonds formation have improved chemical inertness and exhibited significant efficiency in gas absorption and storage. Nevertheless, the C–N-linked COFs reported in the literature usually have moderate specific surface areas and mesoporous pores, as well as relatively low uptake capacities. These characteristics hinder their applicability in gas adsorption under practical conditions.

COFs have been promoted as a novel category of crystalline materials with porosities for suitable gas adsorption and storage applications due to their large intrinsic surface area, high stability, very low densities, and tuneable pore dimension. Pore engineering of COFs is essential for the development of COF materials, serving as a fundamental focus for studying the structure–property relationship and investigating potential applications. The pore engineering of COFs can be categorized into two levels. The first one is to design the chemical environment within the pores. The chemical composition of pores is the main focus at this level. This constitution depends on the selection of linkages and the nature of the building unit skeletons. The connections, which are primarily made up of dynamic covalent bonds like B–O and C=N bonds, have a direct impact on the stability and crystallization of COFs (Geng et al., 2020). Another common approach to adjusting the properties of COFs for additional uses is to functionalize the COF skeletons by adding substituent groups, metal clusters, or heteroatoms. This can be done through either premodification or postfunctionalization processes. This type of pore engineering is known as channel surface engineering. The second task of pore engineering for COFs focuses on the topological structures of the frameworks. This involves designing the shape, size, and arrangement of pores by merging building blocks of different sizes and specific symmetries.

At present, the increasing capability of COFs to uptake lightweight gases, such as hydrogen, carbon dioxide, ammonia, and methane, at room temperature is a key challenge. Along with increasing pore volume and surface area and fine-tuning the pore width, an effective approach for enhancing small gas adsorption capacity is to introduce specific active sites—such as heteroatoms and unsaturated metals, into the polymer networks, thereby boosting the interactions between adsorbent and adsorbate. An azine-based COF, ACOF-1, has been designed through a mild condensation reaction and exhibits significant performance in the storage of small gases (Li et al., 2014). This is mainly due to the relatively small pore width resulting from the azine unit of short structural length. The gas storage potential of COFs for methane, hydrogen, carbon dioxide, and ammonia has been a growing interest in the past few decades. This chapter focuses on the gas adsorption and storage applications of COFs, considering the necessity of environmental remediation and the vast potential of COFs as materials with diversified functionalities in this area.

13.2 APPLICATIONS OF COFs IN GAS STORAGE

13.2.1 Hydrogen Storage

It is widely recognized that the world's petroleum reserves are depleting and that, in order to avoid pollution, a clean, ecologically friendly energy source needs to be discovered. Hydrogen is considered an exceptional candidate since it is simple to produce and has a greater energy per unit mass than all other fuels. Furthermore, it is environmentally benign because water is the primary by-product of its burning process. However, the challenge of storage has been one of the primary issues that have prevented commercial uses thus far. The development of appropriate materials for hydrogen hosting is necessary. Various solid materials have been designed to store hydrogen gas through chemisorption or physisorption. While molecular hydrogen disintegrates into atomic hydrogen during chemisorption, it is stored in molecular form through the physisorption method. Various metal hydrides have been reported as a source of hydrogen. These materials maintain potentially good volumetric and gravimetric capacities, but the optimal hydrolysis needed to achieve 100% conversion to hydrogen limits their capacities. In the physisorption process, the hydrogen molecule reacts with the storage material's surface atoms through weak dispersive and electrostatic interactions. It was anticipated from the beginning that a huge surface area would encourage storage of hydrogen gas by physisorption. Considering this concept, the majority of the research work was devoted to design materials with porosity leading to large surface areas. Due to their advanced structural architecture constructed by lightweight elements, COFs (3D-COFs) have been thought to be as hydrogen storage materials upon their discovery. A wide range of shapes can be achieved by tailoring the reticular structure of COFs. Rather than the presence of improved interaction sites at the pore surface, the unique properties of the materials—like low density, enhanced surface area, topology of framework, and pore width—are the primary cause of their rapid hydrogen uptake. Initially, various complex multiscale theoretical techniques were used to assess the possible uses of COFs for hydrogen storage. Superior gravimetric H_2 uptake for 3D COFs at 77 K was predicted by the majority of these studies. To determine hydrogen interacting sites and the interaction potential curves, Froudakis and colleagues employed density functional theory and second-order Møller–Plesset perturbation theory (MP2) calculations (Furukawa and Yaghi, 2009). The studies revealed that phenyl rings and B_3O_3 or C_2BO rings with binding energies less than 1 kcal mol^{-1} are the most preferred locations for H_2 adsorption (Figure 13.1). Gas chromatography-mass spectroscopic simulations were reported on both 2D and 3D COFs by Yaghi et al. As anticipated, because of large surface area and free volume, 3D COFs have demonstrated a two to three times greater potential to store hydrogen than 2D COFs. Furukawa and Yaghi investigated the capability of several 2D and 3D COFs for hydrogen uptake at 77 K. H_2 uptake of COF-102 was found to be comparable to that of MOF-177 and MOF-5 (Furukawa and Yaghi, 2009).

It was noted that COF's H_2-storage capacities under ambient circumstances remain inadequate and are insufficient to achieve energy targets. The interaction energy of neat COF surfaces with H_2 molecules is weak. It was proposed that applying metal nanoparticles, metal complexes, or positively charged metal ions to the surfaces of COF could boost its room-temperature storage capabilities through the spillover process (Li and Yang, 2006). The term "spillover" refers to the chemisorption of H_2 molecules on a metal-based catalyst after dissociation, which is followed by the migration of monoatomic hydrogen into pores of COF acting as the host. It was discovered that the spillover mechanism can enhance the uptake of H_2 gas by up to six or seven times when COFs are coupled with transition metal-based catalysts

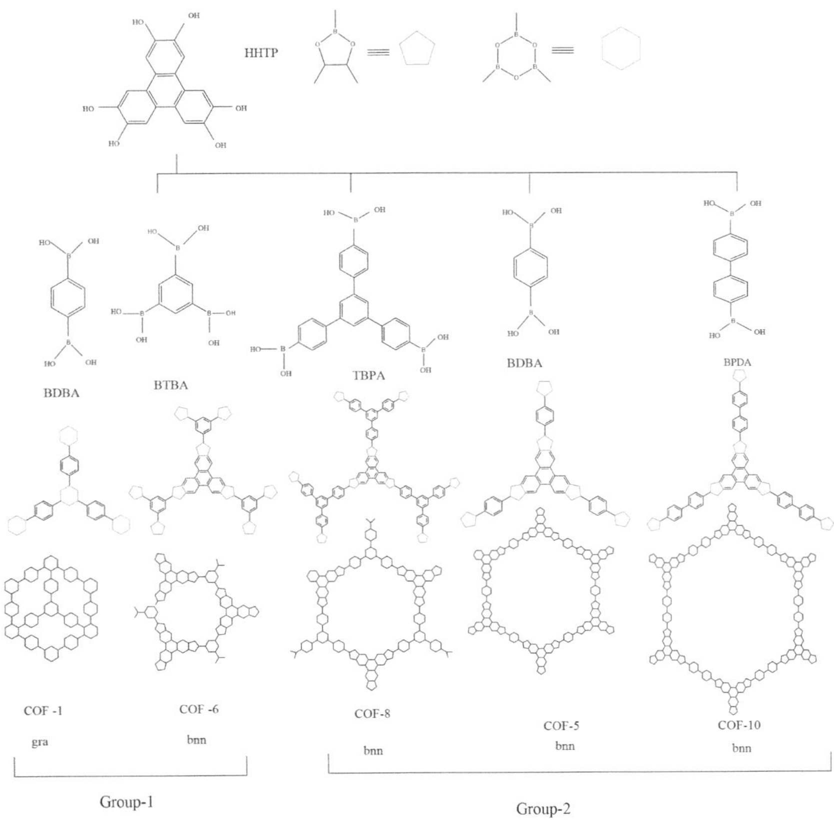

FIGURE 13.1 Condensation reactions involving HHTP and boronic acids employed to produce 2D and 3D COFs with high crystallinity. (Reprinted (adapted) with permission from Furukawa and Yaghi (2009) Copyright {2025} American Chemical Society.)

like Pt/C (Li and Yang, 2006). In this case, the primary spillover source is activated carbon, while COF acts as a secondary one.

A significant factor in determining the total H_2 uptakes is the contact time between the sources of primary and secondary spillover. A carbon bridge can be created by carbonizing a hydrocarbon precursor, like sucrose, to increase contact. In the example of IRMOF-8, Li and Yang found that the uptake of H_2 at ambient conditions may be raised from 0.5 to 4 wt % at 10 MPa pressure by employing a bridge approach (Li and Yang, 2006). Li and Yang reported that the spillover approach can improve COF-1's room-temperature H_2 absorption by two to three and a half times (Li and Yang, 2008). The catalyst for H_2 dissociation in these studies was 5 wt % Pt on activated carbon.

In systems like metal/MOFs or metal/COFs, the idea of spillover does not appear to be a very effective way to boost H_2 uptakes. The hydrogen characteristics of the Pd@COF-102 hybrid material were reported earlier (Kalidindi et al., 2012). The electron tomography approach revealed that the Pd nanoparticles were uniformly dispersed throughout the COF-102 crystals. Powder XRD and NMR spectroscopy did not reveal any discernible degradation of the framework after incorporation of PdNPs into COF-102. Following Pd nanoparticle loading, the hydrogen uptake dropped from 5.3 to 4.6 wt % at 77 K. This is because the specific surface area and volume of the pore decrease after the addition of Pd. According to room temperature hydrogen storage measurements, Pd inclusion inside the COF-102 pores increased H_2 uptakes by a factor of two to three. But in this instance, rather than overflow, the increased H_2 capture was attributed to Pd hydride production and hydrogenation of leftover organic components on the surface of Pd nanoparticles.

Systems where nanocrystalline porous organic polymers are used to stabilize metal nanoparticles have also been found to exhibit the spillover mechanism in hydrogen storage. Liang et al. have presented the synthesis of micro-porous hyper-crosslinked polymers incorporated with well-dispersed Pt nanoparticles (Liang et al., 2012). The capacity of hydrogen storage of hyper-crosslinked polymers is increased to 0.21 wt % at ambient condition by adding 2 wt % Pt nanoparticles. In comparison to materials devoid of Pt nanoparticles, the amount of H_2 gas adsorption has increased by a ratio of 1.75. These methods can aid in the focused synthesis of novel COF materials, which are expected to exhibit even greater performance in terms of their capacity to store hydrogen.

13.2.2 Methane Storage

Methane, the primary component of natural gas, is the least used fossil fuel on Earth even though it makes up around two-thirds of all fossil fuels. There is now a lot of interest in expanding the use of methane as a vehicle fuel because it is readily available and produces fewer carbon emissions than petroleum. Natural gas, which is mostly methane, will be a key transitional fuel in the move away from the present hydrocarbon-based fuels like coal and oil and toward renewable sources. Since it has a high heat of combustion and more hydrogen than any other hydrocarbon fuel, its carbon dioxide emissions are also favourable when compared to the energy generated by combustion (Furukawa et al., 2014). Compared to coal, methane produces about twice as much energy with carbon dioxide emissions. Methane accomplishes this without releasing mercury or producing ash that is high in uranium and thorium, in contrast to coal. But, in order to use this CH_4, low-cost storage and transportation methods are needed. Methane is not liquefiable at room temperature, which raises the cost of transportation (Menon and Komarneni, 1998). Attempts to get around this drawback include the following:

- Preserving methane as compressed natural gas (at about 200 bar) or liquefied natural gas (at about 112 K);
- Turning methane into higher hydrocarbons like ethane or oxygenates like methanol
- Storing in porous materials.

We think that the storage of methane through adsorption on materials with high porosities is the most favourable option among these options because it can function at a suitable temperature (7–298 K) and pressure (1–300 bar) and does not require additional input of energy for conversion to methanol or higher hydrocarbons. Strong covalent connections between light elements like H, Si, B, C, and O hold COFs together. They have the lowest densities of any known material with high crystallinity, huge surface areas, and enormous pore volumes—all of which are necessary for significant methane uptake. In theory, a vast variety of COFs may be created and tested for methane adsorption utilizing different construction components, as well as different numbers of sites of extension and functionalization to achieve different topologies. Although these empirical methods have proven successful, an alternative approach has been investigated in which the most promising candidates are predicted using theory and computation, and only the most promising examples are then synthesized and characterized experimentally. Naturally, this is only feasible if the theory and computation results are trustworthy enough to allow for the confident rejection of low-performance systems without the need for experimentation.

For various two-dimensional and three-dimensional COFs, Han et al. calculated methane uptake at 298 K and 1–100 bar pressure (Han et al.). For all COFs, the uptake of CH_4 was estimated with the help of force fields (ff)-based grand canonical Monte Carlo (GCMC) simulations constructed to suit precise quantum mechanics. Isotherm of methane adsorption for COF-8 and COF-5 predicted by these calculations was found to correlate well with the experiment.

Zhao and Yan created a series of modified COFs by substituting different substituent groups, such as –Cl, –I, $-CF_3$, –CN, $-NH_2$, $-CH_3$, and $-OCH_3$, for some of the H atoms on benzene rings (Zhao and yan, 2014). GCMC simulation was used to examine the substituents' adsorption characteristics in order to investigate their effects on methane storage. At the start, they all retain a strong linearity. All of the substituted COFs were shown to be superior to the original COFs below 20 bar due to their higher interactions with CH_4. COFs possessing imine linkages were created by Jiang et al. to have reactive, accessible, and content-tuneable ethynyl groups on the side walls of one-dimensional (1D) pores (Huang et al., 2015). These COFs provide the perfect framework for pore-wall surface functionalization, which aims to incorporate different types of functional groups with controllable loading amounts, ranging from basic to acidic units and hydrophobic to hydrophilic moieties. The findings showed that COFs with pore walls incorporating amino groups had the highest affinity for CO_2; these groups have shown an increase in the potential to absorb CO_2 (Figure 13.2).

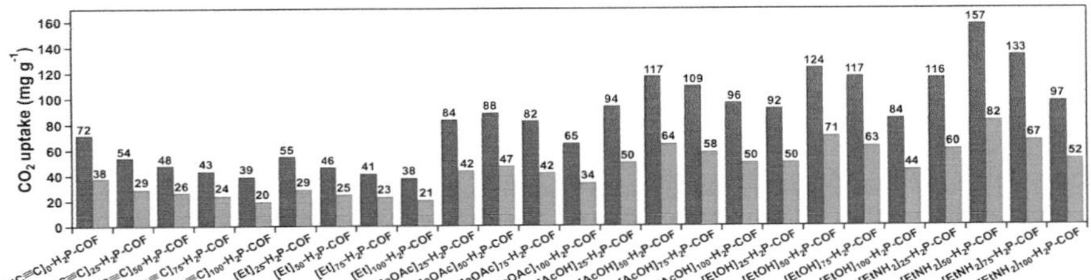

FIGURE 13.2 Adsorption potential of the COFs at 273 K (red bars), 298 K (blue bars), and 1 bar. (Reprinted (adapted) with permission from [Huang et al., 2015] Copyright {2025} American Chemical Society.)

COFs doped with metal ions have also been studied to increase their methane storage capacity. Cao et al. looked at how Li doping affected the COFs' ability to store methane (Lan et al., 2010). Calculations of first-principles demonstrate that Li cation-incorporated COFs can significantly increase the interaction of methane with the material due to the induced dipole interaction and London dispersion forces, which result from the Li cation's strong affinity for methane molecules.

13.2.3 CO_2 Capture

Because of the greenhouse effect, the widespread use of fossil fuels causes a significant amount of CO_2 to be released into the atmosphere, which is a serious environmental concern. The harmful consequences of CO_2 in the atmosphere include rising sea levels and global climate warming. It also directly affects the growth of plants. Conversely, CO_2 serves as a useful reaction medium as well as reactant in the production of various fine chemical compounds and polymers, such as cyclic carbonates, polycarbonates, acrylic acids, and ethers. The design of new porous materials that can capture substantial amounts of CO_2 with high storage capacity will have applications in both industry and academia. In recent years, various COFs have been utilized as crystalline materials with porosity for CO_2 adsorption, demonstrating a great potential for increased CO_2 capture capacity.

Prior to the practical investigation, a theoretical calculation of COFs' gas capture capacity is conducted. Gas storage represents a promising use for two-dimensional COF materials owing to their high density of randomly organized channels and pores that can be functionalized, and post functionalization disturbs the structural framework of COF by breaking of bond, hence it is preferred to develop synthetic methods that generate COF with built-in functionalities. The application of various COFs as CO_2 capture materials was first reported in 2009 (Furukawa and yaghi, 2009). Isotherm for adsorption of CO_2 by COF at low pressure was studied. The results demonstrated that at low pressure, CO_2 adsorption capacity of COFs mainly depends upon the pore size rather than the volume of the pore of COFs.

COFs have been constructed using the reversible B–O bonds formation (boroxine and boronate). Nevertheless, the reversible nature of B–O bonds renders boron-based COFs hydrolytically unstable in moist environments, significantly restricting their potential uses. Mao and group have reported the stabilization of boron-based COFs by incorporating Lewis base and Bronsted acid defect sites, incorporating N– donor groups and Si–O bonds into the structural architecture of COF-1 (Du et al., 2015). COF-1 showed enhanced stability in air for a long time after incorporation of a dual functional (3-aminopropyl) triethoxysilane (APTES) reagent, APTES-COF-1 (Figure 13.3). Furthermore, the adsorption isotherm for CO_2 capture of APTES-COF-1 was studied in detail, and it was found to be identical to that of COF-1 at various temperature ranges. The boron-based COFs have been proven to have several advantages such as good crystallinity, low density, great thermal stability, and structural variation. Nonetheless, their industrial application for CO_2 capture is greatly limited by their comparatively low CO_2 storage capacity at low pressure and their low stability in moisture or water. One possible alternative route could be to enhance their stability in water and their applicability for separating CO_2 under

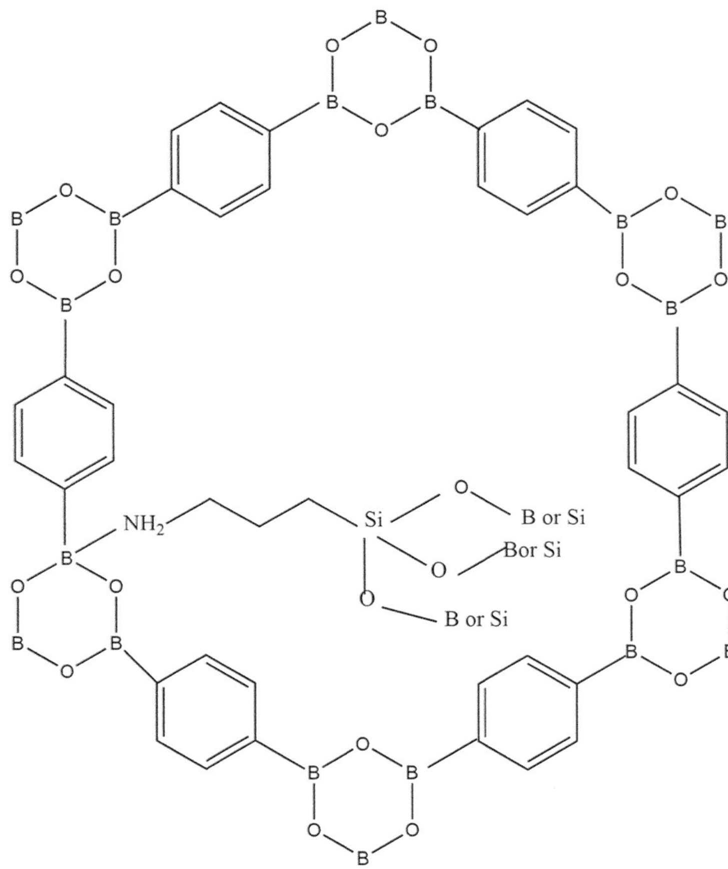

FIGURE 13.3 (3-aminopropyl) triethoxysilane APTES-COF-1.

high pressure by introducing or adding desirable functional groups to boron-based COFs (Du et al., 2015).

Jiang et al. reported that COF showed a significant increase in the CO_2 capture potential after incorporation of alkyl or carboxylic functional groups in their internal pore walls because of strong interactions with CO_2 gas molecules (Huang et al., 2015). Triazine-based COFs (CTFs) have also been extensively studied for CO_2 capture. A covalent triazine-based framework was incorporated with fluorine functional groups (FCTF-1) to study its potential for capturing CO_2 molecules. The addition of fluorine groups improved the framework's capacity for CO_2 adsorption and separation in a number of ways. Highly polar C–F bonds enabled CO_2 adsorption thermodynamically through electrostatic interactions, particularly at low pressures. Furthermore, because the C–F linkages are hydrophobic, FCTF-1 demonstrated water tolerance and maintained its superior CO_2 capture efficiency in the presence of moisture in the gas mixture. CTFs have shown higher chemical and thermal stability in comparison to boron-based COFs, regardless of their low crystallinity, lower structural integrity, and low porosities. Nonetheless, the CTFs exhibit a high CO_2

capacity and maintain their effective CO_2 capture performance after multiple regenerations, even when exposed to water vapour. This makes them excellent candidates for CO_2 capture. Since only a small number of CTFs show some degree of crystallinity and their performance is highly contingent on the synthesis conditions, the aforementioned drawbacks are unavoidable challenges in their further development.

CO_2 absorption potential of imine-based COFs was also studied extensively. Nitrogen-incorporated COFs are found to be good candidates for CO_2 storage due to their increased interaction affinity of nitrogen atoms for CO_2. Murugavel et al. synthesized four COFs from C3-symmetric amine-based building units 1,3,5-tris (4′-amino-3′,5′-isopropylphenyl)benzene (iPrTAPB)) and (1,3,5-tris(4-aminophenyl) benzene (TAPB) and aldehyde components 1,3,5-triformyl-phluroglucinol (TFP)) and (tris(4′-formylphenyl)benzene via Schiff base catalysed co-condensation (Kaleeswaran et al., 2013). These COFs exhibited a 2D honeycomb-like network. TAPB–TFP showed the highest CO_2 uptake potential among all studied COFs. The variation in CO_2 uptake potential of these materials is attributed to the steric hindrances imparted by isopropyl groups and amine groups of higher basicity relative to that of C=N groups.

El-Kaderi et al. reported a mesoporous COF (ILCOF-1) possessing imine linkages with a large surface area designed by condensation of *P*-phenylenediamine in 1,4-dioxane and 1,3,6,8-tetrakis(4′-formylphenyl)-pyrene in 1,4-dioxane and acetic acid at 120°C (Rabbani et al., 2013). The system exhibited a significant CO_2 capture capacity. The presence of nitrogen-based functional groups in the pore walls of ILCOF-1 contributes to the good performance by enhancing CO_2 binding through N·CO_2 interactions. A novel reaction strategy has been introduced for the construction of COFs that relies on the formation of two different types of covalent bonds. Using the orthogonal reaction strategy, a ternary COF (NTU-COF-2) was constructed that includes two types of covalent bonds formed by imine group creation and the formation of a C_2O_2B boronate ring from TAPB, 4-formylphenylboronic acid and 2,3,6,7,10,11-hexahydroxytriphenylene (HHTP) (Zeng et al., 2015). NTU-COF-2 is

TABLE 13.1
Details of Porosities and CO_2 Uptake of Few Selected COFs

COFs	Type of Covalent Bond	Pore Width (nm)	CO_2 Uptake Capacities at Low Pressure	CO_2 Uptake Capacities at High Pressure	Reference
TDCOF-5	Boronate	2.6	47		Kahveci et al. (2013)
COF-1	Boroxine	0.9	51	117	Furukawa and Yaghi (2009)
COF-5	Boronate	2.7	3	443	
COF-6	Boronate	0.9	85	158	
COF-8	Boronate	1.6	33	321	
COF-10	Boronate	3.2	27	514	
COF-102	Boroxine	1.2	34	611	
COF-103	Boroxine	1.2	38	606	

(Continued)

TABLE 13.1 (*Continued*)
Details of Porosities and CO_2 Uptake of Few Selected COFs

COFs	Type of Covalent Bond	Pore Width (nm)	CO_2 Uptake Capacities at Low Pressure	CO_2 Uptake Capacities at High Pressure	Reference
TpPa-1	Imine	1.3	78		Kahveci et al. (2013)
TpBD	Imine	1.7	43		Biswal et al. (2013)
TpPA-NO_2	Imine	1.4	63		Chandra et al. (2013)
TpPa-F_4	Imine	1.4	35		
TpBD-$(NO_2)_2$	Imine	1.7	52		
TpBDMe_2	Imine	1.4	37		
TpBD$(OMe)_2$	Imine	1.4	27		
$[OH]_{25\%}$-H_2P-COF	Imine	2.5	27		Huang et al. (2015)
$[OH]_{50\%}$-H_2P-COF	Imine	2.5	23		
$[OH]_{75\%}$-H_2P-COF	Imine	2.5	26		
$[OH]_{100\%}$-H_2P-COF	Imine	2.5	32		
$[HO_2C]_{25\%}$-H_2P-COF	Imine	2.2	49		
$[HO_2C]_{50\%}$-H_2P-COF	Imine	1.9	68		
$[HO_2C]_{75\%}$-H_2P-COF	Imine	1.7	80		
$[HO_2C]_{100\%}$-H_2P-COF	Imine	1.4	89		
TRITER-1	Imine	1.7		300	Gomes et al. (2015)
ACOF-1	Imine	0.94	90		Li et al. (2014)
COF-LJU2	Imine	0.96	110		Li et al. (2015)
TpPa-COF(MW)	Imine	1.3	111		Kandambeth et al. (2012)
TpPa-2	Imine	1.4	65		Kandambeth et al. (2012)
$[C.5]_0$-H_2P-COF	Imine	2.5	37		Huang et al. (2015)
$[C.5]_{25}$-H_2P-COF	Imine	2.3	27		
$[C.5]_{50}$-H_2P-COF	Imine	2.1	24		
$[C.5]_{75}$-H_2P-COF	Imine	1.9	22		
$[C.5]_{100}$-H_2P-COF	Imine	1.6	20		
$[Et]_{25}$-H_2P-COF	Imine	2.2	28		
$[Et]_{75}$-H_2P-COF	Imine	1.6	21		
$[Et]_{50}$-H_2P-COF	Imine	1.9	23		
$[Et]_{100}$-H_2P-COF	Imine	1.5	19		
FCTF-1	Triazine	0.46, 0.54	105		Huang et al. (2015)
FCTF-1-600	Triazine	0.46, 0.59	124		
CTF-1	Triazine	0.54	55		

the first material of a multiple-component COF reported so far. Isotherms of absorption of CO_2 for NTU-COF-2 were studied in detail, demonstrating that this material can capture up to 10.2 wt % of CO_2 gas at 1.0 bar and 273 K.

Considering the experimental findings for various classes of COFs, triazine- and imine-based COFs display a significantly high CO_2 uptake potential in comparison to boron-based systems, attributing that the presence of nitrogen-based functional groups in COFs could be advantageous for CO_2 uptake to a certain extent. COFs with 3D architecture do not show better CO_2 adsorption capability than 2D COFs, suggesting that a meticulous search for 3D COFs may not be essential for CO_2 capture at low pressure.

13.2.4 Ammonia Storage

Ammonia finds extensive use in industrial applications, including the manufacturing of nitrogen-based functional groups incorporated fertilizers. Compressed liquefied ammonia is often a necessity for transport and use at the commercial level. Nevertheless, the handling of liquid ammonia poses challenges owing to its toxic and corrosive nature. Thus, adsorbents provide a viable solution to this problem by efficient scalable storage of ammonia. Various members of the COF family including both two-dimensional and three-dimensional systems have been screened to evaluate their potential for reversible ammonia uptake and release. According to Ding and colleagues, the boron-containing COF-10 exhibits an ammonia uptake that is remarkably greater than that of other porous materials (Ding and Wang, 2013). COF-10 can release and re-adsorb ammonia in a reversible way, with only a minor decrease in the uptake capacity. During the adsorption cycles, the COF-10 with layered morphology was found to be disturbed; however, its periodicity and atomic connectivity remained intact. The ability of COF-10 to capture a large amount of ammonia is attributed to the formation of a coordinate bond between ammonia and boron.

13.2.5 Adsorption of Volatile Pollutants

With respect to volatile pollutants, greater focus was directed at volatile radionuclides (such as iodine), which are of great concern as environmental contaminants. A computational screening study was performed on 187 types of COFs to investigate the adsorption of gaseous I_2 and CH_3I at 423 K and 1 bar (Lan et al., 2017). The simulated results indicated that 3D-COFs outperformed 2D-COFs in capturing I_2 and CH_3I. Of these COFs, 3D-Py-COF demonstrated the highest I_2 capture of 16.7 g g^{-1}, while COF-103 exhibited a CH_3I capture of 2.8 g g^{-1}. 3D-COFs with pores of 0.9 nm size demonstrated a significantly greater uptake capacity than other COFs. A study on intermolecular bonding between I_2 and the void fraction of COFs was noted, but a consistent relationship between CH_3I and COFs was not found. Most studies have built porous structures with entangled pores and specific pore environments to boost iodine capture. However, Jiang's research team departed from this approach by creating 2D COFs with 1D open channels that lack specific binding sites (Lan et al., 2017). They also revealed the mechanism responsible for the extremely high iodine uptake capacity. The high capture abilities of 1D channelled COFs, which were nearly equivalent to their theoretical values, can be attributed to the near-total occupation of the pore space by iodine, regardless of the pore topology and size. Heteroporous 2D COFs (SIOC-COF-7) featuring hierarchical porosities were employed for the capture of iodine (Yin et al., 2017). It demonstrated an adsorption efficiency of 481 wt % for volatile iodine and 127 wt % for soluble iodine in hexane. The incorporation of solvent molecules in the void structure of COFs resulted in a reduction of the pore volume for iodine, which accounts for the variations in adsorption efficiency.

13.3 CONCLUSION

The chemistry of COFs and their applications have bloomed over the past few years. The COFs have unique properties such as flexible design for molecular structures, persistent porosity, varying degrees of crystallinity, low density, tuneable pore sizes, and various available building blocks, which make them a potential candidate for structural design. This implies that they are promising materials for applications in diversified fields.

The porosities and surface area are the key parameters for applications related to gas adsorption and storage. The key factor to improve the potential of COF materials in these areas is designing the structure of COF, which can enhance porosities and surface areas. In the case of 2D COF, increasing the interlayer spaces would enhance the surface area. In the case of 2D COF, increasing the interlayer spaces would enhance the surface area. The creation of 3D COF possessing high crystallinity would result in a very high BET surface area. Their further applications in this area have a promising path ahead.

REFERENCES

Biswal, B. P., Chandra, S., Kandambeth, S., Lukose, B., Heine, T., and Banerjee, R. (2013). Mechanochemical synthesis of chemically stable isoreticular covalent organic frameworks. *Journal of the American Chemical Society*, 135, 5328–5331.

Chandra, S., Kandambeth, S., Biswal, B. P., Lukose, B., Kunjir, S. M., Chaudhary, M., Babarao, R., Heine, T., and Banerjee, R. (2013). Chemically stable multilayered covalent organic nanosheets from covalent organic frameworks via mechanical delamination. *Journal of the American Chemical Society*, *135*(47), 17853–17861.

Chavez, A. D., Smith, B. J., Smith, M. K., Beaucage, P. A., Northrop, B. H., and Dichtel, W. R. (2016). Discrete, hexagonal boronate ester-linked macrocycles related to two-dimensional covalent organic frameworks. *Chemistry of Materials*, *28*(14), 4884–4888. https://doi.org/10.1021/acs.chemmater.6b01831

Ding, S. Y. and Wang, W. (2013). Covalent organic frameworks (COFs): From design to applications. *Chemical Society Reviews*, *42*(2), 548–568. https://doi.org/10.1039/c2cs35072f

Du, Y., Calabro, D., Wooler, B., Kortunov, P., Li, Q., Cundy, S., and Mao, K. (2015). One step facile synthesis of amine-functionalized cof-1 with enhanced hydrostability. *Chemistry of Materials*, *27*(5), 1445–1447. https://doi.org/10.1021/cm5032317

Furukawa, H. and Yaghi, O. M. (2009). Storage of hydrogen, methane, and carbon dioxide in highly porous covalent organic frameworks for clean energy applications. *Journal of American Chemical Society*, *131*, 8875–8883.

Furukawa, H., Lee, S., and Yaghi, O. M. (2014). High methane storage capacity in aluminum metal–organic.pdf. *Journal of American Chemical Society*, *136*, 5271–5274.

Geng, K., He, T., Liu, R., Dalapati, S., Tan, K. T., Li, Z., Tao, S., Gong, Y., Jiang, Q., and Jiang, D. (2020). Covalent organic frameworks: Design, synthesis, and functions. *Chemical Reviews*, *120*(16), 8814–8933. https://doi.org/10.1021/acs.chemrev.9b00550

Geng, T.-M., Fang, X.-C., Wang, F.-Q., and Zhu, F. (2022). Azine- and azo-based flexible covalent organic frameworks for fluorescence sensing nitro-aromatic compounds and iodine and adsorbing iodine. *Reactive and Functional Polymers*, *176*, 105309. https://doi.org/10.1016/j.reactfunctpolym.2022.105309

Gomes, R., Bhanja, P., and Bhaumik, A. (2015). A triazine-based covalent organic polymer for efficient CO_2 adsorption. *Chemical Communications*, *51*, 10050–10053. https://doi.org/10.1039/C5CC02147B

Huang, N., Chen, X., Krishna, R., and Jiang, D. (2015). Two-dimensional covalent organic frameworks for carbon dioxide capture through channel-wall functionalization. *Angewandte Chemie – International Edition*, *54*, 1–6. https://doi.org/10.1002/anie.201411262

Huang, N., Krishna, R., Jiang, D., Huang, N., Krishna, R., and Jiang, D. (2015). Tailor-made pore surface engineering in covalent organic frameworks : Systematic functionalization for performance screening. *Journal of the American Chemical Society*, *137*(22), 7079–7082. https://doi.org/10.1021/jacs.5b04300

Hunt, J. R., Doonan, C. J., LeVangie, J. D., Côté, A. P., and Yaghi, O. M. (2008). Reticular synthesis of covalent organic borosilicate frameworks. *Journal of the American Chemical Society*, *130*(36), 11872–11873. https://doi.org/10.1021/ja805064f

Kandambeth, S., Mallick, A., Lukose, B., Mane, M. V., Heine, T., and Banerjee, R. (2012). Construction of crystalline 2D covalent organic frameworks with remarkable chemical (acid/base) stability via a combined reversible and irreversible route. *Journal of the American Chemical Society*, *134*, 19524–19527.

Kahveci, Z., Islamoglu, T., Shar, G. A., Ding, R., and El-Kaderi, H. M. (2013). Targeted synthesis of a mesoporous triptycene-derived covalent organic framework. *CrystEngComm*, *15*, 1524–1527. https://doi.org/10.1039/c2ce26487k

Kaleeswaran, D., Vishnoi, P., and Murugavel, R. (2013). [3 + 3] Imine and β-ketoenamine tethered fluorescent covalent-organic frameworks for CO_2 uptake and nitroaromatic sensing. *Journal of Materials Chemistry C*, *3*, 7159–7170. https://doi.org/10.1039/C5TC00670H

Kalidindi, S. B., Oh, H., Hirscher, M., Esken, D., Wiktor, C., Turner, S., Van Tendeloo, G., and Fischer, R. A. (2012). Metal@COFs: Covalent organic frameworks as templates for Pd nanoparticles and hydrogen storage properties of Pd@COF-102 hybrid material. *Chemistry – A European Journal*, *18*(35), 10848–10856. https://doi.org/10.1002/chem.201201340

Lan, J., Cao, D., and Wang, W. (2010). High uptakes of methane in Li-doped 3D covalent organic frameworks. *Langmuir*, *26*(17), 220–226. https://doi.org/10.1021/la9020383

Lan, Y., Tong, M., Yang, Q., and Zhong, C. (2017). Computational screening of covalent organic frameworks for the capture of radioactive iodine and methyl iodide. *CrystEngComm*, *19*(33), 4920–4926. https://doi.org/10.1039/c7ce00118e

Li, B., Huang, X., Gong, R., Ma, M., Yang, X., Liang, L., and Tan, B. (2012). Catalyzed hydrogen spillover for hydrogen storage on microporous organic polymers. *International Journal of Hydrogen Energy*, *37*(17), 12813–12820. https://doi.org/10.1016/j.ijhydene.2012.05.106

Li, R., Sun, F., Liu, Z., Shi, Y., He, S., and Chen, J. (2024). Research progress and prospect of covalent organic frameworks (COFs) and composites: From synthesis to application in water contaminants. *Journal of Environmental Chemical Engineering*, *12*(5), 113944. https://doi.org/10.1016/j.jece.2024.113944

Li, Y. and Yang, R. T. (2006). Significantly enhanced hydrogen storage in metal-organic frameworks via spillover. *Journal of the American Chemical Society*, *128*, 726–727.

Li, Y. and Yang, R. T. (2008). Hydrogen storage in metal-organic and covalent-organic frameworks by spillover. *AIChE Journal*, *54*(1), 269–279. https://doi.org/10.1002/aic.11362

Li, Z., Feng, X., Zou, Y., Zhang, Y., Xia, H., Liu, X., and Mu, Y. (2014). A 2D azine-linked covalent organic framework for gas storage applications†. *Chemical Communications*, *50*, 13825–13828. https://doi.org/10.1039/C4CC05665E

Li, Z., Zhi, Y., Feng, X., Ding, X., Zou, Y., Liu, X., and Mu, Y. (2015). An azine-linked covalent organic framework: Synthesis, characterization and efficient gas storage. *Chemistry – A European Journal*, *21*(34), 12079–12084. https://doi.org/10.1002/chem.201501206

Menon, V. C. and Komarneni, S. (1998). Porous adsorbents for vehicular natural gas storage: A review. *Journal of Porous Materials*, *5*, 43–58. https://link.springer.com/article/10.1023/A:1009673830619

Qian, Y., Liu, Z., Song, H., Yin, R., Yang, H., Li, S., Xiong, W., Yuan, S., Zhang, J., and Pang, H. (2024). Imide-based covalent organic framework with excellent cyclability as an anode material for lithium-ion battery. *Chinese Chemical Letters*, *35*(6), 108785. https://doi.org/10.1016/j.cclet.2023.108785

Rabbani, M. G., Sekizkardes, A. K., Kahveci, Z., Reich, T. E., Ding, R., and El-Kaderi, H. M. (2013). A 2D mesoporous imine-linked covalent organic framework for high pressure gas storage applications. *Chemistry: A European Journal*, *13*, 3324–3328. https://doi.org/10.1002/chem.201203753

Raghav, S., Raghav, J., Muhammad, M.A., Singh, M. B., Jain, P., Kumar D (2025). Recent advancements in synthesis and applications of covalent organic frameworks. *Applied Materials Today 44*, 102694. doi:10.1016/j.apmt.2025.102694

Rochelle, G. T. (2024). Air pollution impacts of amine scrubbing for CO_2 capture. *Carbon Capture Science and Technology*, *11*, 100192. https://doi.org/10.1016/j.ccst.2024.100192

Vitaku, E. and Dichtel, W. R. (2017). Synthesis of 2D imine-linked covalent organic frameworks through formal transimination reactions. *Journal of the American Chemical Society*, *139*(37), 12911–12914. https://doi.org/10.1021/jacs.7b06913

Wang, S., Li, M., Yan, G., Yang, Z., Guo, Y., Sun, X., Wang, Y., Feng, Y., Ding, H., and Zhang, X. (2023). Squaraine-linked zwitterionic COF modified LLZTO nanoparticles for high performance polymer composite electrolytes in Li–S batteries. *Nanoscale*, *15*(31), 12961–12971. https://doi.org/10.1039/D3NR01942J

Wei, S., Zhang, W., Qiang, P., Yu, K., Fu, X., Wu, D., Bi, S., and Zhang, F. (2019). Semiconducting 2D triazine-cored covalent organic frameworks with unsubstituted olefin linkages. *Journal of the American Chemical Society*, *141*(36), 14272–14279. https://doi.org/10.1021/jacs.9b06219

Yin, Z. J., Xu, S. Q., Zhan, T. G., Qi, Q. Y., Wu, Z. Q., and Zhao, X. (2017). Ultrahigh volatile iodine uptake by hollow microspheres formed from a heteropore covalent organic framework. *Chemical Communications*, *53*(53), 7266–7269. https://doi.org/10.1039/c7cc01045a

Zeng, Y., Zou, R., Luo, Z., Zhang, H., Yao, X., Ma, X., Zou, R., and Zhao, Y. (2015). Covalent organic frameworks formed with two types of covalent bonds based on orthogonal reactions. *Journal of the American Chemical Society*, *137*(3), 1020–1023. https://doi.org/10.1021/ja510926w

Zhao, J. and Yan, T. (2014). Effects of substituent groups on methane adsorption in covalent organic frameworks. *RSC Advances*, *4*, 15542–15551. https://doi.org/10.1039/c3ra47429a

Zhuang, H., Guo, C., Huang, J., Wang, L., Zheng, Z., Wang, H., Chen, Y., and Lan, Y. (2024). Hydrazone-linked covalent organic frameworks. *Angewandte Chemie International Edition*, *63*(31), e202404941. https://doi.org/10.1002/anie.202404941

14 Applications of Covalent Organic Frameworks in Degradation of Volatile Organic Compounds

Sushmita Gajurel and Tarnveer Kaur

14.1 INTRODUCTION

Volatile organic compounds (VOCs) are a class of synthetic chemicals, typically in gaseous form, released into the environment through natural and human activities. Naturally occurring VOCs prevail via plant emissions, as tropical regions are the largest contributors on a global scale. These regions release over a 1,000 Tg of carbon annually, primarily as isoprene, monoterpene, and reactive hydrocarbons, which account for 44%, 11%, and 22.5% of emissions, respectively. Although natural sources play a significant role, human activities remain the primary contributors to VOC emissions in indoor and outdoor habitats (David and Carolina Niculescu, 2021, Montero-Montoya et al., 2018, CMM Group, 2023a, Xihe et al., 2023).

VOCs significantly contribute to the formation of ozone, acid rain, and photochemical smog, as well as to the effects of global warming. These organic compounds are classified into three main categories based on their volatility, which refers to their ability to evaporate and enter the air as gas (Figure 14.1). Highly volatile organic compounds include examples such as methyl chloride, propane, and butane. Other commonly used laboratory chemicals, classified as VOCs, include hexanal, formaldehyde, acetone, and toluene. On the other hand, the pesticide is also classified as a semi-volatile organic compound. On the other hand, measuring certain VOCs, such as alcohols, carbonyl compounds, and amines, is quite challenging due to their gaseous state, which makes them difficult to detect as atmospheric components. These compounds are called hazardous air pollutants (U.S. Environmental Protection Agency, CMM Group, 2023b, Wikipedia).

People spending daily hours in offices, residences, and schools are liable to be exposed to indoor VOCs released via paints and furnished materials that cause severe health hazards, mainly in cold weather with limited ventilation. In terms of VOC emissions, isoprenoid (chemical compound) serves as a significant contributor by reacting with the pollutants and deteriorating the air quality. Some of the major health risks caused by these conditions include breathing or respiratory issues, dizziness, memory loss, and skin-related issues. Food, plastic, pharmaceutical, cosmetic, etc., industries are also the primary contributors of VOCs. Studies show that

DOI: 10.1201/9781003605386-14

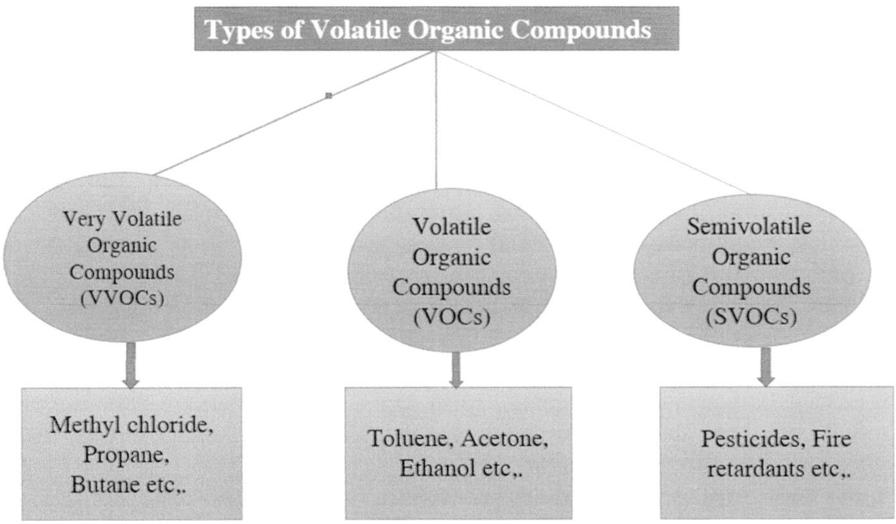

FIGURE 14.1 Types of volatile organic compounds.

TABLE 14.1
Some Examples of VOCs and Their Risks to Human Health and Environment

Volatile Organic Compounds	Risks to Environmental and Human Safety
Benzene	Cancer causing
Chloroform	Kidney failure, dermal infection, mental distress, dizziness
Phenol	Foul smell and poisonous nature
Toluene	Headache, dizziness
Acetaldehyde, acetone	Breathing irritation and eye discomfort
Epoxides	Cancer causing, hazardous
Styrene, ethylene	Atmospheric ozone depletion
Xylene	Depressant effect, liver damage, dermal effect
Amines	Unpleasant odor, carcinogenic

school-going children are the worst sufferers of VOC emissions being daily exposed to traffic exhaust pollution, which causes respiratory disorders and bronchitis (Xihe et al., 2023, Mendell, 2007, Peder et al., 2006, Krassi et al., 2004, Paolin and Matija, 2024, Lu et al., 2024, Batty et al., 2025, Xiao et al., 2024, Haick, 2024, Rehman et al., 2025, Yang et al., 2024, Maojun et al., 2024). Table 14.1 shows some examples of VOCs and their risks on human health and the ecosystem.

Therefore, it is a necessary concern to address the issue of VOC emissions where there is a growing risk each second for health and the environment.

Nowadays, researchers have developed various technologies for VOC removal, and some of them include adsorption, absorption, condensation, membrane separation, catalytic oxidation, photocatalysis, and biodegradation. Although each of these techniques has its advantages, it goes through some major disadvantages such as low selectivity in product formation, low efficient product regeneration, produces secondary pollutants, and expensive methodology (Zou et al., 2018, Yang et al., 2019). However, to resolve this issue, some of the materials are in great demand these days being newly designed and possessing high applicability. Among them, the groundbreaking discovery of covalent organic frameworks (COFs) has gained immense interest in the removal of VOCs. COFs are composed of light elements such as C, N, O, and B (Freund et al., 2021). The prominent feature of COFs includes the predesignable structure, which can be controlled via synthetic strategies and can be managed functionally (Geng et al., 2019). COFs possess a permanent porous network, strong light absorption capacity, and exhibit high chemical stability. These features ensure the correct fit of the VOCs into the active catalytic sites of the COFs. Furthermore, various functionalities prevalent in COFs offer a strong interaction of COFs with VOCs enhancing the removal efficiency (Rejali et al., 2023).

14.2 PROPERTIES OF COF CORRESPONDING TO VOC REMOVAL

The COF represents a novel class of crystalline materials with their high porosity and extreme photostability. Below are some important characteristics that show how COFs are important in VOC removal (Figure 14.2).

14.2.1 High Crystallinity and Porosity

COFs possess high porosity, corresponding to a high surface area. Furthermore, the interlayer connection held by the network within the frame of COFs enables the VOC to be easily adsorbed in substantial amounts. Although zeolites and activated carbon were utilized as traditional methods in adsorbing the VOCs, COFs have significantly served as excellent candidates in adsorbing VOCs even at lower concentrations. Due to their large surface area, rapid-adsorption kinetics is another phenomenon exhibited by COFs in VOC removal (Cui et al., 2021).

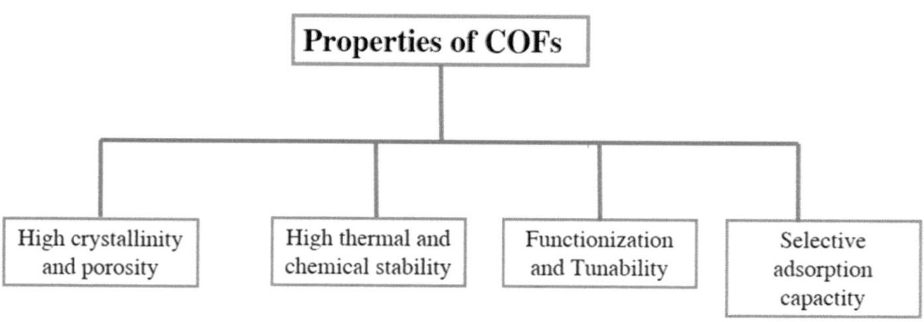

FIGURE 14.2 Properties of COFs.

14.2.2 Functionalization and Tunability

Another unique feature possessed by COFs is their flexible and adaptable structure. Tuning the pore size of COFs depends on the basis of linkers and knots used from organic moieties. While the structural design is so flexible and functionalizable to develop the pore size of necessary importance; therefore, smaller pore sizes are meant to adsorb smaller molecules such as formaldehyde and larger pore sizes adsorb bulkier molecules such as toluene and xylene.

COFs can be functionalized with a broad range of functional groups, which allows an improved interaction ability with various VOCs increasing the selectivity. The introduction of varying functional groups, such as amino, hydroxyl, and carboxyl groups, enhances adsorption via hydrogen bonding, dipole interactions, etc. This feature encourages researchers to develop COFs with diverse structural importance, leading to an efficient VOC removal process (Geng et al., 2019).

14.2.3 Thermal and Chemical Stability

Some of the traditional processes involved in the removal of VOCs, such as catalytic oxidation or regeneration of the adsorbent, demand high temperatures, which might lead to a loss of their structural properties leading to degradation of the catalyst. On the other hand, COFs are resistant to elevated temperature, which shows their survival under harsh conditions. Furthermore, COFs can withstand their structural integrity under acidic as well as basic conditions. These important conditions enable COFs to maintain their heterogeneous and reusable behavior without losing their catalytic activity even after reusing. Hence, the remarkable stability of COFs makes them resistant to degradation on exposure to various harmful VOCs (Cui et al., 2021).

14.2.4 Selective Adsorption Capacity

The selectivity of a catalyst is an important factor in the removal of VOCs. In the case of VOCs being present in a mixture form, COFs can easily adsorb the toxic pollutants by ignoring the nontoxic ones. Moreover, based on their tunable structure and incorporation of the functional groups, COFs can accelerate some chemical reactions that can convert VOCs into nontoxic by-products such as CO_2 and H_2O. This dual characteristic of adsorption and degradation exhibited by COFs makes them an ideal material of catalytic importance (Ge et al., 2024).

14.3 A Mechanistic Study on the Degradation of Volatile Organic Compounds Using COFs

14.3.1 Adsorption Mechanism and Degradation of VOCs

The adsorption method describes the removal of VOCs from the environment by allowing van der Waals or chemical interaction between the adsorbent and the VOCs. Adsorption phenomenon is classified into physical and chemical adsorption.

Physical adsorption is nonspecific, reversible, and favored at low temperatures. On the other hand, chemical adsorption involves a strong chemical bond between the VOCs and the COFs, making it suitable and irreversible for the permanent removal of VOCs. Its high removal efficiency effectively captures VOCs even at minimal concentrations. Furthermore, the process is simple and requires low energy consumption (Khan and Ghoshal, 2000, Gendy et al., 2022).

14.3.2 Photocatalytic Degradation of VOCs

COFs have been widely studied for their strong covalent bond being used as a metal-free heterogeneous photocatalyst. Unlike conventional semiconductors, COFs exhibit shared characteristics and unique advantages in photocatalytic applications. These include their strong covalent bonding and periodic π-conjugated aromatic structures, which enhance substrate adsorption and diffusion, visible-light absorption, and photoelectron transfer efficiency.

To ensure the photocatalytic performance of COFs, it is essential to design catalysts with high photoelectric activity and well-organized band structures. Notably, COFs with hollow structures have gained widespread attention in photocatalysis due to various features.

1. Compared to their solid counterparts, hollow COFs exhibit a larger specific surface area, which improves substrate adsorption and provides more active sites for photocatalytic reactions.
2. The hollow cavity allows multiple reflections and light scattering, facilitating light absorption and generating more photoelectrons.
3. A hollow structure with modified porosity reduces the distance for charge carriers to travel from the inner walls of the framework to the outer surface, thereby minimizing electron–hole recombination (Zhang et al., 2020).

14.3.3 Surface Chemistry and Role of Functional Groups in VOC Degradation

Surface chemistry refers to the general chemical properties of COFs. These properties include hydrophobicity, hydrophilicity, polarity, and chemical stability. Depending on the nature of compounds, VOCs might be polar or nonpolar. When COFs are precisely modified according to their polarity, it can interact with VOCs via dipole–dipole, van der Waals, or hydrophobic interaction. COFs designed with polar functional groups, i.e., hydrophilic COFs, effectively adsorb polar VOCs. On the other hand, COFs with nonpolar functional groups, i.e., hydrophobic COFs, adsorb nonpolar VOCs. Functional groups such as –OH and $-NH_2$ are capable of forming hydrogen bonding with the VOCs facilitating adsorption. Furthermore, polarizable VOCs can be easily attracted by COFs consisting of functional groups such as –COOH or $-SO_3H$. The functional group with their aromatic moieties attached in COFs can easily attract aromatic compounds such as toluene and benzene (Gendy et al., 2022).

COFs are structurally framed with strong covalent bonds, including various linkages such as imine, amide, imide, azine, hydrazone, triazine, boronate, alkene, and β-ketoamine. In the context of photocatalysis, COFs with boronate, imine, and alkene linkages are among the most extensively studied and reported.

Aldehydes and amines on condensation produce an imine linkage with high reversibility that results in the superior crystallinity of COFs. The imine linkage, formed through the condensation of aldehydes and amines, stands out for its high degree of reversibility, resulting in COFs with superior crystallinity. Additionally, imine-based COFs exhibit enhanced chemical stability when compared to their boron-containing counterparts. However, under prolonged exposure to aqueous environments, imine linkages remain susceptible to hydrolysis and degradation. As a result, recent research efforts have been directed toward synthesizing even more robust COF structures. Highly stable COFs, such as those with sp^2 C=C linkages, β-ketoamine bonds, or quinoline linkages, have demonstrated remarkable photocatalytic performance. This exceptional behavior is largely attributed to their fully conjugated electronic structures and extensive π-electron delocalization (Geng et al., 2019, Yusran et al., 2020, Raghav et al., 2025).

14.4 SYNTHETIC METHODOLOGIES FOR COFs

Various synthetic methodologies have been developed for developing COFs with modified structures. Out of them, some commonly used synthetic methods are as follows:

14.4.1 Solvothermal Method

Solvothermal synthesis is a widely used method for preparing crystalline porous COFs. The process is affected by monomer solubility, reaction reversibility, and specific conditions including temperature, reaction time, solvent choice, and catalyst concentration. Typically, the procedure involves mixing suitable monomers, solvents, and a catalyst in a Pyrex tube. This mixture is then sonicated, degassed, sealed, and heated for a predetermined duration. After cooling, the resulting precipitate is isolated through centrifugation or filtration and washed to eliminate impurities. The final product is vacuum-dried and stored under an inert gas atmosphere. This method enables the large-scale production of COFs. For example, TPT-COF-1 is synthesized from the monomers TPT-NH_2 and TPT-CHO, yielding a BET surface area of 1,589 $m^2\ g^{-1}$ and demonstrating high crystallinity (Xu et al., 2016).

14.4.2 Microwave Synthesis

The microwave method is trending as a fast alternative to solvothermal synthesis for producing crystalline porous COFs, significantly reducing reaction times. Boronate-ester-linked COF-5 and COF-102, as well as imine-linked TpPa-COF, were successfully synthesized employing the microwave synthesis method.

Typically, monomers were added in a suitable solvent in a sealed atmosphere using a microwave tube under nitrogen or vacuum and heated with stirring at 100°C for 60 minutes. For boron-based COFs, the crude product undergoes solvent extraction with acetone at 65°C, followed by filtration and vacuum drying, enhancing porosity and oligomer removal. Imine-linked COFs are filtered, washed with mesitylene and acetone, Soxhlet-extracted with THF, and dried at 100°C under vacuum (Campbell et al., 2009, Ritchie et al., 2010, Wei et al., 2015).

14.4.3 Mechanochemical Synthesis

Both solvothermal and microwave syntheses involve complex conditions, such as sealed reactions, inert atmospheres, and precise control of solvents and temperatures. To overcome these challenges, mechanochemical synthesis provides a simpler, cost-effective, and environmentally friendly alternative. This method involves grinding monomers together in a mortar and pestle at room temperature to produce COFs, including TpPa-1, TpPa-2, TpPa-NO_2, TpPa-F_4, TpBD, TpBD-$(NO_2)_2$, TpBD-Me_2, and TpBD-$(OMe)_2$ (Wei et al., 2015, Chandra et al., 2013, Biswal et al., 2013). To enhance this process, liquid-assisted grinding has been introduced, where a small amount of catalyst solution is added during the grinding process. This addition improves the homogeneity of the reactants, makes the reaction rapid, and enhances the crystallinity of the products (Shinde et al., 2016, Peng et al., 2016).

14.4.4 Azine and Imide Linkage COFs

The azine linkage is formed by the shortest hydrazine monomer, which connects two aldehyde molecules, leading to polygonal skeletons that create some of the smallest pores among various linkage types. Several types of knots have been explored for this purpose, including substituted benzene, triphenyl benzene, triphenyl triazine, and pyrene monomers. These knots are very helpful in the synthesis of hexagonal, rhombic, and trigonal COFs. These frameworks have been designed and synthesized in various reaction media at different temperatures, depending on the solubility and reactivity of the π-knots.

In the synthesis of imide linkages, a reaction occurs between amine derivatives and acetic anhydride. This reaction, due to its less reversibility nature, demands high temperatures, reaching up to 250°C (Geng et al., 2019, Zhu et al., 2015, Li et al., 2015).

14.5 COMPARISON OF COFS WITH TRADITIONAL METHODS FOR VOC REMOVAL

Although zeolites and activated carbon have emerged as important candidates for VOC removal, they also suffer from some other drawbacks, which are discussed below. Therefore, COFs with their unique feature have emerged to resolve these existing drawbacks.

14.5.1 Role of Zeolites in VOC Removal

Zeolites are a class of crystalline porous materials termed aluminosilicates. They possess microporous and well-defined structures with high adsorption efficiency. VOCs can easily interact with their well-defined structure via electrostatic interaction. However, the pore size of zeolites being less than 2 nm is unable to adsorb some of the larger VOCs such as toluene. Therefore, it leads researchers to look for better designs that can resolve this issue (John and Doe, 2023).

14.5.2 Role of Activated Carbon in VOC Removal

Activated carbon is another class of important material in VOC removal. Some of the significant characteristics possessed by this material are high surface area and various pore sizes available throughout the structure. On the other hand, although it has varied pore sizes, its pore sizes are nonuniform, and it lacks chemical specificity. It is nonpolar and adsorbs VOCs of nonpolar compounds such as benzene and toluene. Therefore, due to its nonuniform porosity, this method demands some other chemical alternatives (Wee Kong et al., 2018).

14.5.3 Role of COFs as Compared to Zeolites and Activated Carbon in VOC Removal

The development of COFs serves as a boon to the researchers for various chemical applications including VOC removal. As discussed above, this material's structural and chemical properties, tunability, and adsorption capacity are very unique compared to zeolites and activated carbon. Therefore, it can adsorb every small, medium, or large VOC depending on the tuned structure and presence of linkages.

14.6 EXPERIMENTAL STUDIES ON VOC REMOVAL USING COFs

Various experimental studies have been found for the removal of VOCs through advanced mechanisms. Some of the processes are illustrated below.

14.6.1 Functional Organic Ligands in COFs

The efficiency of COFs in removing VOCs can be enhanced by functionalizing COFs with the organic linkers. Chen et al. performed an aldol condensation reaction for the synthesis of vinylene-linked COFs. The synthetic procedure involved the use of 2,4,6-trimethyl-1,3,5-triazine (TMT), which showed excellent activity in the use of diacetylene COF (TMT-BDA) for the photocatalytic degradation of phenol. This method creates triazine-based linkers and provides more room for the delocalization of π-electrons. Interestingly, this procedure provided 96% efficiency within 15 minutes using visible light (Maojun et al., 2024).

Cai and co-workers reported another study in which mesoporous imine-linked COFs were synthesized by the unification of triazine-derived aldehydes and nitrogen-rich building blocks. These COFs proved to be very efficient in degrading phenol, showing a 79% phenol degradation efficiency. The degradation was accomplished within just 30 minutes (He et al., 2018). This shows the significance and advantage of designing and selecting a suitable functional group that improves the interaction of functional groups present in the framework with the VOCs.

14.6.2 Donor–Acceptor (D–A) COFs

The considerable and significant potential shown by donor–acceptor (D–A) COFs in the degradation of VOCs is highly noteworthy. These materials incorporate both electron-donating and electron-accepting components to enhance charge separation and light absorption. For instance, CTF-5Th, a framework linked by triazine, achieved nearly 100% removal efficiency for bisphenol A (BPA) when exposed to visible light. This outstanding performance can be attributed to the effective generation of reactive oxygen species (ROS), such as hydroxyl radicals and singlet oxygen, which is made possible by the well-aligned donor–acceptor structure (Li et al., 2022).

14.6.3 COF Hybrids

The hybridization of COFs with other nanomaterials/semiconductors, such as MOFs or Cds, is emerging as an important area of research for VOC removal. A superior photocatalytic activity is observed using these hybrid COFs. An important characteristic of COF hybridization involves the efficiency of charge separation and extended light absorption range making it a good photocatalyst. In a study, a hybrid photocatalyst has been developed by the combination of a COF with TiO_2 nanoparticles. This provided excellent activity showing the degradation efficiency of phenol by 99.9% within 20 minutes. This significant result is shown due to the excellent synergistic interaction between the COF and TiO_2 nanoparticles. Another experiment was performed for the degradation of acetaldehyde under visible light. In this mechanism, MIL-101-NH_2 was framed with triazine-based COFs; these hybrid COFs triggered the generation of ROS and provided 99% degradation of acetaldehyde (Maojun et al., 2024, Prakash et al., 2024). The significant advantage of combining both materials offers some advantages such as high adsorption capacity from the MOFs and efficient charge transfer from the COFs. Here are a few examples of the photocatalytic degradation of BPA and phenols using COFs illustrated in Table 14.2 (Maojun et al., 2024).

14.7 COMPARISON OF MOFs, COFs, AND CTFs AS PHOTOCATALYSTS FOR VOC REMOVAL

Metal-organic frameworks (MOFs), COFs, and covalent triazine frameworks (CTFs) are undoubtedly very important photocatalysts for the removal of VOCs. Although COFs, MOFs, and CTFs have structural similarities, they differ in their chemical and photophysical characteristics. MOFs are highly ordered cage-like structures that possess a specific surface area of 4,000 $m^2\ g^{-1}$ and a pore size of 2.8–3.4 nm. The aromatic linkers in MOFs improve MOFs' interaction ability with VOCs. On the other hand, several types of VOCs in the presence of polar solvents get attracted to the organic

TABLE 14.2
Examples of COFs in the Degradation of VOCs and Their Degradation Efficiency

Entry	COFs	VOCs	Time (min)	Degradation Efficiency (%)
1.	TMT-BDA (functional organic linkers)	Phenol	15	96
2.	TMT-EDA (functional organic linkers)	Phenol	15	~100
3.	COF_{A+C} (functional organic linkers)	Phenol	30	79
4.	COF-PRD/PMS (functional organic linkers)	BPA	150	100
5.	CTF-5Th (D-A COFs)	BPA	75	~100
6.	COF-TD1 (D-A COFs)	BPA	120	97
7.	$CTFS_5$ (COF hybrids)	BPA	360	84.7
8.	CTF-SD_2 (COF hybrids)	BPA	90	100

Source: Maojun et al. (2024).

linkers while ionic compounds have a strong interaction with the defective sites of the MOFs. These limitations state that it might lead to higher bandgaps resulting in the inactivation of visible lights. On the other hand, the bandgap can be tuned in COFs and CTFs and has more charge mobility. But, COFs possess structural defects compared to MOFs, which is a drawback of COFs, leading to trapping of charge carriers generated by light. When the charge carriers are trapped, it is difficult to be activated by visible light. The major advantages of COFs are their highly conjugated structures where electrons and holes (generated by light) can move with ease within the material. The structure of COFs is highly adjustable by the use of various linkers and building blocks where appropriate bandgaps are generated (Maojun et al., 2024).

14.8 CONCLUSIONS AND FUTURE PERSPECTIVE OF COFs IN VOC DEGRADATION

COFs possess poor crystallinity, which is why it is difficult to study and improve the structure of COFs. It is very challenging to characterize the structure of COFs. Moreover, COFs that are modified and anchored with various metals and elements such as S create defect sites, making it even more difficult to characterize. Therefore, it requires even more sophisticated technology to characterize COFs. Moreover, it is difficult to synthesize COFs on a large scale. Therefore, it is of great importance to think of some alternative methodologies that improve the preparation of COFs on a large scale. Some technologies, such as AI and data mining, might help in the development of methodologies. Although COFs have made large advancements, it is still difficult to understand the mechanisms of VOC degradation by COFs. A thorough understanding is required to evaluate how charges move on COFs and how reactions occur on their surface. Lastly, the development of COFs that are reusable is very challenging in real-world applications. Therefore, catalyst deactivation mechanisms and regeneration methods must be studied effectively for a better understanding.

REFERENCES

Batty, C. A., Victoria K., Pearson, K. O.-F., and Geraint, M. 2025. Volatile organic compounds (VOCs) in terrestrial extreme environments: Implications for life detection beyond earth. *Natural Product Reports* 42(1): 93.

Biswal, B. P., Chandra, S., Kandambeth, S., Lukose, B., Heine, T., and Banerjee, R. 2013. Mechanochemical synthesis of chemically stable isoreticular covalent organic frameworks. *Journal of the American Chemical Society* 135(14): 5328–5331.

Campbell, N. L., Clowes, R., Ritchie, L. K., and Cooper, A. I. 2009. Rapid microwave synthesis and purification of porous covalent organic frameworks. *Chemistry of Materials* 21(2): 204–206.

Chandra, S., Kandambeth, S., Biswal, B. P., Lukose, B., Kunjir, S. M., Chaudhary, M., Babarao, R., Heine, T., and Banerjee, R. 2013. Chemically stable multilayered covalent organic nanosheets from covalent organic frameworks via mechanical delamination. *Journal of the American Chemical Society* 135(47): 17853–17861.

Cui, Y., Zhang, L., and Li, J. 2021. Covalent organic frameworks: Synthesis, properties and applications – An overview. *Polymers* 13(6): 970.

David, E. and Carolina Niculescu, V. 2021. Volatile organic compounds (VOCs) as environmental pollutants. Occurrence and mitigation using nanomaterials. *International Journal of Environmental Research and Public Health* 18(24): 13147.

Freund, R., Zaremba, O., Arnauts, G., Ameloot, R., Skorupskii, G., Dinca, M., Bavykina, A., Gascon, J., Ejsmont, A., Goscianska, J., Kalmutzki, M., Lachelt, U., Ploetz, E., Diercks, C. S., and Wuttke, S. 2021. The current status of MOF and COF applications. *Angewandte Chemie International Edition* 60: 23975–24001.

Ge, S., Wei, K., Peng, W., Huang, R., Akinlabi, E., Xia, H., Shahzad, M. W., Zhang, X., Xu, B. B., and Jiang, J. 2024. A comprehensive review of covalent organic frameworks (COFs) and their derivatives in environmental pollution control. *Chemical Society Reviews* 53(23): 11259–11302.

Gendy, E. A., Oyekunle, D. T., Ifthikar, J., Jawad, A., and Chen, Z. 2022. A review on the adsorption mechanism of different organic contaminants by covalent organic framework (COF) from the aquatic environment. *Environmental Science and Pollution Research* 29(22): 32566–32593.

Geng, K., He, T., Liu, R., Tian Tan, K., Li, Z., Tao, S., Gong, Y., Jiang, Q., and Jiang, D. 2019. Covalent organic frameworks: Design, synthesis, and functions. *Chemical Reviews* 120(16): 8814–8933.

Haick, H. 2024. Advances in volatile organic compounds detection: From fundamental research to real-world applications. *Applied Physics Reviews* 11(4): 040401.

He, S., Yin, B., Niu, H., and Cai, Y. 2018. Targeted synthesis of visible-light-driven covalent organic framework photocatalyst via molecular design and precise construction. *Applied Catalysis B: Environmental* 239: 147–153.

John, S. and Doe, J., (2023). Zeolites as adsorbents for volatile organic compounds: Challenges and opportunities. *Journal of Environmental Science and Technology* 25(3): 123–134.

Khan, F. I. and Ghoshal, A. K. 2000. Removal of volatile organic compounds from polluted air. *Journal of Loss Prevention in the Process Industries* 13(6): 527–545.

Krassi, R., Spickett, J., Bulsara, M., Phillips, M., and Stick, S. 2004. Association of domestic exposure to volatile organic compounds with asthma in young children. *Thorax* 59(9): 746–751.

Li, S., Zhiquan, J., Xinming, J., Jie, Y., Yashuang, W., Sijia, J., Haiyan, Z., Shuang, S., and Tao, Z. 2022. Thiophene-incorporated CTFs with boosted intramolecular charge transfer and defect-abundant networks for potent photosensitized oxidation of organics. *Chemical Engineering Journal* 431: 133900.

Li, Z., Zhi, Y., Feng, X., Ding, X., Zou, Y., Liu, X., and Mu, Y. 2015. An azine-linked covalent organic framework: Synthesis, characterization, and efficient gas storage. *Chemistry – A European Journal* 21(34): 12079–12084.

Lu, S., Xin, Y., Yue, C., and Ling, Z. 2024. Recent developments and challenges for volatile organic compounds control by the synergistic of adsorption and photocatalysis. *Applied Catalysis O: Open* 193: 206975.

Maojun, D., Sun, J., Chakraborty, J., Zhou, M., and Van Der Voort, P. 2024. From design to applications: A comprehensive review on porous frameworks for photocatalytic volatile organic compounds (VOCs) removal. *ChemCatChem* 16(7): e202300783.

Mendell, M. J. 2007. Indoor residential chemical emissions as risk factors for respiratory and allergic effects in children: A review. *Indoor Air* 17(4): 259–277.

Montero-Montoya, R., Lopez-Vargas, R., and Arellano-Aguilar, O. 2018. Volatile organic compounds in air: Sources, distribution, exposure, and associated illnesses. *Annals of Global Health* 84(2): 225–238.

Paolin, E. and Matija, S. 2024. Volatile organic compounds (VOCs) in heritage environments and their analysis: A review. *Applied Sciences* 14(11): 4620.

Peder, W., Wilkins, C. K., Clausen, P. A., and Gunnar, D. N. 2006. Organic compounds in office environments-sensory irritation, odor, measurements, and the role of reactive chemistry. *Indoor Air* 16(1): 7–19.

Peng, Y., Xu, G., Hu, Z., Cheng, Y., Chi, C., Yuan, D., Cheng, H., and Zhao, D. 2016. Mechano assisted synthesis of sulfonated covalent organic frameworks with high intrinsic proton conductivity. *ACS Applied Materials and Interfaces* 8(18): 18505–18512.

Prakash, K., Deka, R., and Mobin, S. M. 2024. A review on covalent organic frameworks: Exploration of their growing potential as porous materials in photocatalytic applications. *Inorganic Chemistry Frontiers* 11: 6711–6752.

Raghav, S., Raghav, J., Muhammad, M. A., Singh, M. B., Jain, P., Kumar, D. 2025. Recent advancements in synthesis and applications of covalent organic frameworks. *Applied Materials Today*, 44(June): 102694.

Rehman, S., Zheng, Z., Aujla, M. I., and Mehmood, T. 2025. Recent advances in adsorptive removal of hazardous VOCs by metal-organic-framework-based materials. *Chemical Engineering Journal* 505: 159257.

Rejali, N. A., Dinari, M., and Wang, Y. 2023. Post-synthetic modifications of covalent organic frameworks (COFs) for diverse applications. *Chemical Communications* 59: 11631–11647.

Ritchie, L. K., Trewin, A., Reguera-Galan, A., Hasell, T., and Cooper, A. I. 2010. Synthesis of COF-5 using microwave irradiation and conventional solvothermal routes. *Microporous and Mesoporous Materials* 132(1): 132–136.

Shinde, D. B., Aiyappa, H. B., Bhadra, M., Biswal, B. P., Wadge, P., Kandambeth, S., Garai, B., Kundu, T., Kurungot, S., and Banerjee, R. 2016. A mechanochemically synthesized covalent organic framework as a proton-conducting solid electrolyte. *Journal of Materials Chemistry A* 4(9): 2682–2690.

The CMM Group. 2023a. Volatile Organic Compounds Environmental Impact. *The CMM Group Blog*. https://www.thecmmgroup.com/vocs-impact-environment.

The CMM Group. 2023b. Volatile Organic Compounds Environmental Impact. *Environmental Perspectives*.

U.S. Environmental Protection Agency. 2025. Technical Overview of Volatile Organic Compounds. *EPA.gov*. https://www.epa.gov/indoor-air-quality-iaq/technical-overview-volatile-organic-compounds.

Wee Kong, P., Yusoff, R., and Aroua, M. K. 2018. A review on activated carbon adsorption for volatile organic compounds (VOCs). *Reviews in Chemical Engineering* 35(5): 649–668.

Wei, H., Chai, S. Z., Hu, N. T., Yang, Z., Wei, L. M., and Wang, L. 2015. The microwave-assisted solvothermal synthesis of a crystalline two-dimensional covalent organic framework with high CO_2 capacity. *Chemical Communications* 51(59): 12178–12181.

Xiao, Z., Xuerui, Y., Hongming, G., Jialiang, H., Tongguang, Z., Jianian, C., Xukang, P., Guangli, X., Mingyue, L., and Wei, Z. 2024. Characterization and sources of volatile organic compounds (VOCs) during 2022 summer ozone pollution control in Shanghai, China. *Atmospheric Environment* 327: 120464.

Xihe, Z., Zhou, X., Wang, C., and Zhou, H. 2023. Environmental and human health impacts of volatile organic compounds: A perspective review. *Chemosphere* 313: 137489.

Xu, L. Q., Ding, S. Y., Liu, J. M., Sun, J. L., Wang, W., and Zheng, Q. Y. 2016. Highly crystalline covalent organic frameworks from flexible building blocks. *Chemical Communications* 52(28): 4706–4709.

Yang, C., Miao, G., Pi, Y., Xia, Q., Wu, J., Li, Z., and Xiao, J. 2019. Abatement of various types of VOCs by adsorption/catalytic oxidation: A review. *Chemical Engineering Journal* 370: 1128–1153.

Yang, Y., Xianhong, M., Qiang, C., Qiang, X., Lina, W., Jian, S., Wenkai, G., Huijie, T., Lili, Y., and Chen, F. 2024. Characteristics of volatile organic compounds under different operating conditions in a petrochemical industrial zone and their effects on ozone formation. *Environmental Pollution* 363: 125254.

Yusran, Y., Li, H., Guan, X., Fang, Q., and Qiu, S. 2020. Covalent organic frameworks for catalysis. *EnergyChem* 2(3): 100035.

Zhang, X., Li, Z., Chen, Z., Li, Q., Xu, J., and Chen, J. 2020. Covalent organic framework photocatalysts: Structures and applications. *Chemical Society Reviews* 49(24): 4135–4165.

Zhu, Y., Wan, S., Jin, Y., and Zhang, W. 2015. Desymmetrized vertex design for the synthesis of covalent organic frameworks with periodically heterogeneous pore structures. *Journal of the American Chemical Society* 137(41): 13772–13775.

Zou, W., Gao, B., Ok, Y. S., and Dong, L. 2018. Integrated adsorption and photocatalytic degradation of volatile organic compounds (VOCs) using carbon-based nanocomposites: A critical review. *Chemosphere* 218: 845–859.

15 Challenges, Future Perspectives, and Emerging Trends in COF's Research

Rahul Badru, Khushboo Verma, and Avtar Singh

15.1 INTRODUCTION

Covalent organic frameworks (COFs) are a class of porous organic frameworks recognized for their potent use in gas separation and storage, metal ion sensing, energy storage, catalysis, and environment remediation (Han et al. 2023). COFs have several advantages over metal-organic frameworks (MOFs), such as the potential to undergo chemical changes, extensive network and tunable pore size, and significant structural diversity (Mohan et al. 2023; Raghav et al. 2025). Unlike MOFs, COFs are made up of lighter elements including carbon, hydrogen, oxygen, nitrogen, and boron. They exhibit greater energy densities and lower molecular weights (Ampong et al. 2024). Furthermore, COFs are popular for their exceptional structural stability, because the extensive framework is supported by strong covalent bonding (Nguyen 2021). This is contrary to MOFs, which face structural stability issues, that majorly rely on metal-oxygen bonds (Cao et al. 2019). Covalent bonding in organic molecules underwent a significant shift in the 20th century, from empirical techniques to more systematic and rational strategies for the synthesis of organic compounds (Liu et al. 2016). In this era, the field of COFs has seen numerous developments, with key developments marking the emergence of these materials (Diercks and Yaghi 2017). A significant advancement in materials science was made in 2005 when Yaghi and his colleagues developed COF for the first time (Cote et al. 2005). This structure was formed by dehydrating benzenediboronic acid (BDBA), resulting in COF-1 with two-dimensional (2D) layers composed of benzene and B_3O_3 rings linked through covalent bonding between the boronic acid groups. The flexibility between the layers was attained since they were weakly bounded together by van der Waals forces. A first-time reticular synthesis was accomplished by the process of creating porous materials with precisely controlled structures using organic building blocks. The unique structure of COF-1, which linked benzene rings through boron and oxygen atoms, became the foundation for numerous other inventions (Cote et al. 2005). A new class of materials with exceptional structural variety and tunability has emerged as a result of this discovery. By 2007, an array of COFs were designed with a wide range of possible organic linkers (El-Kaderi et al. 2007). In 2009, Furukawa

DOI: 10.1201/9781003605386-15

and his group achieved a significant breakthrough by forming boronate-based COFs, namely COF-5, that showed potential for separation and gas storage applications (Furukawa and Yaghi 2009). COF-5 was synthesized by the polymerization of hexahydroxytriphenylene and 1,4-phenylenebis (boronic acid) in the presence of dioxane and mesitylene solvent and heated up to 100°C for 72 hours (Han et al. 2023). COF-5 has shown its use in supercapacitors as an electrode material, exhibiting a specific capacitance of 85 F g^{-1} with a scan rate of 2 mV s^{-1} (Dubey et al. 2024). The high crystallinity and electrical conductivity up to 2.35×10^{-2} S cm^{-1} for the structure of COF-5 was achieved using microwave-assisted synthesis, which is suitable for its role in supercapacitors. The development of covalent triazine frameworks in 2011 was another remarkable advancement that garnered interest because of their exceptional thermal stability and potential uses in industries including gas storage, capture, and catalysis (Zhang et al. 2013). The field kept on expanding in 2013 by Yaghi and his group, following the successful synthesis of a further class of COFs, known as extended π-conjugated COFs. The COF crystals were synthesized through an imine condensation reaction between 4,4′-biphenyldialdehyde and tetra-(4-anilyl) methane in 1,4-dioxane at 120°C, resulting in the formation of a highly porous diamond-like network with nine interwoven folds (Zhu et al. 2020). In 2016, another major development occurred with the development of dynamic COFs, which were capable of undergoing reversible structural transformations (Hu et al. 2020). The adaptable materials allowed for the design of frameworks that could be customized to meet specific requirements (Evans et al. 2018). Another significant advancement was made in 2018 with the production of single-crystal COF and thin films, which obtained precise crystal structure and orientation at the micrometer scale (Yang et al. 2022). A single crystal was achieved using a two-step procedure, which transforms boronate ester-linked 2D COFs from nanoscale seeds to large-size single crystals, with the use of a solvent that hinders the generation of additional nanoparticles, which often leads to amorphous powders. The resulting crystalline materials demonstrated superior charge transport properties compared to traditional powders, enhancing the potential applications of COFs in various industries, particularly in electronics (Evans et al. 2018, Yang et al. 2022). The development of redox-active COFs in 2019, capable of storing and releasing charge, opened up new opportunities for applications in electrochemistry and energy storage. These redox-active COFs have shown great potential for use in sustainable energy technology (Royuela et al. 2019). In 2021, the emergence of high-valency COF structures was observed that provided an improved connection between the building units, marking a significant advancement in COF synthesis. These novel structures surpassed the existing valency constraints by using multivalent polycubane linkers, which had been capped at 3 and 4 because of the sp^2 and sp^3 hybridization of carbon, by using multivalent polycubane linkers (Gropp et al. 2020). Numerous kinds of COFs, including imine-based COFs (Ding et al. 2011), borate COFs (Wang et al. 2021), vinyl-linked COFs (Bi et al. 2022), and azine-type COFs (Li et al. 2015), have been synthesized by researchers in recent years. Among these, imine-based COFs have gained significant attention in research owing to tunable functionality, high surface area, and chemical diversity. However, challenges remain in the synthesis methods and the production of high-quality COFs.

COFs can be categorized into 1D, 2D, and 3D frameworks. 1D COFs have shown potential as building blocks for non-centrosymmetric crystals, because of their high degree of flexibility in molecular packing (Ding and Wang 2013, Huang et al. 2016). Usually, covalent bonds formed in 1D COFs are limited to just one dimension. Therefore, the synthesis of 1D COFs is challenging, nevertheless, because of the highly anisotropic character of these structures and the ability of organic chains to adopt random packing as a result of entropy effects (Xu et al. 2020). It is clear that creating a well-ordered 1D COF is difficult when using linear polymers as an example, which have mostly been amorphous or weakly crystalline since the formation of Bakelite in 1907. Two primary goals need to be addressed in order to attain this:

i. To create a structure-defined framework, rather than a mere dense organic polymer.

   While creating a 1D COF, the main objective is to create a framework instead of a mere random assembly of polymer chains. In order to ensure that the organic building blocks are capable of developing a stable and periodic structure, a thorough design is required. The framework needs to be made up of covalent connections, which enable the desired dimensionality while maintaining the structural integrity (Xu et al. 2020).

ii. To maintain a precise control over the packing of organic chains to form a single-crystal structure.

   The periodic packing of the organic chains must be controlled in order to achieve a single-crystalline structure. This entails being understanding of how the chains interact and how to organize them in an ordered manner. The COF and the structured arrangement of the chains are stabilized by noncovalent interactions such as hydrogen bonding and π–π stacking (Dou et al. 2014).

Although topochemical polymerization has been used to convert pre-packed molecular crystals into single-crystalline polymers, its uses have not been explored yet (Dou et al. 2014, Kissel et al. 2012). Consequently, developing a direct crystallization approach from solution is still a major goal. Additionally, 1D COF films, like COF-4,4′-methylenedianiline and COF-4,4′-oxydianiline, have demonstrated multilevel resistive switching, which is important for neuromorphic computing (Zhou et al. 2024). TiO_2 films are included to improve the electric field at COF interfaces, which allows for greater control over synaptic weights.

Flat, planar units are used to build two-dimensional COFs, which can be layered to generate multilayer structures (Zhai et al. 2017). These planar monomers are arranged to produce a well-ordered polymeric structure, which in turn creates 2D atomic layers with unique topologies (Huang et al. 2017). These 2D layers are mostly held together by π–π stacking interactions, which are specified by the alignment of π-conjugated segments that are related to the atomic planes (Mandal et al. 2017). Strong covalent linkage within each layer maintains the structural integrity of the COF in this arrangement, while non-covalent forces between layers support the stability of the framework (Dubey et al. 2024).

Three-dimensional (3D COFs are formed from nonplanar, tetrahedral building units, which resemble large, porous extended networks like MOFs. This synthesis involves multidirectional growth of polymeric backbone, incorporating chain folding and layer-overlapping techniques. This extension is both functional and spatial, giving these materials dimensional flexibility that enables their use in various technical fields (Li et al. 2015). 3D COFs are well known for having significantly large surface areas (more than 5,000 $m^2 g^{-1}$) and low densities (less than 0.13 g cc^{-1}), which make them suitable for various applications (Baldwin et al. 2016). According to studies, 3D COFs' low density, high porosity, and huge surface areas make them ideal candidates to store more gas than many MOFs, particularly for hydrogen gas (Zeng et al. 2016). The high porosity, huge surface areas, and low densities of these 3D COFs make them very useful for gas storage, with hydrogen capacity exceeding that of 2D COFs (Zeng et al. 2016). Methane and carbon dioxide storage capacities of boronate-ester-based COFs (COF-102 and COF-103) have been shown in theoretical calculations to be superior to those of standard MOFs and other nanoporous materials (Ma and Scott 2018).

3D COFs have exceptional adsorbent qualities in addition to their storage capacity (Yaghi et al. 2019). Numerous linkages, including boron-oxygen, imine, hydrazone, imide, and azine, can be found in these frameworks (Yaghi et al. 2019, DeBlase and Dichtel 2016). Early 3D COF instances used boron-oxygen bonds, which led to the crystals with low densities, as COF-108, which has a large BET surface area and a density of 0.17 g cm^{-3} (El-Kaderi et al. 2007). Metallocenes have been used to investigate the host–guest relationship of COF, and recent metalation of 3D COFs with dehydrobenzoannulenes has exhibited a notable retention of surface area (Baldwin et al. 2016). Even being microwave-assisted synthesis of 3D COFs is reported, it is still not well understood (El-Kaderi et al. 2007).

15.2 SYNTHETIC CHALLENGES

15.2.1 Synthetic Challenges in 1D COFs

The maintenance of structural regularity and porosity is a significant problem in the synthesis of one-dimensional COFs (Ding and Wang 2013). The root cause of these challenges is the complexity of creating highly ordered, stable, and functional materials that are appropriate for pioneering applications like memristors for neuromorphic computing and other electrical devices (Zhou et al. 2024). The synthesis of 1D COFs is affected by many important challenges. The potential of COFs to create crystalline structures with long-range order is one of their primary characteristics (Ma et al. 2018). However, it is especially difficult to achieve this feature in 1D COFs. The arrangement of organic building blocks during synthesis must be precisely controlled in order to provide the linear, extended structure needed for 1D COFs. Any defects in the 1D structure or crystallinity can have a detrimental effect on overall performance and electronic conductivity, as well as cause irregularity in material characteristics, which are necessary for devices such as memristors (Zhou et al. 2024). Subsequently, these imperfections in the crystal structure of COFs lead to the variations in the electrical and mechanical properties of the material, which makes its use unfit for electronic devices (Wan et al. 2024). The stability of 1D COFs depends on a strong covalent bonding present between organic molecules, and to preserve the material's structural integrity, exact control over these connections is required (Ma et al. 2018). Furthermore, synthetic methods, temperatures, choice of solvent, or impurity can cause changes in bond formation of COFs, which ultimately lead to incomplete or uneven COF networks, reducing the mechanical and electrical strength and limiting their potential in device applications (Wan et al. 2024).

15.2.2 Synthetic Challenges in 2D COFs

The complex nature of COF structures, their sensitivity to reaction conditions, and scalability constraints are the main causes of these challenges. 2D and 3D COFs have recently been modified by adding aliphatic monomers, improving their potential in some important applications like drug delivery, gas separation, luminescence, and catalysis (Ziang et al. 2019, Mandal et al. 2017, Keller and Bein 2021). However, to retain framework firmness and foster interlayer π–π stacking, aliphatic monomers must be coupled with aromatic ones (Ziang et al. 2019). This combination often results in a partial loss of crystallinity, affecting material performance and complicating structural analysis. Additionally, the limited availability of aliphatic monomers, their differing reactivity, and crystallization kinetics pose challenges in synthesis (Yang et al. 2023).

These issues are addressed via a new approach, which was developed to create fully saturated 3D COFs using only aliphatic building units (Ren et al. 2024). The methodology involves using 1,3,5,7-tetraaminoadamantane (TAA) and bicyclo[2.2.2]octane-1,4-dicarbaldehyde (BODA) to form an imine-based COF via a Schiff base reaction. The synthesis optimization was carried out using a Lewis acid modulator to facilitate the condensation between TAA and BODA. After obtaining the aliphatic COF, its fully saturated form can be achieved by reducing the imine bonds (Qian et al. 2022). Below is a more detailed discussion of the synthesis challenges faced in COF production.

15.2.2.1 Synthesis Complexity and Control

The synthesis of COF involves the formation of strong covalent bonds between organic monomers under specific conditions. However, ensuring the correct geometry and orientation of these monomers during synthesis is challenging. The process often requires precise control over temperature, solvent choice, and reaction time. Any variation can lead to the formation of unwanted side products or incomplete reactions, resulting in poorly defined frameworks. The synthesis of COFs with high crystallinity is often difficult because the coordination between the organic monomers needs to be exact. The reaction conditions for creating COFs, such as solvothermal conditions or mild conditions for some COFs (like dynamic covalent chemistry), require careful monitoring to avoid defects in the framework. Another approach to enhancing the crystallinity of COFs involves designing monomers that naturally promote the formation of ordered structures (Ascherl et al. 2016). Bein and colleagues utilized monomers shaped like a screw that stacked in a specific, uniform arrangement. This design helps avoid the degenerative offsets observed in other 2D COFs, potentially aiding in the merging of the boundaries (Chen et al. 2013).

Similarly, Jiang *et al.* were able to improve the crystallinity by modifying the electronic interactions among the monomers (Jiang et al. 2013). The authors demonstrated that using a combination of terephthalaldehyde and its perfluorinated equivalent enhanced the interlayer stacking energy, leading to 2D COFs with superior crystallinity and larger surface area compared to those formed from each monomer individually. Another challenge is with the choice of suitable linkers that will create highly ordered, yet stable, COFs that maintain their structure across a range of applications. The specific organic linkers used to create the network structures of COFs also play a crucial role. Designing these linkers to have both the required functional groups and the stability needed for robust framework formation is a complex task (Lyle et al. 2019).

One of the major challenges in the synthesis of COF is the solubility of the building blocks or precursors. Organic linkers and monomers must dissolve in suitable solvents to react effectively. However, many COF precursors are either poorly soluble or insoluble in DMF, MeOH, acidic and basic aqueous solutions, making it difficult to achieve high reaction rates and uniform synthesis of the framework (Kumar et al. 2021). One of the primary challenges has been identified, particularly when dealing with the large complex molecules (Kumar et al. 2021). COFs are often quite insoluble in common organic solvents such as THF, DMSO, and methanol, even though they may be synthesized successfully. This aspect causes an issue for postsynthesis processing and modifications, such as purification or adding more functional groups in the structure of COFs. Additionally, more idealistic approaches are needed to solve this problem, including fabricating linker designs that increase solubility without affecting the framework's integrity (Smith et al. 2016). The formation of imperfections in the crystalline structures of COFs during the polymerization process is a major obstacle in their synthesis. These defects may result from the competing side reactions or incomplete reactions, which may cause disordered or low-crystallinity frameworks. A research paper emphasizes that achieving a high degree of crystallinity is crucial for the material's performance in applications such as gas storage, catalysis, and molecular sieving. Research done in COFs has also shown that these defects can be minimized by optimizing reaction parameters like solvent polarity, temperature, and reaction time, but even then, perfect crystalline COFs are challenging to obtain (Daliran et al. 2023).

After attaining the successful synthesis, solubility, and crystallinity, the hurdles continue in the large-scale production of COFs. COFs have been successfully synthesized in laboratory-scale experiments, but scaling up their production to industrial levels presents a lot of difficulties. The issue of reproducibility arises, as small variations in the synthesis environment or raw materials can result in large differences in the final product. A study by Vardhan et al. states that reproducibility at a larger scale remains a significant challenge. The specific reaction conditions used in COF synthesis may not easily translate to industrial production settings without losing key properties such as structural integrity or functionality (Vardhan et al. 2023).

15.2.2.2 Challenges in Scalability and Reproducibility of 2D COF Synthesis

2D and 3D COFs possess extraordinary crystallinity and permanent porosity. To remove the guest molecules from their pores is quite challenging owing to the extreme capillary forces existing between the host–guest molecules (Roy and Evans 2024).

Furthermore, the economic and energy input methods typically used for COF synthesis, such as solvothermal or microwave-assisted synthesis, can be energy-intensive and costly. These processes may not be sustainable or cost-effective on a large scale, presenting challenges for the commercialization of COFs. Abuzeid et al. suggested that innovative synthesis strategies, like room temperature synthesis or greener solvents, are being explored to overcome these cost and energy barriers (Abuzeid et al. 2021).

The exact mechanism behind the formation of COFs is still not very clear. So, the researchers have to put in so many trials in order to optimize the reaction conditions for synthesizing the desired COF. A deeper understanding of the thermodynamics and kinetics of the formation of COF might be helpful to improve the reliability and predictability of the method of synthesis of COFs. Deeper research is required to gain insights into the formation and seeding of COF at the molecular level, by using sophisticated characterization tools including XRD and electron microscopy (Chen et al. 2025).

The COFs involving imine linkages have been the choice of mechanistic research for a long time, pertaining to the unique and peculiar set of properties associated with them (Xiao et al. 2023). In an attempt to disclose the mechanistic details of the method of formation of imine-based COFs, Smith et al. synthesized imine-linked COF by treating tris(4-aminophenyl)benzene with terephthaldehyde (Smith et al. 2016). They discovered that the crystallization pattern varies for imine-linked and boron-linked COFs. In the imine-linked COF, the material transitions occurred from an amorphous polymeric state with low surface area to a crystalline state through a two-stage process. Initially, a high precipitation rate leads to the formation of an amorphous polymer, which slowly undergoes imine exchange over time liberating the crystalline 2D form of COF under the same experimental conditions (Xiao et al. 2023). In contrast, no such transition was observed for borate COFs; as they directly form crystalline network polymers within minutes, and their crystallinity remains unchanged over time. Water and acetic acid are essential in promoting crystallinity and improving COF yields. These findings offer new theoretical insights and guidance for synthesizing COFs in any combination of solvent systems. Subsequent research also confirmed the rebuilding process underlying the formation of crystalline imine COFs, where they transit from amorphous to crystalline forms (Wang et al. 2017, Gao et al. 2015). Further investigations into the formation mechanisms of COFs led to the proposal of a revised model for imine-linked 2D COFs (Feriante et al. 2020). Using TAPB-PDA-COF again as the model system, it was observed that initially, the imine-based COFs rapidly organized into crystalline 2D nanosheets, which then reorganized over several hours into a piled structure, as demonstrated by recording in situ X-ray diffraction measurements at different reaction times. This new research suggests that the previous theory, which claimed that amorphous polymeric materials could create a porous framework by simultaneous bond cleavage and reformation, may not be universally applicable (Xiao et al. 2023).

15.2.2.3 Performance and Stability

After the synthesis, COFs can exhibit varying levels of stability depending on their chemical composition. For example, some COFs may degrade when exposed to moisture, acids, or bases. This instability limits their practical use in certain applications where environmental resistance is critical (Haase and Lotsch 2020).

The COFs exhibited hydrolytic instability when exposed to humid conditions, which is a key limitation for their use in applications like gas storage or sensing in humid environments (Yang et al. 2022). The documented boron-based COFs possess advantages including low density, excellent crystallinity, extensive surface area, high structural divergence, and thermal stability. But, their relatively low CO_2 adsorption capacity and structural sensitivity under moist conditions render their practical utility in CO_2 capture (Zeng et al. 2016). The introduction of special functional groups into their structure during their formation or postmodification may improve their water sensitivity and their ability in gas adsorption (Sharma et al. 2020, Bunck and Dichtel 2013).

15.2.3 Synthetic Challenges in 3D COFs

Early 3D COFs often relied on B–O linkages, but these linkages are susceptible to hydrolysis and oxidation, which limits their practical use (Ma et al. 2018). For instance, COF-1 and COF-2, which are based on boroxine and boronate ester linkages, respectively, demonstrate poor stability in aqueous environments due to the reversible dehydration reactions. As a result, nucleophilic attack on the boron sites in these linkages leads to decomposition. This has prompted a shift toward the use of more stable nitrogen-based linkages in the design of COFs. Currently, imine linkages, which are formed from multifunctional precursors containing amine and aldehyde groups, are widely used in 3D COFs due to their extraordinary chemical and thermal stability. In particular, imine-linked COFs are known for their excellent water stability, which may be assigned to their electron-rich structural framework (Uribe-Romo et al. 2009). For example, the imine-linked 3D COF-300, despite having limited porosity and low crystallinity, demonstrates improved aqueous stability, aligning with the principles of hydrazine-, triazine-, and azine-linked COFs (Uribe-Romo et al. 2009, Segura et al. 2016). The stability of COFs toward water has been demonstrated to be enhanced by intramolecular H-bonding (Segura et al. 2016). In addition to this, the aqueous stability of the COFs can also be enhanced by modifying the reactive functionalities into nonreactive functionalities like imine to amide, and the latter have been found to possess enhanced stability under acidic as well as basic conditions (Wan et al. 2009). Alternatively, the aqueous stability of the COFs can also be enhanced by uniting the COFs with some waterproof material (Gao et al. 2021). As a case, the conjunction of the COF with GO resulted in the formation of water-stable composites.

15.2.3.1 Complexity of 3D Network Design

The challenges that are associated with the design and synthesis of 3D-COFs are found to be complex and multifaceted in comparison to other categories of COFs. A few of these challenges include the accuracy in monomer unit selection, achieving a long-range order while keeping the COF rigid and flexible. A detailed account of these three key challenges is described below:

One of the key challenges pertaining to the construction of 3D-COFs is the selection of appropriate monomers and maintaining control over the geometry of the resulting 3D framework. Monomers typically utilized in COF synthesis are organic building units with varied functionalities like amines, boronic acids, or aldehydes (Jackson et al. 2012, Du et al. 2016). The dioxin-linked COFs have been generated by condensing catechol with electron-deficient 1,2-difluoroarenes. Nitrogen linkages, including imines, azines, hydrazones, squaraines, imides, and imidazoles, are formed through reactions such as the condensation of aldehydes with amines, hydrazines, hydrazides, squaric acid, anhydrides, and diamines, respectively (Ranjeesh et al. 2019, Das et al. 2014). Phenazine linkages are synthesized by condensing quinones with aromatic 1,2-diamines. The process of creating β-ketoenamine linkages involves reacting amine linkers with a 1,3,5-triformylphloroglucinol unit (Jin et al. 2022).

C=C bond linkages are used to synthesize sp^2 carbon-based COFs (sp^2c-COFs) through Wittig, Knoevenagel, and Horner–Wadsworth-Emmons reactions (Zhang et al. 2018). Due to the absence or poor reversible nature of C=C bond formation, the resultant framework typically exhibits poor porosity and crystallinity.

Even with considerable tuning of reaction conditions, direct polycondensation processes for some monomers do not produce crystalline and porous sp^2c-COFs. To get over this problem, a templated synthesis approach is reported, wherein imine-linked COFs act as templates for the creation of a sequence of sp^2c-COFs, and the imine-linked templates are then hydrolyzed out (Jin et al. 2020). This strategy is versatile and allows the design of sp^2c-COFs with various topologies. In a 3D COF, the geometrical arrangement of the monomers is particularly crucial. The linkers or monomers must be positioned in a way that allows for efficient bonding in three dimensions, which is harder to achieve compared to 2D COFs. For example, some frameworks require monomers with specific bond angles (such as tetrahedral or octahedral coordination) to maintain the desired 3D topology. 3D COFs are generally formed from highly connected building units with specific spatial orientations, such as nodes with tetrahedral, triangular prism, or cubic symmetries (Gropp et al. 2020, Zhu et al. 2023). However, synthesizing polyhedral organic building blocks with high valency (greater than 5) presents a significant challenge, as it relies on the organic chemistry of sp^2 and sp^3 hybridization, which typically results in valencies of 3 and 4 (Gropp et al. 2020). Another approach to constructing 3D COFs involves controlling the conformational flexibility of the linkages to achieve desired topologies (Zhu et al. 2023). This method has been used to create 3D COFs with nbo21, srs22, and rra23 topologies, where rigid linkages such as tetrahedral SiO_4 or BO_4 were designed to form the required networks. Despite these advances, the construction of 3D COFs remains difficult due to the limited availability of flexible linkages. As a result, there is a strong need to further develop 3D COFs and expand the library of building blocks (Qian et al. 2023).

15.2.3.2 Synthesis Challenges Related to Maintaining Long-Range Connectivity in Three Dimensions

In 3D COFs, maintaining **long-range connectivity,** i.e., the continuous bonding of monomers across large distances presents another significant challenge. Unlike 2D

COFs, where the network is flat and easier to maintain, 3D networks need to connect through multiple, spatially distinct directions, which complicates the synthesis process (Lu et al. 2024). The covalent bonds in a 3D COF must extend uniformly in three dimensions, forming a continuous and rigid network throughout the material. Achieving this long-range connectivity without defects is particularly difficult because any disruption (e.g., an incomplete bond or a wrong angle between monomers) could break the framework's continuity, leading to a loss of structural integrity and material performance. This challenge becomes more pronounced when the network is being formed in solvothermal or solvent-assisted conditions, where the monomers may not align properly, or the reaction may produce unwanted side products (Lu et al. 2024).

Recent research highlighted that altering the solvent, catalyst, and temperature can alter the geometry and thus the properties of the synthesized COF. Zhao et al. synthesized two different COF structures in two different solvent combinations, which were found to exhibit different adsorption capacities and chemical stabilities (Kang et al. 2023). In another report, Deng et al. introduced the usage of Lewis acids, such as tris(pentafluorophenyl)borane, to synthesize a wide range of imine-linked COFs, in contrast with traditional methods that use Brønsted acids or inorganic Lewis acids as catalysts (Shi et al. 2022). Wang and coworkers successfully synthesized two COF variants, 2D-PATB-COF and 3D-PATB-COF, through a solvothermal process that differs in fluorescence intensities (Mercado et al. 2018). However, this method is typically limited to imine-linked COFs synthesized under solvothermal conditions, which require high temperatures, large quantities of organic solvents, and prolonged reaction times. As a result, there is a strong demand for simpler and more scalable synthesis methods for COFs to be used as active materials. To address the limitations of the solvothermal method, alternative techniques are being explored for producing polycrystalline COFs (Hu et al. 2023), which typically exist as bulk powders or thin films.

The kinetics of the polymerization process must be finely tuned to allow for the continuous formation of bonds between the monomers while maintaining the desired network structure. In 3D COFs, the polymerization reactions may lead to side reactions or incomplete bond formation, which can disrupt the long-range connectivity. Achieving high **crystallinity** is particularly difficult because the 3D arrangement of monomers is more prone to defects compared to simpler 2D materials (Bisbey and Dichtel 2017).

15.2.3.3 Scaling Up to Industry

Synthetic strategies such as **solvothermal synthesis** or **microwave-assisted methods** are often employed for the synthesis of COF, but **scaling these methods** to industrial production remains challenging, as small fluctuations in the reaction environment can disrupt the desired connectivity. Furthermore, achieving uniformity in large-scale synthesis is difficult because maintaining **consistent temperature** and **solvent conditions** across a larger volume of reactants is often not feasible. The nucleation and framework formation of the COF are mimicked via molecular modeling to gain better insights into the COF structure and energy changes involved in their synthesis. These theoretical studies exclude the need of hit and trials required otherwise for achieving the desired optimal conditions for the construction of the COF (Xiao et al. 2023).

15.2.3.4 Synthesis Challenges Related to Balancing Structural Rigidity with Flexibility

Another major challenge faced while designing 3D-COFs is to create a rigid framework with sufficient flexibility. Rigidity does provide stability and strength to the framework, whereas flexibility is required for enhancing properties like molecular diffusion, gas storage, and catalysis. Keeping the right balance among these two is quite difficult as these two vary with the choice of monomer and covalent framework which is attained. A rigid network ensures the stability of the COF under harsh experimental conditions. Rigidity is usually attained by using rigid linkers involving benzene or naphthalene moieties, which resist the deformation of the framework. The rigidity of the framework is important for applications in gas sensing, gas storage, and molecular sieving. Whereas flexibility permits the COF framework to contract or expand for accommodating guest molecules of varied sizes and thus enhances the COFs ability in metal ion sensing, gas storage, and catalysis. It also enables the COF to change its geometry in response to conditions of temperature and pressure. The flexibility allows COFs to maintain dynamic interactions allowing easy diffusion of gases or guest molecules into the environment. Thus, both aspects of the COFs need to be balanced in the constructed framework. If the COF is too flexible, its structure can be altered easily over passage of time or under extreme conditions of temperature and pressure, and if it is too rigid, it impairs the catalytic and gas-storage properties of the COF. The paper discusses strategies for addressing this challenge, such as the use of mixed-linker approaches, where some linkers are rigid while others are flexible, or the incorporation of soft bonds that can provide some degree of flexibility without compromising stability. One potential solution involves creating hybrid frameworks that incorporate both rigid and flexible components, providing a tunable balance between mechanical strength and adaptability. For example, linkers that are partially flexible or have rotatable groups may allow the framework to adjust to external conditions while still maintaining the overall structural rigidity (Xiao et al. 2023, Zhang et al. 2017). While only a handful of 3D COFs could be successfully synthesized into single crystals, the majority of COFs are typically grown into polycrystalline forms. Thus, the preparation of single-crystal COFs remains challenging due to the nature of polycondensation reactions and crystallization. Most polycondensation reactions do not involve a nucleation process, making it difficult to achieve crystallization. A third mono-functionalized monomer has been added to the reaction system to create imine-linked COFs as single crystals for 3D COFs. This additional component helps repair structural defects, leading to the formation of a well-defined lattice. In contrast, boronate-ester COFs have demonstrated the potential for single-crystal formation upon the addition of monomers in multiple stages during the polymerization process. While the synthesis of single crystals for 2D COFs is still being explored, the dipole effect on linker sites has been successfully employed to produce single crystals of 2D COFs (Jiang 2020).

Crystalline lattices can be targeted by carefully choosing appropriate fundamental monomeric systems. Latest developments in the polymerization and processing of 3D COFs have led to established methods for producing materials with a wide range of linkages (Yang et al. 2023). These approaches often result in highly cross-linked materials with high density and poor solubility. Even with the use of multi-topic linkers which can undergo desired polymerization, it remains challenging to consistently produce crystalline, highly porous materials. The fact that dimensional polymerization occasionally fails to generate long-range ordered crystalline structures has remained a key focus while studying crystalline organic structures. Several theories have been proposed regarding the formation of the amorphous materials in such cases. A few suggest that a disordered framework arises due to poor control over the kinetics of polymerization (Yang et al. 2023, Zhang et al. 2022). Others propose that a random formation of linearly aligned polymers may result in the formation of several unreacted dangling chemical functionalities (Zhu et al. 2021). Another possibility is that the desired crystalline lattice exists but is distorted on a nanoscale level due to interpenetration or interlayer stacking, making the appearance of the material chaotic when analyzed using traditional techniques (60). All these scenarios are likely to be system-dependent and cannot be rectified even with the sophisticated analytical tools (Castano et al. 2021, Robinson et al. 2021).

The discovery of alternative methods for producing a crystalline form of 3D COFs complicates the situation further. As a case, mechanochemical and reconstruction approaches are recognized as effective for producing high-quality materials, although their mechanisms may differ from those used in solution-based condensation syntheses (Zhang et al. 2022). Additionally, new synthetic techniques have enabled the creation of crystalline materials in various forms, including single crystals, polycrystalline materials, monoliths, and thin films (Huang et al. 2023, Guan et al. 2018). While there have been notable advancements in material quality over the past two decades, avoiding amorphous forms by the synthesis process optimization remains largely an empirical challenge. A significant reason for this is the ongoing difficulty in achieving reproducible postsynthetic processing of these materials.

The synthesis of 3D COFs is generally carried out at elevated temperatures, often around 120°C but sometimes as high as 160°C (Fang et al. 2015). This is usually done using the solvothermal method over long periods of time or through quick microwave heating, which provides extra energy to facilitate covalent bond rearrangements and enables the COF building units to attain the necessary connectivity. Although this approach is able to address the crystallization challenge but imposes restrictions on the practical utility of 3D COFs. The prolonged reaction conditions render the commercialization of this methodology, and the elevated temperatures can reduce the compatibility of functional groups and thus limiting their applications. Thus, the development of methods that can quickly produce crystalline materials under milder conditions will be crucial for expanding the potent applications of 3D COFs. Recently, Guan et al. synthesized highly crystalline, imine-based 3D COFs at room temperature in the catalytic presence of an ionic liquid (IL) (Guan et al. 2018). While this ionothermal method shows great potential, several aspects of the approach still require optimization. The 3D COFs produced in the present study trapped up to 12% of IL within their pores, which may have contributed to the small internal surface areas. The current practice of screening polymerizations using powder X-ray diffraction is often biased toward identifying conditions where nucleation is not well-controlled, leading to rapid precipitation and small crystalline domains. There is limited understanding of the optimal polymerization conditions or the necessary error correction rates to improve the quality of materials. This disparity stands in contrast to the more well-established discipline of controlled radical polymerization, where well-defined polymers are produced from irreversible processes by carefully balancing chain growth and deactivation rates.

15.3 EMERGING TRENDS IN COVALENT ORGANIC FRAMEWORKS

15.3.1 Advancements in Structural Features of COFs

Advancements in the structural features of COFs have made them highly versatile for various applications. However, the preparation and development of COFs is a distinct field of research in itself, with many groups focusing exclusively on synthesizing and optimizing these frameworks. Meanwhile, other researchers are dedicated to exploring their use in specific applications, such as detection. The ability to control pore sizes, pore volume, surface area, and functionalization enables the researchers to construct unique COF design for utilization in particular application (Liu et al. 2021). The COFs of specific choice can be generated for utilization in gas storage, selective metal ion absorption, or catalytic applications (Akinnawo 2024). In addition to this, the introduction of functionalities like hydroxy, amino, or carboxylic serves to improve their reactivity further in catalysis, drug delivery, or molecular recognition. Section 15.2 of this chapter highlights the structural improvements and advancements made so far in 1D, 2D, and 3D COFs.

15.3.2 Advancements in the Applicability of COFs

High pore volume and structural stability are two key parameters responsible for the increasing demand of COFs in divergent applications (Figure 15.1). The absence of metal ions in their structure and the extended covalent framework are the reasons behind their utmost stability under strong acidic, basic, or harsh conditions (Bagheri et al. 2025, Abuzeid et al. 2021). That is the reason why COFs are resistant to solvolysis or oxidation under normal conditions. Owing to their large surface area and porosity, they find extensive usage in adsorption of gases like CO_2 and H_2 (Agwupuye et al. 2025). Lu and coworkers explored the potential usage of three polymeric porous COF structures in methane, CO_2, and hydrogen storage (Lu et al. 2010). Out of the three, one showed maximum hydrogen uptake of 4.2 wt%, while the other showed the highest uptake of methane and CO_2 at 77 K. The COF structures have been found to possess high thermal stability up to a temperature of 400°C in addition to their large pore volume and surface area (Zhu et al. 2025). Cote et al. synthesized COF-1 and COF-5 through the condensation of hexahydroxytriphenylene and phenyldiboronic acid. The XRD results of synthesized COFs show extended layers of porous graphite with staggered (COF-1, P63/mmc) and eclipsed (COF-5, P6/mmm) crystal structures (Cote et al. 2005).

The ability to functionalize COFs has significantly broadened their practical applications. Researchers have shown that COFs can be modified with various functional groups, enhancing their electrochemical properties and making them appropriate for energy storage devices like supercapacitors and batteries, where charge–discharge

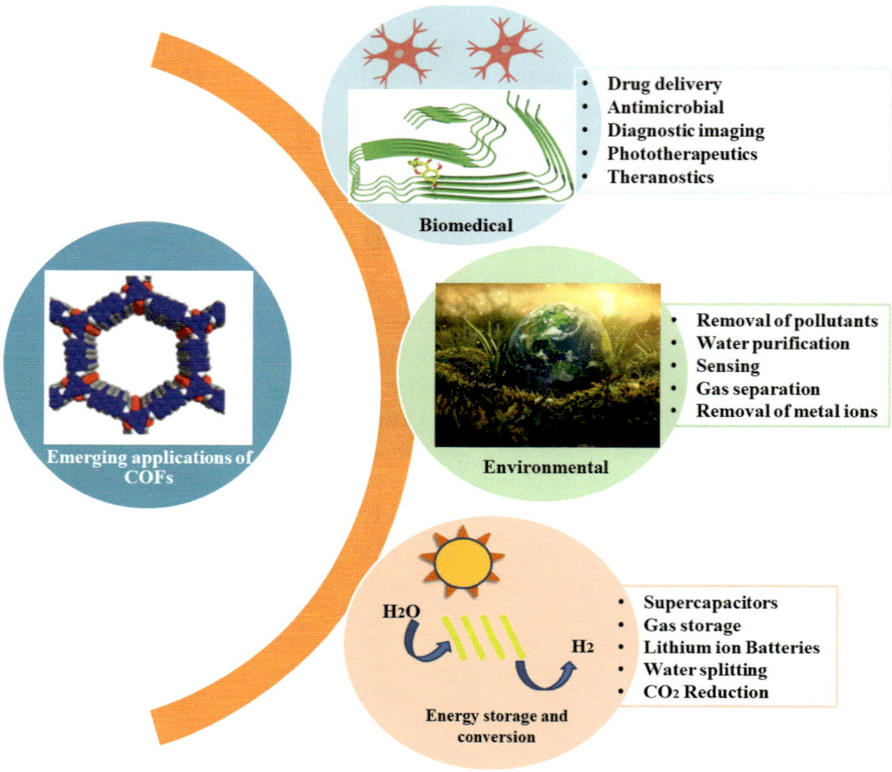

FIGURE 15.1 Emerging applications of covalent organic frameworks.

processes can be used to store energy (Chafiq et al. 2023). In addition, COFs can be engineered to have specific electronic characteristics, such as electron mobility and conductivity, allowing their use in electronic devices like field-effect transistors, photodetectors, and sensors (Yang et al. 2022). Their tunable structures also make them ideal for flexible electronics, where lightweight, flexible devices are essential. Some researchers also discussed the use of functionalized COFs for food sample preparation in detecting contaminants. It covers various strategies, such as modifying COFs with organic groups, Fe_3O_4, MOFs, and molecularly imprinted polymers, to improve extraction selectivity and efficiency. Liu et al. developed flexible, free-standing COF membranes using a liquid–liquid interface-confined reaction at room temperature, demonstrating tunable aperture sizes and high solvent permeances, with imine-linked COFs showing superior nonpolar solvent permeance, making them ideal for precise molecular sieving and membrane-based separation applications (Wang et al. 2023, Liu et al. 2020).

15.3.3 Biomedical Applications of COFs

COFs can be precisely engineered at the molecular level, making them ideal platforms for various biomedical applications, such as drug delivery, phototherapeutics, antimicrobial treatments, and diagnostic imaging (Esrafili et al. 2021, Zhao et al. 2017). These advantages stem from COFs' excellent biocompatibility, high surface area, tunable porosity, and ease of modification, making them promising candidates for advancing medical treatments and therapies.

COFs have shown significant potential in drug delivery and theranostics because of their porous nature and high surface area, which enable efficient encapsulation of therapeutic agents like small molecules, proteins, and nucleic acids (Kaur et al. 2024). This allows for controlled and sustained release of drugs, enhancing therapeutic efficacy while minimizing side effects by targeting specific areas such as tumors or inflamed tissues. As a case, Hassan and coworkers constructed nine COFs and evaluated their potential in drug delivery (Hassan et al. 2024). The authors found that the one loaded with paclitaxel showed maximum proliferation in ovarian cancer cells.

COFs have shown great promise in phototherapeutics, especially in photodynamic therapy (PDT) and photothermal therapy (PTT). The authors reviewed the potential of COFs in cancer phototherapy, highlighting their use in photodynamic and photothermal therapies, and combination treatments inducing immunogenic cell death. They addressed challenges like toxicity and stability, proposing surface modifications of COFs, and outlined future research directions to facilitate their clinical adoption (Singh et al. 2024). In PDT, COFs can be engineered with photoactive components that absorb light and generate reactive oxygen species (ROS), which induce oxidative stress to kill cancer cells. This enables highly targeted therapy, minimizing damage to healthy tissues. In PTT, COFs absorb light and convert it into heat, causing localized hyperthermia to disrupt cancerous tissues. Some COFs are designed to combine both PDT and PTT, enhancing therapeutic efficacy in cancer treatment, making them effective and multifunctional tools for tumor ablation. Researchers created a multifunctional therapeutic agent (H–CeO_2@Ce_6@PDA@GOx NPs) that combines photothermal, photodynamic, chemodynamic, and starvation therapies, significantly enhancing glioblastoma treatment by promoting cell death through synergistic actions without additional drugs (Akdogan et al. 2024).

COFs are also demonstrating potential in combating antibiotic resistance through their antimicrobial properties (Wang et al. 2024). Their high surface area, porous structure, and functionalization versatility enable COFs to interact effectively with bacterial membranes, providing a platform for antimicrobial activity. COFs functionalized with antibiotics or antimicrobial peptides, such as silver-based COFs, have shown the ability to neutralize bacterial pathogens, including multidrug-resistant strains like *Escherichia coli and Staphylococcus aureus* (Bedair et al. 2024). Moreover, COFs can be engineered for the controlled release of the antimicrobial agents, ensuring targeted and sustained delivery to infection sites, minimizing resistance (Mehvari et al. 2024). This makes COFs ideal for wound-healing applications, where they prevent infection while promoting tissue repair. Dong and colleagues encapsulated ibuprofen in a porphyrin-based COF (IBU@DhaTph) using interfacial polymerization, achieving controlled drug release and photodynamic sterilization under visible light. This COF membrane, when used in band-aid dressings, demonstrated significant anti-infectivity and enhanced wound-healing properties (Ding et al. 2021).

COFs have also shown significant promise in biosensing and bioimaging applications due to their tunable porosity and conjugated structures (Karthik et al. 2024). The COFs are able to selectively bind bio-macromolecules including nucleic acids, proteins, and microbes, enabling their use in the detection of pathogens and biomarkers. For instance, in a work by Azzouz et al., porous organic frameworks (POFs) were used to detect the cancer biomarker carcinoembryonic antigen (Azzouz et al. 2025). The conjugation of linkers with COF structure helps them to exhibit fluorescence properties that enable their use in bio-imaging (Karthik et al. 2024). They have been employed for real-time in vivo imaging of cancerous cells. Thus, they are highly promising in the field of precise medicine and drug delivery. Additionally, being metal free they are counted biocompatible and thus find a place in biomedical applications in a long run.

15.3.4 COFs in Environmental Pollution Remediation

COFs are a class of emergent materials which have found immense potential in environmental remediation since their discovery (Paz et al. 2023). Their peculiar properties, including high thermal stability, high porosity, low density, and tunability, enable their usage in the detection and removal of a broad range of pollutants from water, air, and soil (Rakkesh et al. 2024). COFs are able to selectively adsorb a number of toxic effluents on their functionalized surface with high adsorption capacity. COFs functionalized with desired groups helped in achieving the removal of toxic pollutants including gaseous contaminants, heavy metal ions, and organic materials from the environment. The COFs have been reported to selectively adsorb gases like methane and CO_2 owing to their large surface area (Wang et al. 2023), and thus serve in improving the quality of air. The COFs exhibiting free carboxyl and amine groups have been found to remove lead and mercury ions selectively from the aqueous environment. A study by Batool et al. highlighted the role of terephthaldehyde-linked COFs in the adsorptive removal of pharmaceutical compounds like erythromycin, dexamethasone, and azithromycin (Batool et al. 2024). The molecular dynamic studies were employed to establish the H-bonding and Π electronic interactions between the COF and the pharmaceutical entities. COFs with pore-based structures exhibited better pollutant removal capacity, with global minima values at 298 K showing DEG/F-COFs = –665.81, AZM/F-COFs = –638.53, EMC/F-COFs = –566.31, and CMC/F-COFs = –326.75 KJ mol^{-1} (Aksu and Keskin 2024). On a similar note, the COFs have been used in the removal of microplastics and pharmaceutical residue from water (Batool et al. 2024). These divergent examples spotlight the versatility of COF structures in air and water purification and thus environmental remediation.

Ding and coworkers developed a hydrazone-based COF with the incorporation of thioether functionality for the detection and removal of Hg^{2+} ions. COF-LZU8 demonstrated significant fluorescence properties, with a maximum emission at 460 nm. The fluorescence was highly sensitive to the presence of Hg^{2+} ions, showcasing a strong quenching effect upon binding. The fluorescence results showed that detection limit for Hg^{2+} was determined to be 25.0 ppb, indicating high sensitivity and potential for real-time monitoring of mercury levels in different environments (Ding et al. 2016). These hybrid substances improve the conductivity of COFs while maintaining their structural integrity, and thus pave the way toward highly efficient and scalable energy solutions. Huang et al. designed a stable COF with tailored structural features—such as pore size, skeleton, and sulfide-functionalized pore walls—for efficient and selective Hg(II) removal, establishing a benchmark system for water purification (Huang et al. 2017). In another study, the authors developed a COF adsorbent, TFPOTDB-SO_3H, for efficient Pb(II) removal from water, achieving high adsorption capacity, selectivity, and reusability with a 99.75% removal efficiency and excellent performance across multiple cycles (Khosravani et al. 2024).

15.3.5 COFs in Energy Storage and Conversion

COFs are increasingly utilized as electrode materials for supercapacitors and batteries (Wan et al. 2024). A notable example is a novel polyimide COF with *N,N,N,N*-tetraphenylphenylenediamine (TPPDA) units, which was developed for lithium-ion batteries, achieving a high discharge voltage of 3.6 V by utilizing TPPDA as redox-active sites, offering a new strategy for high-voltage organic electrodes in energy storage (Gu et al. 2022). Another innovative system involves COFs doped with redox-active units, such as porphyrins or quinones, which enhance charge storage through faradaic reactions (Wang et al. 2023). Hybrid systems, such as COFs combined with polyaniline (PANI), have also shown promise, with capacitance values exceeding 600 F g^{-1}, owing to the synergistic combination of COFs' porosity and PANI's electrical conductivity. A novel composite of COF and PANI was developed, showing improved performance in supercapacitors with 83% capacity retention after 30,000 cycles, demonstrating potential for faradaic supercapacitors and other electrochemical devices (Liu et al. 2019).

COFs are pivotal in the development of post-lithium energy storage technologies. For instance, Pang and colleagues (2024) introduce a 3D CN-COF material as a promising anode for potassium-ion batteries. By utilizing a high-concentration THF-based electrolyte, the material achieved improved potassium storage performance, including a high reversible capacity of 385.8 mAh g^{-1} at 50 mA g^{-1} and stable cycling with 95.3 mAh g^{-1} after 1,500 cycles at 500 mA g^{-1}. The design optimizes the formation of a stable solid-electrolyte interphase, enhancing cycling stability and reaction kinetics, offering a significant advancement for K-storage materials. In aluminum-ion batteries (AIBs), Lu and coworkers (2020) demonstrate the use of COFs as cathode materials. The COFs enable fast anion diffusion and intercalation without structural collapse, thanks to their robust frameworks and a large surface area of 1,794 m^2 g^{-1} (Lu et al. 2020). The AIBs show impressive performance, with a reversible discharge capacity of 150 mAh g^{-1} after 13,000 cycles at 2 A g^{-1} and a rate capability of 113 mAh g^{-1} at 5 A g^{-1}, highlighting the potential of COFs for storing large-sized ions like chloroaluminate.

COFs have exhibited exceptional efficiency in photocatalytic applications, particularly in water splitting and CO_2 reduction (Yang et al. 2020). For example, two nanoscale COFs (nano-COFs) exhibit a hydrogen evolution rate of 392.0 mmol g^{-1} h^{-1}, one of the highest for organic photocatalysts. Downsizing improves water dispersibility and light-harvesting, with a reverse concentration-dependent photocatalytic effect and molecule-like excitonic properties (Zhao et al. 2024).

Similarly, a Ni-based COF (Ni-TpBpy) efficiently reduces CO_2 to CO with 4,057 µmol g^{-1} CO in 5 hours and 96% selectivity over H_2 evolution. Even at low CO_2 pressure (0.1 atm), CO selectivity remains at 76%. The synergistic effects of single Ni catalytic sites and TpBpy contribute to enhanced CO_2 activation and selective CO production (Zhong et al. 2019). These examples highlight the tunability of COFs for specific photocatalytic energy conversion reactions.

Hybrid systems combining COFs with graphene derivatives or metal oxides have led to significant advancements in energy storage and conversion (Chu et al. 2021). A hybrid hydrogel made from 2D COF and reduced graphene oxide (rGO) nanosheets was developed as an anode material for microbial fuel cells (MFCs) (Liu et al. 2024). This composite enhances extracellular electron transfer by preventing rGO stacking, providing high surface area (313.2 m^2 g^{-1}), mesopores/macropores, and electrical conductivity (492 S m^{-1}). The COF introduces redox-active quinone groups, boosting bacterial growth and biofilm formation. MFCs with COF/rGO anodes achieved a maximum power density of 905.1 mW m^{-2} and operated for 600 hours, also facilitating organic contamination degradation. Another work highlights the potential of 2D nanoarchitectures in improving MFC performance. In another study (Rangappa et al. 2023), the researchers introduce a strategy for integrating TiO_2 nanodots (≈1.82 nm) into COFs using a pore-wall modification and site-specific nucleation approach. The method stabilizes TiO_2 nanodots and promotes their controlled growth within COF pores, improving active site accessibility and charge transfer. The resulting TiO_2-functionalized COFs show enhanced catalytic activity due to strong N–Ti–O bonds. Density functional theory analysis confirms that this modification strengthens interactions between COF and TiO_2, improving energy transfer, and highlighting the potential for COF-metal oxide heterostructures in artificial photosynthesis for carbon fuel production. These hybrid materials leverage the structural advantages of COFs while addressing their conductivity limitations, paving the way for highly efficient and scalable energy solutions.

15.4 CONCLUSION

The development of 3D COFs faces several challenges and issues over the 2D COFs. A bunch of limited covalent linkages, like hydrazone and azines, have been explored so far for synthesizing 3D COFs. Much more advancement has been made in the synthesis and application of 2D COFs in comparison to 3D COFs. One of the main hurdles is that the usage of dynamic covalent bonding, including hydrazones and azines, in the formation of 3D COFs is limited (Dubey et al. 2024). Many research reports on 3D COFs focus mostly on the development and characterization of novel structures, and even their potential applications are still unexplored. The synthesis, modification, and applications of 2D COFs have been extensively studied so far. Diverse COF structures can be effortlessly assembled through the use of dynamic covalent connections, which allow for the formation of numerous structural configurations through reversible bonds. This structural flexibility plays a crucial role in fabricating materials of specific characteristics and functionalities. This reversible property of linkages also provides insights into the mechanism for error-free synthesis by enabling the formation of thermodynamically stable structures instead of kinetically trapped ones. As a result, superior quality of COF materials is produced with more stability. Since many COFs have poor synthetic accessibility because of the entrapment of amorphous networks, the researchers are able to regulate the crystallization process by using dynamic covalent processes. The stability of COF also gets impacted by these dynamic linkages. In contrast to boron-oxygen couplings, which are more susceptible to oxidation and hydrolysis, imine linkages, a form of dynamic linkage, provide better thermal and chemical stability. The ability to include dynamic links creates new opportunities for COFs with special qualities, such as responsive materials. It shows an intriguing field of research in materials science that can change its structure or characteristics in response to an external stimulus. The structure of 3D COFs can be better understood by the fabrication of single crystals. However, the diversity of their frameworks is restricted by a limited number of building units and challenges caused by interpenetration in 3D COFs, which affects both the area of surface and pore volume. Another significant hurdle is crystallization that arises from the requirement to create stable, crystalline frameworks using strong covalent linkages in the fabrication of 3D COFs (Ma et al. 2018). The necessity of reversible linkages to allow for error correction during synthesis is as important as the requirement of stable and crystalline frameworks. Amorphous networks are frequently kinetically trapped by the bulk structure and rigid nature of monomer precursors, which prevents them from getting crystallized. 3D COFs rely entirely on covalent interactions, which are more challenging to control, in contrast to 2D COFs, where reaction sites are readily accessible and van der Waals interactions support layered structures. Regardless of such difficulties, the formation of stoichiometric amounts of water during the condensation reaction of boronic acid and imine attains the reaction equilibrium in 3D COF synthesis. To overcome some crystallization difficulties, the use of water-immiscible solvents or modification of pressure in the reaction vessel can be accomplished (Ma et al. 2018). Kinetic entrapment can be minimized by optimizing reaction conditions along with controlling monomer concentration. Common solvothermal methods for COF synthesis operate at high temperatures, which may limit the compatibility of functional groups and slow down reaction rates. To expand the potential applications of 3D COFs, milder reaction conditions and better catalysts, such as metal triflates, are needed. Additionally, the use of ILs in room temperature synthesis of 3D COFs may affect the internal surface area (Kumar et al. 2021).

15.5 FUTURE PROSPECTS

There is a significant opportunity to synthesize 2D and 3D COFs with desired structural features by wisely selecting appropriate monomers with correct geometry and sufficient functionality or by precise control over temperature, solvent choice, and reaction time. Any variation can lead to the formation of poorly defined frameworks or amorphous structures as ensuring the correct geometry and orientation of these monomers during synthesis is quite challenging. There is a great opportunity to investigate strategies for controlling the folding and interpenetration of COFs, which could enable the predesign of COFs, much like the progress made in MOFs. This can be achieved with the help of molecular modeling, which can really help access the structure–activity relationship for COFs. Understanding the kinetics and thermodynamics of polymerization and crystallization processes is further crucial to further explore the applicability of COFs and to design COFs specifically for a particular application.

Nearly every application of COFs, including adsorption, photocatalysis, drug delivery, and environment remediation, is related to their large surface area, high porosity, and high pore volume, or more simply to their structural framework. Another major challenge is the absence of a universal method for producing high-quality crystallites or single crystals of COFs. Future research should prioritize exploring the crystallization mechanisms, as this is essential for improving the quality of COF materials. An extensive work on the mechanistic understanding of crystallization of COF is desired to track the reaction pathways and gain better insight into how the COFs form and evolve at the molecular level. Gathering data on typical COF systems will deepen our understanding of how these factors influence the final properties of COFs. Investigating defect formation in COFs is also critical for linking their structures to their functions and properties. While current research focuses mainly on polycrystalline COFs, more emphasis should be placed on single crystals to better understand the role of defects. These defects mostly arise because of incomplete linkages, leading to frameworks that are disordered or have low crystallinity. Future work should also center on designing and engineering interfaces within COFs, as this could reveal novel functions unique to these materials, opening up new applications.

Another important aspect to look at is scalability of the use of COFs in the industrial level. The research work in laboratories is aimed at enhancing the structural and functional aspects of COFs either by improving their crystallinity by altering the growth parameters or by doping any additional material into them. Only a few of them have accessed their reliability into a particular application on an industrial scale.

From a Physics standpoint, it is important to comprehend interactions at different spatiotemporal scales. Real-time observations and time-resolved experiments could offer insights into how ordered structures interact with photons, excitons, and other particles, which is key to exploring functional properties. COFs have promising applications in fields such as adsorption, separation, storage, and catalysis. However, further research is needed to understand the unique properties that set COFs apart, ensuring their full potential is realized in practical applications. These perspectives underscore the exciting future possibilities for COFs, highlighting the need for interdisciplinary collaboration and innovative strategies to address current challenges. The future prospects of COFs in energy storage applications are promising and multifaceted. COFs are gaining attention in diverse batteries beyond traditional lithium-ion batteries, with potential applications in sodium-ion batteries, lithium-sulfur batteries, and even multivalent ion, metal-oxygen, and metal-halogen batteries. This growing interest highlights the versatility of COFs as materials for energy storage solutions. Their metal-free composition makes them sustainable, and their stability under both acidic and alkaline conditions enhances their suitability for long-term energy storage applications.

The inherent flexibility of COFs also makes them promising for developing flexible energy storage devices, which are crucial for applications in wearable electronics and flexible displays. By engineering COFs at the molecular level, their properties can be tailored, making them ideal candidates for thin and flexible electrode materials. A significant challenge, however, is their low conductivity. Ongoing research aims to improve conductivity by enhancing the interacting areas of COFs and introducing fused π-systems, which is essential for better performance in energy storage applications. Combining COFs with conductive polymers, such as PEDOT or PANI, can further improve their conductivity and mechanical stability, which is vital for creating efficient, flexible energy storage devices. Advancements in COFs for energy storage are poised to make a substantial impact on clean energy technologies, aligning with global sustainability goals for affordable and reliable energy access.

REFERENCES

Abuzeid, H. R.; EL-Mahdy, A. F. M.; Kuo, S.-W. 2021. Covalent Organic Frameworks: Design Principles, Synthetic Strategies, and Diverse Applications. *Giant* 6:100054.

Agwupuye, J. A.; Romanus, E. N.; Ihedioha, J. N.; Amodu, I. O.; Ekpong, B. O.; Charlie, D. E.; Umo, F. E.; Agwupuye, M. C.; Zeeshan, M. 2025. Engineering of Covalent Organic Framework (COF) via Mono-Doping and Co-Doping for the detection of CO_2 Gas Pollutant. *Next Mater.* 7:100460.

Akdogan, Ç. Z. S.; Çetin, E. A.; Onur, M. A.; Önel, S.; Tuncel, A. 2024. In Vitro Synergistic Photodynamic, Photothermal, Chemodynamic, and Starvation Therapy Performance of Chlorine 6 Immobilized, Polydopamine-Coated Hollow, Porous Ceria-Based, Hypoxia-Tolerant Nanozymes Carrying a cascade system. *ACS Appl. Bio Mater.* 7(5):2781–2793.

Akinnawo, S. O. 2024. Covalent organic frameworks: Design, synthesis, characterization, and applications. *ChemPhysMater* 3(1):36–63.

Aksu, G. O.; Keskin, S. 2024. Rapid and Accurate Screening of the COF Space for Natural Gas Purification: COFInformatics. *ACS Appl. Mater. Interfaces* 16(15):19806–19818.

Ampong, D. N.; Effah, E.; Tsiwah, E. A.; Kumar, A.; Agyekum, E.; Doku, E. N. A.; Issaka, O.; Agyemang, F. O.; Darkwa, K. M.; Gupta, R. K. 2024. Advances and challenges in covalent organic frameworks as an emerging class of materials for energy and environmental concerns. *Coord. Chem. Rev.* 519:216121.

Ascherl, L.; Sick, T.; Margraf, J. T.; Lapidus, S. H.; Calik, M.; Hettstedt, C.; Karaghiosoff, K.; Dö blinger, M.; Clark, T.; Chapman, K. W.; Auras, F.; Bein, T. 2016. Molecular docking sites designed for the generation of highly crystalline covalent organic frameworks. *Nat. Chem.* 8:310–316.

Azzouz, A.; Hejji, L.; Kim, K.-H. 2025. Recent advances in porous organic framework-based aptasensors for diagnosis of cancer diseases. *TrAC Trends in Anal. Chem.* 183:118106.

Bagheri, A. R.; Su, H.; Shokrollahi, A.; Samadi, A.; Song, C.; Kong, L.; Yi, S.; Zhao, S. 2025. Covalent organic framework-based materials for ion separation: A review. *Chem. Eng. J.* 504:158745.

Baldwin, L. A.; Crowe, J. W.; Pyles, D. A.; McGrier, P. L. 2016. Metalation of a mesoporous three-dimensional covalent organic framework. *J. Am. Chem. Soc.* 138:15134–15137.

Batool, R.; Sahrish; Ahad, S. A.; Maqsood, Q.; Ali, S. W.; Abbas, S. M. 2024. Redefining Sustainability: Next-gen wastewater treatment breakthroughs. *Clean. Water* 1:100018.

Bedair, H. M.; Hamed, M.; Mansour, F. R. 2024. New emerging materials with potential antibacterial activities. *Appl. Microbiol. Biotechnol.* 108:515.

Bi, S.; Meng, F.; Wu, D.; Zhang, F. 2022. Synthesis of vinylene-linked covalent organic frameworks by monomer self-catalyzed activation of Knoevenagel condensation. *J. Am. Chem. Soc.* 144:3653–3659.

Bisbey, R. P.; Dichtel, W. R. 2017. Covalent organic frameworks as a platform for multidimensional polymerization. *ACS Cent. Sci.* 3:533–543.

Bunck, D. N.; Dichtel, W. R. 2013. Postsynthetic functionalization of 3D covalent organic frameworks. *Chem. Comm.* 49:2457–2459.

Cao, S.; Li, B.; Zhu, R.; Pang, H. 2019. Design and synthesis of covalent organic frameworks towards energy and environment fields. *J. Chem. Eng.* 355:602–623.

Castano, I.; Evans, A. M.; Reis, R. D.; Dravid, V. P.; Gianneschi, N. C.; Dichtel, W. R. 2021. Mapping grains, boundaries, and defects in 2D covalent organic framework thin films. *Chem. Mater.* 33:1341–1352.

Chafiq, M.; Chaouiki, A.; Ko, Y. G. 2023. Advances in COFs for energy storage devices: Harnessing the potential of covalent organic framework materials. *Energy Storage Mater.* 63:103014.

Chen, X.; Addicoat, M.; Irle, S.; Nagai, A.; Jiang, D. 2013. Control of crystallinity and porosity of covalent organic frameworks by managing interlayer interactions based on self-complementary π-electronic force. *J. Am. Chem. Soc.* 135:546–549.

Chen, Y.; Shi, Y.; Song, G.; Yang, B.; Pang, H. 2025. Compositing MXene with organic coordination frameworks (MOFs, COFs, and HOFs) for enhanced electrochemical performance. *Next mater.* 6:100323.

Chu, J.; Wang, Y.; Zhong, F.; Feng, X.; Chen, W.; Ai, X.; Yang, H.; Cao, Y. 2021. Metal/covalent-organic frameworks for electrochemical energy storage applications. *EcoMat* 3(5):e12133.

Cote, A. P.; Benin, A. I.; Ockwig, N. W.; Keeffe, M. O.; Matzger, A. J.; Yaghi, O. M. 2005. Porous, crystalline, covalent organic frameworks. *Science* 310:1166–1170.

Dalapati, S.; Jin, E.; Addicoat, M.; Heine, T.; Jiang, D. 2016. Highly emissive covalent organic frameworks. *J. Am. Chem. Soc.* 138:5797–5800.

Daliran, S.; Blanco, M.; Dhakshinamoorthy, A.; Oveisi, A. R.; Alemán, J.; García, H. 2023. Defects and disorder in covalent organic frameworks for advanced applications. *Ad. Funct. Mater.* 34:2312912.

Das, G.; Balaji Shinde, D.; Kandambeth, S.; Biswal, B. P.; Banerjee, R. 2014. Mechanosynthesis of imine, β-ketoenamine, and hydrogen-bonded imine-linked covalent organic frameworks using liquid-assisted grinding. *Chem. Comm.* 50:2615–12618.

DeBlase, C. R.; Dichtel, W. R. 2016. Moving beyond boron: The emergence of new linkage chemistries in covalent organic frameworks. *Macromol.* 49:5297–5305.

Diercks, C. S.; Yaghi, O. M. 2017. The atom, the molecule, and the covalent organic framework. *Science* 355:6328.

Ding, L.-G.; Wang, S.; Yao, B.-J.; Li, F.; Li, Y.-A.; Zhao, G.-Y.; Dong, Y.-B. 2021. Synergistic antibacterial and anti-inflammatory effects of a drug-loaded self-standing porphyrin-COF membrane for efficient skin wound healing. *Adv. Healthcare Mater.* 10:2001821.

Ding, S. Y.; Wang, W. 2013. Covalent organic frameworks (COFs): From design to applications. *Chem. Soc. Rev.* 42:548–568.

Ding, S.; Gao, J.; Wang, Q.; Zhang, Y.; Song, W.; Su, C.; Wang, W. 2011. Construction of covalent organic framework for catalysis: Pd/COF-LZU1 in Suzuki-Miyaura coupling reaction. *J. Am. Chem. Soc.* 133:19816–19822.

Ding, S.-Y.; Dong, M.; Wang, Y.-W.; Chen, Y.-T.; Wang, H.-Z.; Su, C.-Y.; Wang, W. 2016. Thioether-based fluorescent covalent organic framework for selective detection and facile removal of mercury(II). *J. Am. Chem. Soc.* 138(9):3031–3037.

Dou, L.; Zheng, Y.; Shen, X.; Wu, G.; Fields, K.; Hsu, W.C.; Zhou, H.; Yang, Y.; Wudl, F. 2014. Single-crystal linear polymers through visible light-triggered topochemical quantitative polymerization. *Science* 343:272–277.

Du, Y.; Yang, H.; Whiteley, J. M.; Wan, S.; Jin, Y.; Lee, S. H.; Zhang, W. 2016. Ionic covalent organic frameworks with spiroborate linkage. *Angew. Chem. Int. Ed.* 55:1737–1741.

Dubey, P.; Shrivastav, V.; Boruah, T.; Zoppellaro, G.; Zbořil, R.; Bakandritsos, A.; Sundriyal, S. 2024. Unveiling the potential of covalent organic frameworks for energy storage: Developments, challenges, and future prospects. *Adv. Energy Mater.* 14:2400521.

El-Kaderi, H. M.; Hunt, J. R.; Mendoza-Cortés, J. L.; Côté, A. P.; Taylor, R. E.; Keeffe, M. O.; Yaghi, O. M. 2007. Designed synthesis of 3D covalent organic frameworks. *Science* 316: 268–72.

Esrafili, A.; Wagner, A.; Inamdar, S.; Acharya, A. P. 2021. Covalent organic frameworks for biomedical applications. *Adv.Healthcare Mater.* 10(20):2002090.

Evans, A. M.; Parent, L. R.; Flanders, N. C.; Bisbey, R. P.; Vitaku, E.; Kirschner, M. S.; Schaller, R. D.; Chen, L. X.; Gianneschi, N. C.; Dichtel, W. R. 2018. Seeded growth of single-crystal two-dimensional covalent organic frameworks. *Science* 361:52–57.

Fang, Q.; Wang, J.; Gu, S.; Kaspar, R. B.; Zhuang, Z.; Zheng, J.; Guo, H.; Qiu, S.; Yan, Y. 2015. 3D porous crystalline polyimide covalent organic frameworks for drug delivery. *J. Am. Chem. Soc.* 137:8352–8355.

Feriante, C.; Evans, A. M.; Jhulki, S.; Castano, I.; Strauss, M. J.; Barlow, S.; Dichtel, W. R.; Marder, S. R. 2020. New mechanistic insights into the formation of imine-linked two-dimensional covalent organic frameworks. *J. Am. Chem. Soc.* 142:18637–18644.

Furukawa, H.; Yaghi, O. M. 2009. Storage of hydrogen, methane, and carbon dioxide in highly porous covalent organic frameworks for clean energy applications. *J. Am. Chem. Soc.* 131:8875–8883.

Gao, Q.; Bai, L.; Zeng, Y.; Wang, P.; Zhang, X.; Zou, R.; Zhao, Y. 2015. Reconstruction of covalent organic frameworks by dynamic equilibrium. *Chemistry* 21:16818–16822.

Gao, W.; Cheng, J.; Yuan, X.; Tian, Y. 2021. Covalent organic framework-graphene oxide composite: A superior adsorption material for solid phase microextraction of bisphenol A. *Talanta* 222:121501.

Gropp, C.; Ma, T.; Hanikel, N.; Yaghi, O. M. 2020. Design of higher valency in covalent organic frameworks. *Science* 370:eabd6406.

Gu, S.; Hao, R.; Chen, J.; *et al.*2022. A star-shaped polyimide covalent organic framework for high-voltage lithium-ion batteries. *Mater. Chem. Front.* 6(17):2545–2550.

Guan, X.; Ma, Y.; Li, H.; Yusran, Y.; Xue, M.; Fang, Q.; Yan, Y.; Valtchev, V.; Qiu, S. 2018. Fast, ambient temperature and pressure ionothermal synthesis of three-dimensional covalent organic frameworks. *J. Am. Chem. Soc.* 140:4494–4498.

Haase, F.; Lotsch, B. V. 2020. Solving the COF trilemma: Towards crystalline, stable and functional covalent organic frameworks. *Chem. Soc. Rev.* 49:8469–8500.

Han, X.; Zhang, W.; Chen, Z.; Liu, Y.; Cui, Y. 2023. The future of metal–organic frameworks and covalent organic frameworks: Rational synthesis and customized applications. *Material Horizon* 10:5337–5342.

Hassan, A.; Roy, S.; Das, A.; Wahed, S. A.; Bairagi, A.; Mondal, S.; Chatterjee, N.; Das, N. 2024. Covalent organic frameworks as potential drug carriers and chemotherapeutic agents for ovarian cancers. *ACS Biomater. Sci. Eng.* 10(7):4227–4236.

Hu, J.; Gupta, S. K.; Ozdemir, J.; Beyzavi, H. 2020. Applications of dynamic covalent chemistry concept toward tailored covalent organic framework nanomaterials: A review. *ACS Appl. Nano Mater.* 3:6239.

Hu, J.; Huang, Z.; Liu, Y. 2023. Beyond solvothermal: Alternative synthetic methods for covalent organic frameworks. *Angew. Chem. Int. Ed.* 62:e202306999.

Huang, L.; Yang, J.; Zhao, Y.; Miyata, H.; Han, M.; Shuai, Q.; Yamauchi, Y. 2023. Monolithic covalent organic frameworks with hierarchical architecture: attractive platform for contaminant remediation. *Chem. Mater.* 35:2661–2682.

Huang, N.; Wang, P.; Jiang, D. 2016. Covalent organic frameworks: A materials platform for structural and functional designs. *Nature Rev. Mater.* 1:16068.

Huang, N.; Zhai, L.; Xu, H.; Jiang, D. 2017. Stable covalent organic frameworks for exceptional mercury removal from aqueous solutions. *J. Am. Chem. Soc.* 139:2428.

Jackson, K. T.; Reich, T. E.; El-Kaderi, H. M. 2012. Targeted synthesis of a porous borazine-linked covalent organic framework. *Chem. Comm.* 48:8823–8825.

Jiang, D. 2020. Covalent organic frameworks: An amazing chemistry platform for designing polymers. *Chem* 6:2461–2483.

Jiang, S. Y.; Gan, S. X.; Zhang, X.; Li, H.; Qi, Q. Y.; Cui, F. Z.; Lu, J.; Zhao, X. 2019. Aminal-linked covalent organic frameworks through condensation of secondary amine with aldehyde. *J. Am. Chem. Soc.* 141:14981–14986.

Jin, E.; Geng, K.; Lee, K. H.; Jiang, W.; Li, J.; Jiang, Q.; Irle, S.; Jiang, D. 2020. Topology-templated synthesis of crystalline porous covalent organic frameworks. *Angew. Chem. Int. Ed.* 59:12162–12169.

Jin, F.; Lin, E.; Wang, T.; Geng, S.; Wang, T.; Liu, W.; Xiong, F.; Wang, Z.; Chen, Y.; Cheng, P.; Zhang, Z. 2022. Bottom-up synthesis of 8-connected three-dimensional covalent organic frameworks for highly efficient ethylene/ethane separation. *J. Am. Chem. Soc.* 144:5643–5652.

Kang, C.; Zhang, Z.; Kusaka, S.; Negita, K.; Usadi, A. K.; Calabro, D. C.; Baugh, L. S.; Wang, Y.; Zou, X.; Huang, Z.; Matsuda, R.; Zhao, D. 2023. Covalent organic framework atropisomers with multiple gas-triggered structural flexibilities. *Nature Mater.* 22:636–643.

Karthik, C. S.; Skorjanc, T.; Shetty, D. 2024. Fluorescent covalent organic frameworks – Promising bioimaging materials. *Mater. Horiz.* 11:2077–2094.

Kaur, H.; Siwal, S. S.; Saini, R. V.; Thakur, V. K. 2024. Covalent–organic framework-based materials in theranostic applications: Insights into their advantages and challenges. *ACS Omega* 9(6):6235–6252.

Keller, N.; Bein, T. 2021. Optoelectronic processes in covalent organic frameworks. *Chem. Soc. Rev.* 50:1813–1845.

Khosravani, M.; Ghanatghestani, M. D.; Moeinpour, F.; Parvaresh, H. 2024. Efficient lead removal from aqueous solutions using a new sulfonated covalent organic framework: Synthesis, characterization, and adsorption performance. *Arab. J. Chem.* 17(1):105429.

Kissel, P.; Erni, R.; Schweizer, W. B.; Rossell, M. D.; King, B. T.; Bauer, T.; Götzinger, S.; Schlüter, A.D.; Sakamoto, J. 2012. A two-dimensional polymer prepared by organic synthesis. *Nature Chem.* 4:287–291.

Kumar, S.; Ignacz, G.; Szekely, G. 2021. Synthesis of covalent organic frameworks using sustainable solvents and machine learning. *Green Chem.* 23:8932–8939.

Li, Z.; Zhi, Y.; Feng, X.; Ding, X.; Zou, Y.; Liu, X.; Mu, Y. 2015. An azine-linked covalent organic framework: synthesis, characterization and efficient gas storage. *Chemistry* 21:12079.

Liu, F.; Yang, Y.; Xu, M.; Chen, Y.; Li, W. 2024. 2D covalent organic framework/reduced graphene oxide hybrid hydrogel for simultaneous power generation and organic contamination degradation in microbial fuel cells. *Mater. Today Energy* 39:101464.

Liu, J.; Han, G.; Zhao, D.; Lu, K.; Gao, J.; Chung, T.-S. 2020. Self-standing and flexible covalent organic framework (COF) membranes for molecular separation. *Sci. Adv.* 6:41.

Liu, S.; Yao, L.; Lu, Y.; Hua, X.; Liu, J.; Yang, Z.; Wei, H.; Mai, Y. 2019. All-organic covalent organic framework/polyaniline composites as stable electrodes for high-performance supercapacitors. *Mater. Lett.* 236:354–357.

Liu, X.; Pang, H.; Liu, X.; Li, Q.; Zhang, N.; Mao, L.; Qiu, M.; Hu, B.; Yang, H.; Wang, X. 2021. Orderly porous covalent organic frameworks-based materials: Superior adsorbents for pollutants removal from aqueous solutions. *The Innovation* 2(1):100076.

Liu, Y.; Ma, Y.; Zhao, Y.; Sun, X.; Gándara, F.; Furukawa, H.; Liu, Z.; Zhu, H.; Zhu, C.; Suenaga, K. 2016. Weaving of organic threads into a crystalline covalent organic framework. *Science* 351:365–369.

Lu, H.; Ning, F.; Jin, R.; Teng, C.; Wang, Y.; Xi, K.; Zhou, D.; Xue, G. 2020. Two-dimensional covalent organic frameworks with enhanced aluminum storage properties. *ChemSusChem* 13:3447.

Lu, M.; Zhang, S. B.; Li, R. H.; Dong, L. Z.; Yang, M. Y.; Huang, P.; Liu, Y. F.; Li, Z. H.; Zhang, H.; Zhang, M. S. L.; Lan, Y. Q. 2024. 3D covalent organic frameworks with 16-connectivity for photocatalytic $C(sp^3)$–$C(sp^2)$ cross-coupling. *J. Am. Chem. Soc.*146:25832–25840.

Lu, W.; Yuan, D.; Zhao, D.; Schilling, C. I.; Plietzsch, O.; Muller, T.; Bräse, S.; Guenther, J.; Blümel, J.; Krishna, R.; Li, Z.; Zhou, H.-C. 2010. Porous polymer networks: Synthesis, porosity, and applications in gas storage/separation. *Chem. Mater.* 22(21):5964–5972.

Lyle, S. J.; Waller, P. J.; Yaghi, O. M. 2019. Covalent organic frameworks: Organic chemistry extended into two and three dimensions. *Trends Chem.* 1:172–184.

Ma, T. A.; Kapustin, E. A.; Yin, S. X.; Liang, L.; Zhou, Z.; Niu, J.; Wang, Y.; Su, J.; Li, J.; Wang, X.; Wang, W. D.; Wang, W.; Sun, J.; Yaghi, O. M. 2018. Single-crystal x-ray diffraction structures of covalent organic frameworks. *Science* 361:48–52.

Ma, X.; Scott, T.F. 2018. Approaches and challenges in the synthesis of three-dimensional covalent-organic frameworks. *Comm. Chem.* 1:98.

Mandal, A. K.; Mahmood, J.; Baek, J.-B. 2017. Two-dimensional covalent organic frameworks for optoelectronics and energy storage. *ChemNanoMat* 3:373–391.

Mehvari, F.; Ramezanzade, V.; Asadi, P.; Singh, N.; Kim, J.; Dinari, M.; Kim, J. S. 2024. A panoramic perspective of recent progress in 2D and 3D covalent organic frameworks for drug delivery. *Aggregate* 5(2): e480.

Mercado, R.; Fu, R. S.; Yakutovich, A. V.; Talirz, L.; Haranczyk, M.; Smit, B. 2018. In silico design of 2D and 3D covalent organic frameworks for methane storage applications. *Chem. Mater.* 30:5069–5086.

Mohan, B.; Kumari, R.; Virender; Singh, G.; Singh, K.; Pombeiro, A. J. L.; Yang, X.; Ren, P. 2023. Covalent organic frameworks (COFs) and metal–organic frameworks (MOFs) as electrochemical sensors for the efficient detection of pharmaceutical residues. *Environ. Int.* 175:107928.

Nguyen, H. L. 2021. Reticular design and crystal structure determination of covalent organic frameworks. *Chem. Sci.* 12:8632–8647.

Pang, Y.; Lai, Q.; Xia, H.; Zhang, W.; Chen, H.; Chen, R.; Lin, Z.; Zheng, J. 2024. Covalent-organic-framework enabled efficient three-dimensional K-storage via electrolyte solvation manipulation. *ACS Appl. Mater. & Interfaces* 16:70606–70617.

Paz, R.; Viltres, H.; Gupta, N. K.; Phung, V.; Srinivasan, S.; Rajabzadeh, A. R.; Leyva, C. 2023. Covalent organic frameworks as highly versatile materials for the removal and electrochemical sensing of organic pollutants. *Chemosphere* 342:140145.

Qian, C.; Feng, L.; Teo, W. L.; Liu, J.; Zhou, W.; Wang, D.; Zhao, Y. 2022. Imine and imine-derived linkages in two-dimensional covalent organic frameworks. *Nature Rev. Chem.* 6:881–898.

Qian, C.; Teo, W. L.; Gao, Q.; Wu, H.; Liao, Y.; Zhao, Y. 2023. Polycrystalline covalent organic frameworks. *Mater. Today* 71:91–107.

Raghav, S., Raghav, J., Muhammad, M.A., Singh, M.B., Jain, P. & Kumar, D. 2025. Recent advancements in synthesis and applications of covalent organic frameworks. *Applied Materials Today*, 44(June): 102694.

Rakkesh, R. A.; Naveen, T. B.; Durgalakshmi, D.; Balakumar, S. 2024. Covalent organic frameworks: Pioneering remediation solutions for organic pollutants. *Chemosphere* 346:140655.

Rangappa, A. P.; Kumar, D. P.; Do, K. H.; Wang, J.; Zhang, Y.; Kim, T. K. 2023. Synthesis of Pore-Wall-Modified Stable COF/TiO_2 Heterostructures via site-specific Nucleation for an enhanced Photoreduction of Carbon Dioxide. *Adv. Sci.* 10:2300073.

Ranjeesh, K. C.; Illathvalappil, R.; Veer, S. D.; Peter, J.; Wakchaure, V. C.; Goudappagouda, K.; Raj, K. V.; Kurungot, S.; Babu, S. S.; 2019. Imidazole-linked crystalline two-dimensional polymer with ultrahigh proton conductivity. *J. Am. Chem. Soc.* 141:14950–14954.

Ren, J.; Ji, C.; Du, B.; Liu, Q.; Yu, K.; Ahn, D.; Zhang, Z.; Ye, Y.; Göb, R.; Zhao, D. 2024. A fully saturated covalent organic framework. *J. Am. Chem. Soc.* 146: 30784–30789.

Robison, L.; Gong, X.; Evans, A. M.; Son, F. A.; Wang, X.; Redfern, L. R.; Wasson, M. C.; Syed, Z. H.; Chen, Z.; Idrees, K. B.; Islamoglu, T.; Delferro, M.; Dichtel, W. R.; Coudert, F. X.; Gianneschi, N. C.; Farha, O. K. 2021. Transient catenation in a zirconium-based metalorganic framework and its effect on mechanical stability and sorption properties. *J. Am. Chem. Soc.*143:1503–1512.

Roy, R.; Evans, A. M. 2024. Reproducibility challenges in activating two-dimensional polymers and three-dimensional covalent organic frameworks. *Comm. Mater.* 5:102.

Royuela, S.; Almarza, J.; Mancheño, M.J.; Pérez-Flores, J. C.; Michel, E.G.; Ramos, M.M.; Zamora, F.; Ocón, P.; Segura, J. L. 2019. Synergistic effect of covalent bonding and physical encapsulation of sulfur in the pores of a microporous COF to improve cycling performance in Li–S batteries. *Chem. Eur. J.* 25:12394–12404.

Segura, J. L.; Mancheño, M. J.; Zamora, M. F. 2016. Covalent organic frameworks based on Schiff-base chemistry: Synthesis, properties and potential applications. *Chem. Soc. Rev.* 45:5635–5671.

Sharma, R. K.; Yadav, P.; Yadav, M.; Gupta, R.; Rana, P.; Srivastava, A.; Zbořil, R.; Varma, R. S.; Antonietti, M.; Gawande, M. B. 2020. Recent development of covalent organic frameworks (COFs): Synthesis and catalytic (organic-electro-photo) applications. *Materials Horizons* 7:411–454.

Shi, X.; Yi, L.; Deng, H. 2022. Covalent organic frameworks catalyzed by organic Lewis acid. *Sci. China Chem.* 65:1315–1320.

Singh, N.; Won, M.; An, J.; Yoon, C.; Kim, D.; Lee, S. J.; Kang, H.; Kim, J. S. 2024. Advances in covalent organic frameworks for cancer phototherapy. *Coord. Chem. Rev.* 506:215720.

Smith, B. J.; Overholts, A. C.; Hwang, N.; Dichtel, W. R. 2016. Insight into the crystallization of amorphous imine-linked polymer networks to 2D covalent organic frameworks. *Chem. Comm.* 52:3690–3693.

Uribe-Romo, F. J.; Hunt, J. R.; Furukawa, H.; Klöck, C.; O'Keeffe, M.; Yaghi, O. M. 2009. A crystalline imine-linked 3-D porous covalent organic framework. *J. Am. Chem. Soc.* 131:4570–4571.

Vardhan, H.; Rummer, G.; Deng, A.; Ma, S. 2023. Large-scale synthesis of covalent organic frameworks: Challenges and opportunities. *Membranes* 13:696.

Waller, P. J.; Gándara, F; Yaghi, O. M. 2015. Chemistry of covalent organic frameworks. *Acc. Chem. Res.* 48:3053–3063.

Wan, S.; Guo, J.; Kim, J.; Ihee, H.; Jiang, D. 2009. A photoconductive covalent organic framework: Self-condensed arene cubes composed of eclipsed 2D polypyrene sheets for photocurrent generation. *Angew. Chem. Int. Ed.* 48:5439–5442.

Wan, Y.; Shen, B.; Zhu, X.; Guo, Z. 2024. Application of covalent organic frameworks as electrode materials for supercapacitors. *ChemPhysMater* 3(4):388–414.

Wang, H.; He, B.; Liu, F.; Stevens, C.; Brady, M. A.; Cai, S.; Wang, C.; Russell, T. P.; Tan, T. W.; Liu, Y. 2017. Orientation transitions during the growth of imine covalent organic framework thin films. *J. Mater. Chem. C* 5:5090–5095.

Wang, J.; Feng, J.; Lian, Y.; Sun, X.; Wang, M.; Sun, M. 2023a. Advances of the functionalized covalent organic frameworks for sample preparation in the food field. *Food Chem.* 405:134818.

Wang, K.; Kang, X.; Yuan, C.; Han, X.; Liu, Y.; Cui, Y. 2021. Porous 2D and 3D covalent organic frameworks with dimensionality-dependent photocatalytic activity in promoting radical ring-opening polymerization. *Angew. Chem. Int. Ed.* 60:19466–19476.

Wang, S.; Chen, Z.; Cai, Y.; Wu, X.-L.; Wang, S.; Tang, Z.; Hu, B.; Li, Z.; Wang, X. 2023b. Application of COFs in capture/conversion of CO_2 and elimination of organic/inorganic pollutants. *Environ. Funct. Mater.* 2(1):76–92.

Wang, Z.; Wang, C.; Chen, Y.; Wei, L. 2023c. Covalent organic frameworks for capacitive energy storage: recent progress and technological challenges. *Adv. Mater. Technol.* 8(8):2201828.

Wang, Z.-Q.; Wang, X.; Li, C.; Yang, Y.-W. 2024. Covalent organic frameworks for antibacterial applications. *Cell Rep. Phys. Sci.* 5(3):101845.

Xiao, J.; Chen, J.; Liu, J.; Ihara, H.; Qiu, H. 2023. Synthesis strategies of covalent organic frameworks: An overview from nonconventional heating methods and reaction media. *GEE* 8:1596–1618.

Xu, H. S.; Luo, Y.; Li, X.; See, P. Z.; Chen, Z.; Ma, T.; Liang, L.; Leng, K.; Abdelwahab, I.; Wang, L.; Li, R.; Shi, X.; Zhou, Y.; Lu, X. F.; Zhao, X.; Liu, C.; Sun, J.; Loh, K. P. 2020. Single crystal of a one-dimensional metallo-covalent organic framework. *Nature Comm.* 11:1.

Yaghi, O. M.; Kalmutzki, M. J.; Diercks, C. S. 2019. Introduction to Reticular Chemistry: Metal-Organic Frameworks and Covalent Organic Frameworks. Wiley-VCH Verlag GmbH & Co. KGaA: Weinheim, Germany.

Yang, M.; Hanayama, H.; Fang, L.; Addicoat, M. A.; Guo, Y.; Graf, R.; Harano, K.; Kikkawa, J.; Jin, E.; Narita, A.; Müllen, K. 2023a. Saturated linkers in two-dimensional covalent organic frameworks boost their luminescence. *J. Am. Chem. Soc.* 145:14417–14426.

Yang, Q.; Luo, M.; Liu, K.; Cao, H.; Yan, H. 2020. Covalent organic frameworks for photocatalytic applications. *Appl. Catal. B: Environ.* 276:119174.

Yang, S.; Yang, C.; Dun, C.; Mao, H.; Khoo, R. S. H.; Klivansky, L. M.; Reimer, J. A.; Urban, J. J.; Zhang, J.; Liu, Y. 2022a. Covalent organic frameworks with irreversible linkages via reductive cyclization of imines. *J. Am. Chem. Soc.* 144:9827–9835.

Yang, X.; Gong, L.; Liu, X.; Zhang, P.; Li, B.; Qi, D.; Wang, K.; He, F.; Jiang, J. 2022b. Mesoporous polyimide-linked covalent organic framework with multiple redox-active sites for high-performance cathodic Li storage. *Angew. Chem. Int. Ed.* 61:e202207043.

Yang, Y.; Börjesson, K. 2022. Electroactive covalent organic frameworks: A new choice for organic electronics. *Trends Chem.* 4(1):60–75.

Yang, Y.; Ratsch, M.; Evans, A. M.; Börjesson, K. 2023b. Layered 3D covalent organic framework films based on carbon–carbon bonds. *J. Am. Chem. Soc.* 145:18668–18675.

Yin, Y.; Zhang, Y.; Zhou, X.; Gui, B.; Wang, W.; Jiang, W.; Zhang, Y.-B.; Sun, J.; Wang, C. 2024. Ultrahigh–surface area covalent organic frameworks for methane adsorption. *Science* 386:693–696.

Zeng, Y.; Zou, R.; Zhao, Y. 2016. Covalent organic frameworks for CO_2 capture. *Adv. Mater.* 28:2855–2873.

Zhai, L.; Huang, N.; Xu, H.; Chen, Q.; Jiang, D. 2017. A backbone design principle for covalent organic frameworks: the impact of weakly interacting units on CO_2 adsorption. *Chem. Comm.* 53:4242.

Zhang, J.; Han, X.; Wu, X.; Liu, Y.; Cui, Y. 2017. Multivariate chiral covalent organic frameworks with controlled crystallinity and stability for asymmetric catalysis. *J. Am. Chem. Soc.* 139:8277–8285.

Zhang, Q.; Dai, M.; Shao, H.; Tian, Z.; Lin, Y.; Chen, L.; Zerng, X. C. 2018. Insights into high conductivity of the two-dimensional iodineoxidized sp^2-c-COF. *ACS Appl. Mater. Interfaces* 10:43595–43602.

Zhang, W.; Chen, L.; Dai, S.; Zhao, C.; Wei, C. M. L.; Zhu, M.; Chong, S. Y.; Yang, H.; Liu, L.; Bai, Y.; Yu, M.; Xu, Y.; Zhu, X. W.; Zhu, Q.; An, S.; Sprick, R. S.; Little, M. A.; Wu, X.; Jiang, S.; Wu, Y.; Zhang, Y. B.; Tian, H.; Zhu, W. H.; Cooper, A. I. 2022a. Reconstructed covalent organic frameworks. *Nature* 604:72–79.

Zhang, Y. B.; Su, J.; Furukawa, H.; Yun, Y.; Gandara, F.; Duong, A.; Zou, X.; Yaghi, O. M., 2013. Single-crystal structure of a covalent organic framework. *J. Am. Chem. Soc.* 135:16336.

Zhang, Z.; Zhang, L.; Wang, X.; Wang, T.; Cheng, C.; Liu, X. 2022b. Micro/nano-scaled covalent organic frameworks: Polymerization, crystallization and self-assembly. *ChemNanoMat* 8:e202100345.

Zhao, F.; Liu, H.; Mathe, S. D. R.; Dong, A.; Zhang, J. 2017. Covalent organic frameworks: From materials design to biomedical application. *Nanomaterials* 8(1):15.

Zhao, W.; Luo, L.; Cong, M.; Liu, X.; Zhang, Z.; Bahri, M.; Li, B.; Yang, J.; Yu, M.; Liu, L.; Xia, Y.; Browning, N. D.; Zhu, W.-H.; Zhang, W.; Cooper, A. I. 2024. Nanoscale covalent organic frameworks for enhanced photocatalytic hydrogen production. *Nat. Commun.* 15:6482.

Zhong, W.; Sa, R.; Li, L.; He, Y.; Li, L.; Bi, J.; Zhuang, Z.; Yu, Y.; Zou, Z. 2019. A covalent organic framework bearing single Ni sites as a synergistic photocatalyst for selective photoreduction of CO_2 to CO. *J. Am. Chem. Soc.* 141(18):7615–7621.

Zhou, P. K.; Li, Y.; Zeng, T.; Chee, M. Y.; Huang, Y.; Yu, Z.; Yu, H.; Yu, H.; Huang, W.; Chen, X. 2024. One-dimensional covalent organic framework-based multilevel memristors for neuromorphic computing. *Angew. Chem. Int. Ed.* 63:e202402911.

Zhu, D.; Li, X.; Li, Y.; Barnes, M.; Tseng, C. P.; Khalil, S.; Rahman, M.; Ajayan, P. M.; Verduzco, R. 2021. Transformation of one-dimensional linear polymers into two-dimensional covalent organic frameworks through sequential reversible and irreversible chemistries. *Chem. Mater.* 33:413–419.

Zhu, D.; Zhu, Y.; Chen, Y.; Yan, Q.; Wu, H.; Liu, C. Y.; Wang, X.; Alemany, L. B.; Gao, G.; Senftle, T. P.; Peng, Y.; Wu, X.; Verduzco, R. 2023. Three-dimensional covalent organic frameworks with pto and mhq-z topologies based on Tri- and tetratopic linkers. *Nat. Comm.* 14:2865.

Zhu, Q.; Wang, X.; Clowes, R.; Cui, P.; Chen, L.; Little, M. A.; Cooper, A. I. 2020. 3D cage COFs: A dynamic three-dimensional covalent organic framework with high-connectivity organic cage nodes. *J. Am. Chem. Soc.* 142:16842.

Zhu, Y.; Chen, L.; Pan, J.; Jiang, S.; Wang, J.; Zhang, G.; Zhang, K. 2025. Recent advances in COF-derived carbon materials: synthesis, properties, and applications. *Prog. Mater. Sci.* 148:101373.

Index

N

O

P

Q

R

S

T

U

V

W

X